T0173350

AQA
AS/A-level

Design and Technology

Product Design

Will Potts, Julia Morrison, Ian Granger and Dave Sumpner

HODDER Education

Photo credits for this book can be found on page 366.

Every effort has been made to trace all copyright holders, but if any have been inadvertently overlooked, the Publishers will be pleased to make the necessary arrangements at the first opportunity.

Although every effort has been made to ensure that website addresses are correct at time of going to press, Hodder Education cannot be held responsible for the content of any website mentioned in this book. It is sometimes possible to find a relocated web page by typing in the address of the home page for a website in the URL window of your browser.

Hachette UK's policy is to use papers that are natural, renewable and recyclable products and made from wood grown in well-managed forests and other controlled sources. The logging and manufacturing processes are expected to conform to the environmental regulations of the country of origin.

Orders: please contact Hachette UK Distribution, Hely Hutchinson Centre, Milton Road, Didcot, Oxfordshire, OX11 7HH. Telephone: +44 (0)1235 827827. Email education@hachette.co.uk. Lines are open from 9 a.m. to 5 p.m., Monday to Friday. You can also order through our website: www.hoddereducation.com

ISBN: 9781510414082

First published in 2017 by
Hodder Education,
An Hachette UK Company
Carmelite House
50 Victoria Embankment
London EC4Y 0DZ

www.hoddereducation.co.uk

Impression number 12

Year 2024

Illustrations by Integra Software Services Ltd.

Typeset in India by Integra Software Services Ltd.

Printed by Ashford Colour Press Ltd.

A catalogue record for this title is available from the British Library.

Contents

Introduction to AQA AS/A-level Design and Technology: Product Design

This book has been written to help you fully understand the wide range of technical content and approaches to designing you need for success in AS and A-level Product Design.

The AQA AS and A levels in Product Design require you to develop skills and understanding in both making and designing.

Making products involves working with a range of materials. Only by experiencing working with materials first-hand can you start to understand their properties and the ways in which those materials can be used for a variety of purposes. You will need to know different methods of working with materials in order to understand the right method of production for your product.

When designing, you will need to understand what the user or client requires, and then go on to provide them with a functional prototype, which can be tested so you and your user or client can judge how successful it is.

The course is divided into two main parts:

1 Technical principles
2 Designing and making principles

Both of these parts will be assessed in the final examination. For AS students there will be one paper which will last 1 hour and 30 minutes and will make up 50 per cent of their qualification. For A-level students there will be two papers, one lasting 2 hours and 30 minutes that will cover technical principles and one lasting 1 hour and 30 minutes that will cover designing and making principles. These papers will combine to make up 50 per cent of the A-level qualification. The other 50 per cent for AS and A-level is for the non-examined assessment (NEA) project you will complete during your course.

The NEA project is covered in Section 3 of this book, and the written paper in Section 4.

How to use this book

Features to help you

A range of different features appear throughout the book to help you learn and improve your knowledge and understanding of designing and making.

Learning outcomes

Clear learning objectives for each topic explain what you need to know and understand.

A-level only

This indicates content that is only relevant for students taking the A level course.

Activity

Short activities are included to help you to understand what you have read. Your teacher may ask you to complete these.

Key terms

All important terms are defined.

Key points

Summaries of key points appear at the end of each topic to help you remember the most important aspects of a topic.

Check your knowledge and understanding

These short questions test your knowledge and understanding of each topic.

Further reading

Suggestions for books and weblinks which will help you gain a wider understanding of each topic.

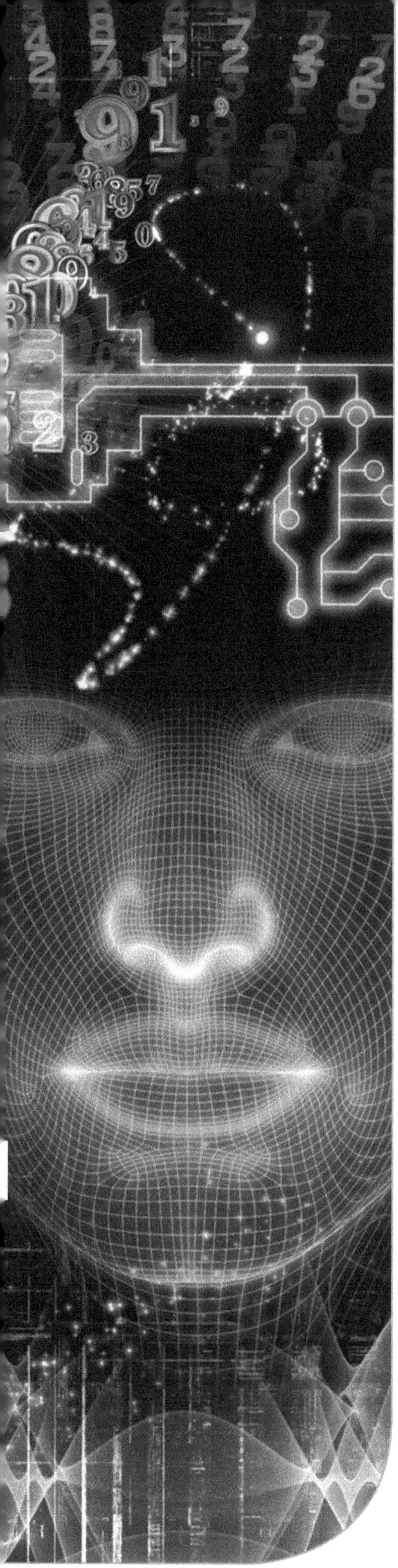

Section 1 Technical principles

In this section you will learn about the following:

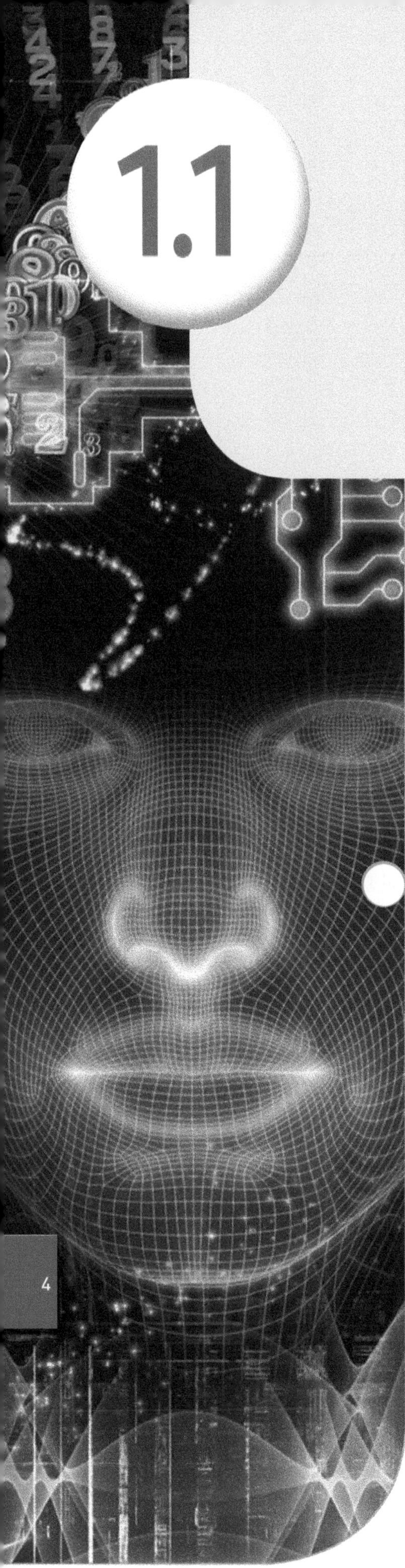

1.1 Materials and their applications

LEARNING OUTCOMES

By the end of this section you should have developed a knowledge and understanding of:

- the physical and mechanical properties and working characteristics of materials
- the names of specific materials
- why specific materials are suitable for given applications based on their physical, mechanical and working properties, product function, aesthetics, cost, manufacture and disposal
- the classification of materials
- practical workshop methods for investigating and testing material properties
- industrial tests for mechanical material properties.

Materials and applications

The ability to choose the best material for product manufacture is an essential skill for both designers and manufacturers. Different materials work in different ways and it is important to have a sound knowledge of material properties in order to make sure the product is a success. Selection is not always easy and often compromises have to be made but the final choice is mostly influenced by product function and the material's properties.

As well as knowledge of material classifications and properties, designers and manufacturers must also consider factors such as aesthetics, cost, manufacturing method and disposal. Some materials are more aesthetically pleasing than others, for example stainless steel is naturally bright grey and is more aesthetically pleasing than the dull coloured low carbon steel. Stainless steel is, however, more expensive than low carbon steel; therefore the designer may select low carbon steel and apply a surface finish to enhance aesthetics. Some woods have a stunning grain pattern but the designer may choose a veneered **manufactured board** to save overall manufacturing costs. The **stock form** of the material can also influence cost. The stock form is the set of standard sizes in which materials are available; for example manufactured boards such as medium density fibreboard (MDF) are available as a standard stock size of 2,440 mm x 1,220 mm (and in predetermined thicknesses, of 25, 18, 15, 12, 9, 6 and 4 mm). This allows for standardisation if designers are designing products to be manufactured in another part of the world.

Metals have a range of stock forms such as square tubing, square bar, round tubing, round bar and hexagonal bar. Generally, for metals, the more complex the stock form, the more expensive the material component will be.

The use of stock forms offers advantages such as uniformity of material sizes across countries, the material form can be transported more easily once converted into stock forms rather than as raw material (for example, a whole tree trunk), standard stock sizes are less expensive than specialist sizes because they are processed in large quantities, and there is less waiting time than for a specialist size as the material will be repeatedly manufactured in the same sizes.

Material costs directly influence product manufacture as well as retail price. Many polymer products are considered 'cheap' but traditional polymers come from crude oil, the cost of which fluctuates due to supply and demand and the impact of geopolitical or weather factors (such as political instability or the impact of a hurricane). As such, polymers are an expensive raw material, but due to fast and economic processing (such as lack of secondary finishing) the cost per product can be significantly reduced compared to a traditional material.

Material properties

There are two principal categories of material properties: **mechanical** and **physical**.

- Mechanical properties are associated with how a material reacts to an external force.
- Physical properties are associated with the actual make-up or structure of the material.

Mechanical properties of materials

Compressive strength: the ability to withstand being crushed or shortened by pushing forces (compression).

Tensile strength: the ability to resist stretching or pulling forces (tension).

Bending strength: the ability to resist forces that may bend the material.

Shear strength: the ability to resist sliding forces on a parallel plane.

Torsional strength: the ability to withstand twisting forces from applied torque or torsion.

Hardness: the ability to resist abrasive wear such as scratching, surface indentation or cutting.

Toughness: the ability to absorb impact force without fracture.

Plasticity: the ability to be permanently deformed (shaped) and retain the deformed shape.

Ductility: the ability to be drawn out under tension, reducing the cross-sectional area without cracking, for example stretching a material into a wire.

Malleability: the ability to withstand deformation by compression without cracking. Malleability increases with a rise in temperature.

Elasticity: the ability to be deformed and then return to the original shape when the force is removed.

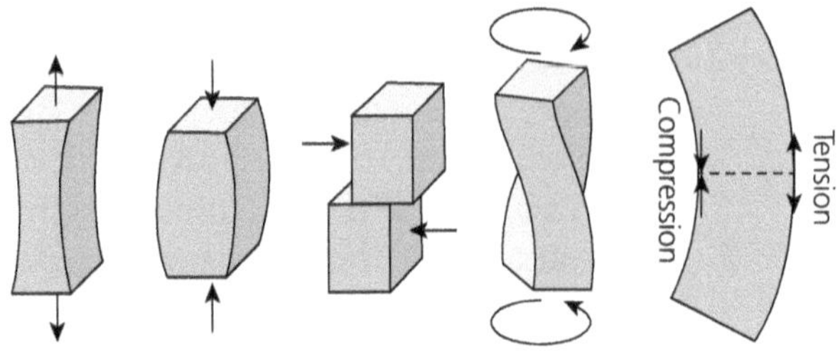

Figure 1.1.1 Tension, compression, shear, torsion, bending.

Physical properties of materials

Electrical properties

- **Electrical conductor:** allows the flow of electrical current through the material. A good conductor gives very little resistance to the flow of charge.
- **Electrical insulator:** does not allow the flow of electricity through the material.

Thermal properties

- **Thermal conductor:** allows the transfer of heat energy through the material. A material with high thermal conductivity allows the transfer of heat to occur quickly across the material.
- **Thermal insulator:** prevents the transfer of heat through the material.
- **Thermal expansion:** the increase in material volume in response to a heat input.

Optical properties

- **Opaque:** prevents light from travelling through.
- **Translucent:** allows light through but diffuses the light so that objects appear blurred. Frosted glass is an example of a translucent material.
- **Transparent:** allows light to pass through easily which means you can see clearly through the material.

Figure 1.1.2 Galvanised low carbon steel watering can.

Figure 1.1.3 High density polyethylene (HDPE) watering can.

Density: the mass of the material in a standard volume of space.

Fusibility: the ability of the material to be fused or converted from a solid to a liquid or molten state, usually by heat. Good fusibility is an essential property for a metal being cast.

Magnetism: the natural force between objects that causes the material to attract iron or steels.

Corrosion/degradation resistance: the ability of the material to withstand environmental attack and decay.

When selecting a suitable material for a watering can, the designer and manufacturer must find a balance between many factors such as material properties, product function, aesthetics and cost. The two watering cans shown both perform the same functions – to hold water, to provide the means of carrying the water and to provide a means of pouring the water onto the required garden area.

ACTIVITY

Figure 1.1.4 Copper wiring.

List the properties that make copper a suitable material for electrical wiring. For each property listed, give a specific reason to explain why the property is suitable.

Table 1.1.1 Comparison of watering cans.

Factors to consider	Galvanised low carbon steel watering can	HDPE watering can
Hardness	Galvanised layer makes the metal more resistant to scratches from gravel if placed on the ground, but a deep scratch will allow water in, which may lead to the low carbon steel rusting.	HDPE may be scratched by the gravel but this will not affect the product finish, function or degradation.
Toughness	Low carbon steel can withstand dropping from a carrying height, but if the galvanised layer is punctured, rust may develop.	HDPE can withstand a drop from carrying height, and will not be damaged by the impact.
Manufacture	Multiple processes involved including cutting, bending, rolling, drilling, MIG (metal insert gas) welding and galvanising for a finish.	Single process such as blow moulding or rotational moulding to produce a one-piece product with integral features such as ergonomic textured grips on the handle.
Aesthetics	Zinc metallic finish applied via the galvanising process.	Pigmented a colour during manufacture. Colour of product can easily be changed by using a different pigment, e.g. green for plant use, red for a watering can used to contain herbicide.
Costs	Low carbon steel is the least expensive metal to use and is less expensive than using a polymer such as HDPE. The manufacture cost is higher as it is a multi-process manufacture.	Polymer is expensive to buy but due to the speed of processing and the fact that it is a single process, the overall unit cost is much lower than that of the metal watering can.

Classification of materials

Materials for product design and engineering are generally split into groups or classifications: **metals**, **woods**, **polymers**, **papers and boards**, **composites**, **smart materials** and **modern materials**.

Some materials have more than one classification, for example metals can be classified as **ferrous**, **non-ferrous** or **alloy**. It is important for the designer and manufacturer to be familiar with each material classification so that they can select the material most suited to a specific application.

Figure 1.1.5 Classification of materials.

Metals

METALS		
Ferrous	**Ferrous:** a metal containing mostly iron and carbon. Ferrous metals are magnetic and will rust.	Low carbon steel, medium carbon steel, cast iron
Non-ferrous	**Non-ferrous:** a metal that does not contain iron. Non-ferrous metals are not magnetic and do not rust.	Aluminium, copper, zinc, silver, gold, titanium, tin
Alloy	**Alloy:** a metal made of two or more metals, or combining two or more elements, one of which must be a metal. Alloys can also be sub classified as ferrous alloys or non-ferrous alloys.	Ferrous alloys: stainless steel, die steel (tool steel) Non-ferrous alloys: bronze, brass, duralumin, pewter

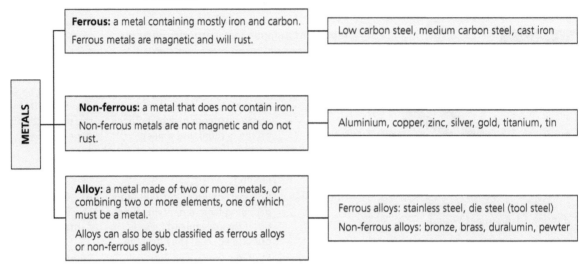

Figure 1.1.6 Flow chart showing different types of metals.

Figure 1.1.7 Aluminium drinks can.

Aluminium is suitable for a drinks can for a number of reasons.

Aluminium is:

- malleable which allows the can to be deep drawn into shape
- lightweight which makes it easier to lift and transport; aluminium adds little to the product weight
- a food safe material, which means the user will not be poisoned when drinking from the can
- non-ferrous so will not rust on contact with the liquid in the can
- very easy to recycle and use again for other products because it has a low melting point (about 660 °C), therefore saving finite resources
- aesthetically pleasing, with a natural silvery colour, which offers a contemporary, clean look to the product.

Woods

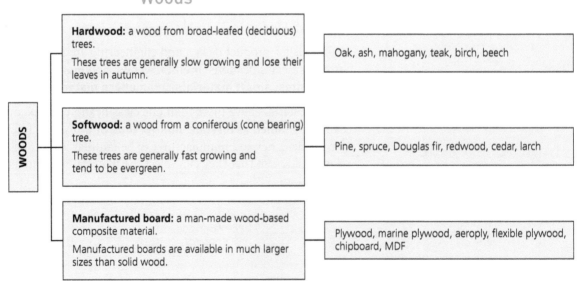

WOODS		
Hardwood	**Hardwood:** a wood from broad-leafed (deciduous) trees. These trees are generally slow growing and lose their leaves in autumn.	Oak, ash, mahogany, teak, birch, beech
Softwood	**Softwood:** a wood from a coniferous (cone bearing) tree. These trees are generally fast growing and tend to be evergreen.	Pine, spruce, Douglas fir, redwood, cedar, larch
Manufactured board	**Manufactured board:** a man-made wood-based composite material. Manufactured boards are available in much larger sizes than solid wood.	Plywood, marine plywood, aeroply, flexible plywood, chipboard, MDF

Figure 1.1.8 Flow chart showing different types of woods.

Figure 1.1.9 Solid wood table top showing the grain and end grain.

Teak is suitable for an outdoor garden bench for many reasons.

Teak:

- contains natural oils resistant to moisture, making it very weather resistant and unlikely to quickly degrade due to the effects of weathering
- contains natural oils which resist acids and alkalis, meaning the bench is unlikely to degrade due to bird droppings or cleaning detergents
- is aesthetically pleasing due to the natural dark brown/red colour and straight grain
- is hard, meaning it will withstand scratches from items such as buttons on people's jeans when they sit down or move around on the bench.

Woods are often used in manufacture because of the natural aesthetics in their grain pattern.

Polymers

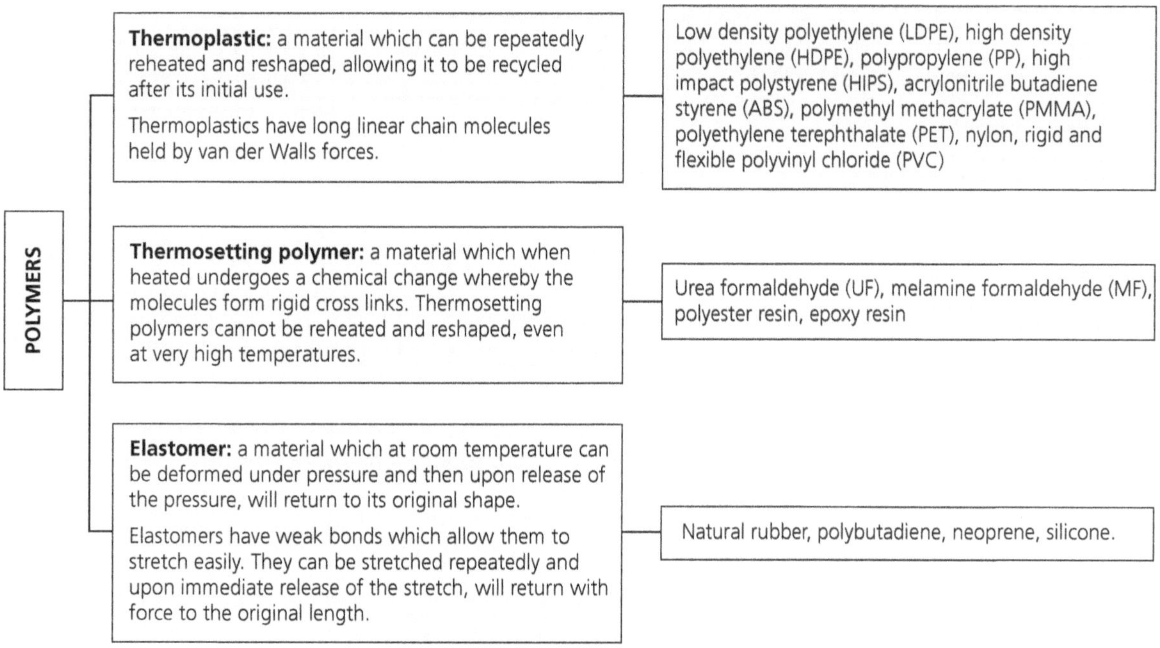

POLYMERS
Thermoplastic: a material which can be repeatedly reheated and reshaped, allowing it to be recycled after its initial use.
Thermoplastics have long linear chain molecules held by van der Walls forces.
Low density polyethylene (LDPE), high density polyethylene (HDPE), polypropylene (PP), high impact polystyrene (HIPS), acrylonitrile butadiene styrene (ABS), polymethyl methacrylate (PMMA), polyethylene terephthalate (PET), nylon, rigid and flexible polyvinyl chloride (PVC)
Thermosetting polymer: a material which when heated undergoes a chemical change whereby the molecules form rigid cross links. Thermosetting polymers cannot be reheated and reshaped, even at very high temperatures.
Urea formaldehyde (UF), melamine formaldehyde (MF), polyester resin, epoxy resin
Elastomer: a material which at room temperature can be deformed under pressure and then upon release of the pressure, will return to its original shape.
Elastomers have weak bonds which allow them to stretch easily. They can be stretched repeatedly and upon immediate release of the stretch, will return with force to the original length.
Natural rubber, polybutadiene, neoprene, silicone.

Figure 1.1.10 Flow chart showing different types of polymers.

Figure 1.1.11 PET bottle.

Polyethylene terephthalate (PET) is suitable for a drinks bottle because it is:

- a **thermoplastic** which allows it to be recycled; this is important for a single use product as it means it will not contribute to landfill
- tough – if the user drops the bottle, the contents will not be released
- available in transparent form, which allows the user to see how much drink is left
- impermeable to carbon dioxide, making it ideal for carbonated drinks.

PET can also be pigmented to give the bottle a colour and identify the drink, for example green pigment for a carbonated water drink. (Please note, PET is not on the specification, but it is a commonly used polymer.)

Figure 1.1.12 Kitchen worktop.

Melamine formaldehyde (MF) is suitable for a kitchen worktop because it:

- is a **thermosetting polymer** with a high melting point, so it will not be affected by hot pans placed on the surface
- is chemical resistant which allows the worktop to be cleaned with detergent

Figure 1.1.13 Wetsuits are made from neoprene.

- is hard, which enables the user to clean the surface with a scouring pad without scratching the surface
- can be pigmented to give the kitchen worktop a specific colour to fit with the kitchen aesthetics.

Neoprene is suitable for a wetsuit because it:

- is an **elastomer** so the wetsuit will stretch and release to fit tightly around the body
- has good degradation resistance so will not be damaged by salt water in the sea
- can be pigmented to given a particular colour choice for the consumer, or to provide a company brand colour option.

Papers and boards

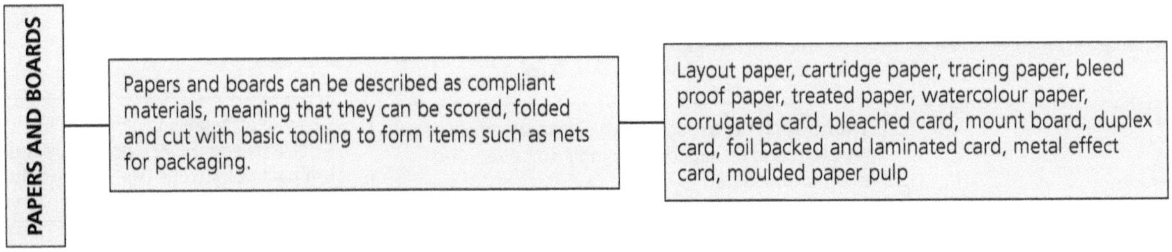

PAPERS AND BOARDS
Papers and boards can be described as compliant materials, meaning that they can be scored, folded and cut with basic tooling to form items such as nets for packaging.
Layout paper, cartridge paper, tracing paper, bleed proof paper, treated paper, watercolour paper, corrugated card, bleached card, mount board, duplex card, foil backed and laminated card, metal effect card, moulded paper pulp

Figure 1.1.14 Flow chart showing different types of papers and boards.

Corrugated card is suitable for a take-away food container such as a pizza box because:

- it is compliant and easy to cut and fold to a box shape using a die cutter
- it is a food safe material – the pizza will not be contaminated by the corrugated card
- it is an insulating material due to the air pockets within the corrugations, which will help to keep the pizza warm
- it is a lightweight material which makes it easy to carry on a delivery bike
- a pizza box is a single use product, and corrugated card is easily recycled, so the box should not contribute to landfill
- it is biodegradable, so will not contribute to landfill issues if it is thrown away and not recycled.

Composites

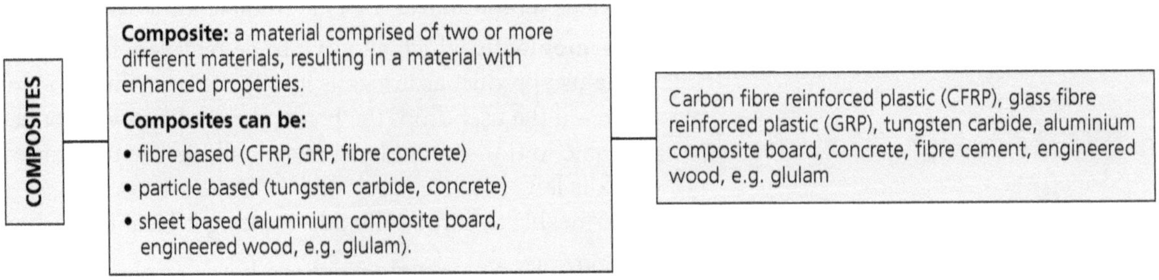

COMPOSITES
Composite: a material comprised of two or more different materials, resulting in a material with enhanced properties.
Composites can be:
• fibre based (CFRP, GRP, fibre concrete)
• particle based (tungsten carbide, concrete)
• sheet based (aluminium composite board, engineered wood, e.g. glulam).
Carbon fibre reinforced plastic (CFRP), glass fibre reinforced plastic (GRP), tungsten carbide, aluminium composite board, concrete, fibre cement, engineered wood, e.g. glulam

Figure 1.1.15 Flow chart showing different types of composites.

GRP (glass fibre reinforced plastic) is suitable for a boat hull because it:

- can be manufactured via the lay-up method, allowing complex 3D shapes such as the hull to be created
- can be pigmented to produce a range of colours for improved aesthetics or corporate branding such as racing team colours

- has chemical resistance so will not corrode or decay when in the salty sea water
- is a tough material and able to withstand minor impact from waves, etc. without damage.

Smart materials

SMART MATERIALS

Smart material: a material whose physical properties change in response to an input or change in the environment, such as electricity, pressure, temperature or light.

Shape memory alloys (SMA), thermochromatic pigment, thermochromatic film, phosphorescent pigment, photochromic pigment, electroluminescent wire, piezo electric material

Figure 1.1.16 Flow chart showing different types of smart materials.

Figure 1.1.17 Emergency exit sign (phosphorescent pigment).

Figure 1.1.18 Thermometer temperature strip (thermochromatic film).

Thermochromatic film is suitable for thermometers because:

- it changes colour in response to temperature change
- the colour change, such as to red for hot, makes it easier to read than small numbers or lines as used on traditional thermometers
- it is a non-toxic material, therefore much safer to use than mercury thermometers, for example for taking the temperature of a young child
- it can be incorporated into a film strip, making it flexible enongh to go on to a forehead when taking a temperature reading.

Modern materials

MODERN MATERIALS

Modern material: a material developed through the invention of new or improved processes, e.g. as a result of man-made materials or human intervention.

Modern materials are not 'smart materials' because they do not react to external change.

Kevlar, precious metal clay (PMC), high density modelling foam, polymorph

Figure 1.1.19 Flow chart showing different types of modern materials. For more an modern materials, see page 37.

Figure 1.1.20 Kevlar body armour.

Material disposal

Material disposal at the end of the product's life is also a major consideration, and not only influences the manufacturer's material choice but also influences consumer choice. Products made from materials that are difficult to recycle may influence the choice of the environmentally aware consumer. For example, when buying a bicycle, the consumer may choose an aluminium bike frame rather than a carbon fibre reinforced plastic (CFRP) frame because, although the CFRP may have performance advantages, it is difficult to recycle, whereas aluminium has a low melting point for a metal, which makes it one of the most commonly recycled materials.

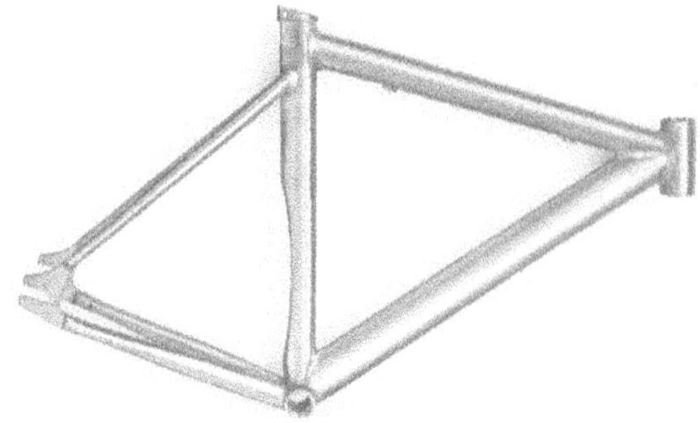

Figure 1.1.21 Aluminium bike frame.

Methods for investigating and testing materials

When selecting the most appropriate material for a specific application it is often necessary to perform specific tests on mechanical properties. These tests come in two main categories: simple workshop tests (comparative) and scientific measured tests (industrial). Whichever test is chosen the results need to be comparable (fair) and therefore we must ensure that all possible variables are removed apart from the material being tested.

The possible variables are:

- size of material sample being tested
- environmental conditions
- testing equipment
- force applied
- method of force application
- position of force application
- individual taking measurements.

Practical workshop tests

These tests can be easily carried out in a workshop using basic tools and equipment.

Tensile testing

Tensile strength is the ability to resist stretching or pulling forces. This can be tested in a workshop by clamping material samples of the same length and thickness into a vice and applying a load such as weights to the unclamped end. This method can be used to see how much the material deflects under the load and/or how much load each material can bear with the least deflection. The less deflection under load, the more tensile strength the material has.

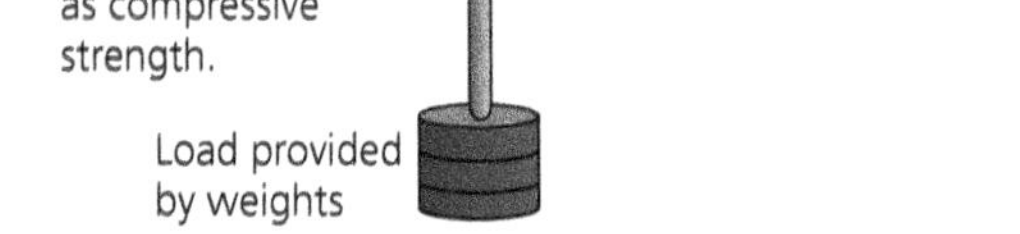

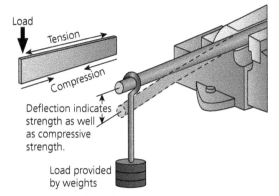

Figure 1.1.22 Tensile strength test.

Toughness testing

This test establishes how much impact force the material can absorb. Material samples are clamped into a vice. Each sample is then hit with the same force, using a hammer. Tough materials will absorb the impact, whereas brittle materials may bend or even shatter.

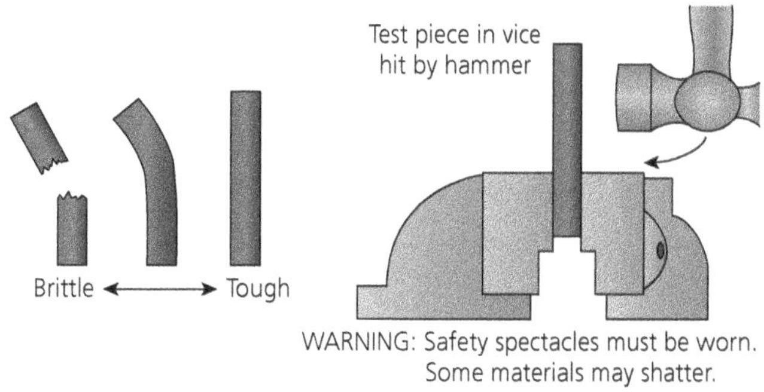

Figure 1.1.23 Testing for toughness.

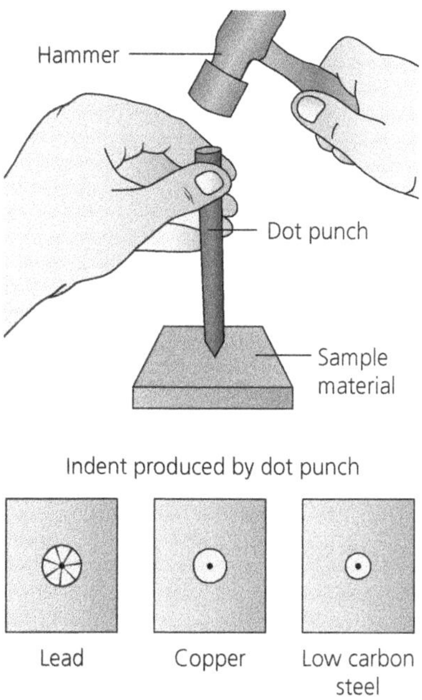

Figure 1.1.24 Testing for hardness.

Hardness testing

There are two aspects to testing hardness: abrasive wear and resistance to surface indentation.

Abrasive wear can easily be tested by running a file over the surface of the material, using the same force for each material sample. The material with the fewest scratches is the hardest.

Resistance to surface indentation can be tested using a dot punch and a hammer. The material sample is fully supported underneath and the dot punch is placed onto the material surface. The dot punch is hit once with a hammer, using the same force for each material sample. The smaller the indent, the harder the material.

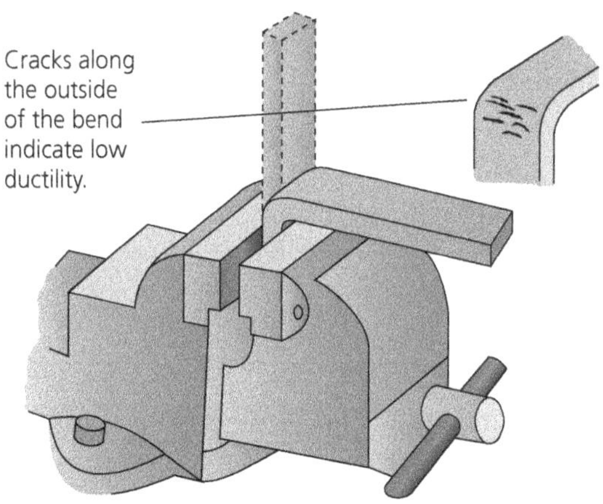

Figure 1.1.25 Testing for ductility.

Malleability and ductility testing

This test can be used to test both ductility and malleability using one test piece.

Secure the test piece in a vice. Try to bend the test piece to 90 degrees. Cracks or surface damage on the outside of the bend indicates a lack of ductility. Cracks or surface damage on the inside of the bend indicates a lack of malleability.

Corrosion testing

This test can be used to determine the rate of corrosion for a material when exposed to environmental effects such as rain or sun.

All test pieces should be the same size. Materials can simply be placed outside in an area exposed to weather effects and left for a certain length of time. Materials can then be visually inspected for surface corrosion. Depending on the intended product function, further material testing such as toughness or hardness could then be carried out.

Conductivity testing

There are two types of conductivity testing: electrical conductivity and thermal conductivity.

Electrical conductivity can be tested using a multimeter. Collect material samples of the same dimensions and, on the top surface, mark a set distance between the two ends of the material sample. The distance is the spacing for each probe of the multimeter. Place the probes on the distance marks and measure the resistance on the multimeter. The higher the resistance, the lower the conductivity.

Thermal conductivity can be tested using a thermometer. Collect material samples of the same dimensions. Measure a set distance from one end of the material on which to place a thermometer. A Bunsen burner is lit under the end of the material. Record the time it takes for the temperature to reach the set point at the other end of the material sample. The shorter the time it takes to reach the set temperature, the higher the thermal conductivity of the material.

Industrial tests

These tests are usually carried out in a lab at a materials testing facility with specific testing machinery, using standardised test pieces of a material.

Tensile testing

A standard test piece is placed into a tensometer machine and held in clamps at each end. One clamp is fixed and the other moves on a worm drive gear mechanism. As the worm drive travels at a constant rate, the test piece is put under tension. As the test piece is stretched, the load and distance travelled is plotted, giving information on elastic limit, yield point, maximum load and final breaking point after 'necking'.

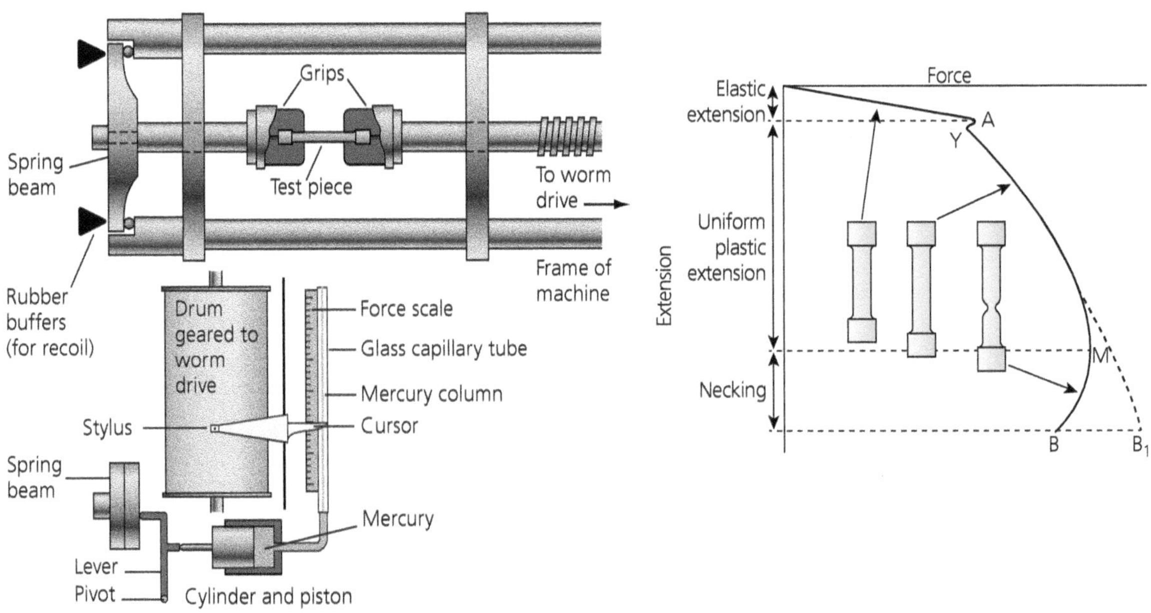

Figure 1.1.26 Testing for tensile strength.

Toughness testing

The Izod impact test is used to test the toughness of a material. A notched test piece is held vertically in the vice of the test machine. A pendulum is released from a set position and swings to strike the test piece. The energy absorbed by the test piece is calculated from the height the pendulum swings to after it hits the test piece. The material that absorbs the most impact is the toughest. This is shown by the distance the pendulum swings after breaking the material. The further it swings, the less energy the sample absorbs and the more brittle the material.

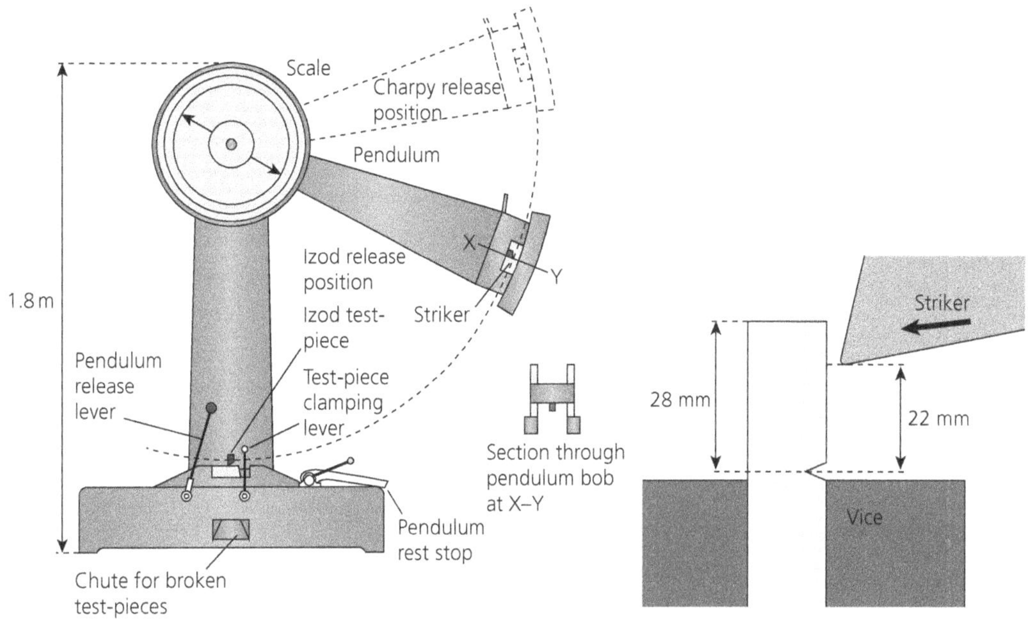

Figure 1.1.27 Testing for toughness.

Hardness testing

There are three common tests for hardness: Rockwell, Brinell and Vickers pyramid.

Rockwell test: a preload is applied to the material sample using a diamond indenter which breaks through just the surface of the material. The preload is the datum or zero reference position. An additional load is then applied to the test material and held for a predetermined length of time (dwell time). The load is then released and the distance between the preload and the applied load is measured, and the hardness of the material calculated. The smaller the indentation depth, the harder the material.

Brinell test: a hardened standard size steel ball is forced into the material's surface using a pre-set load. The diameter of the indent in the surface is measured. The smaller the diameter indent, the harder the material. (See Figure 1.1.28 where D = the diameter of the steel ball and d = the diameter of the indent in the material surface.)

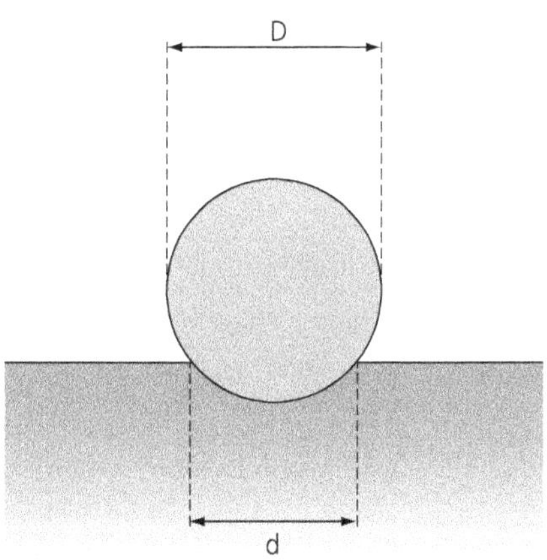

Figure 1.1.28 Brinell test.

Vickers pyramid test: used for very hard materials, this test uses a diamond square-based pyramid to indent the surface of the material. Diamond is used because it will not deform under load. A microscope is used to measure the size of the indent. The smaller the indent, the harder the material.

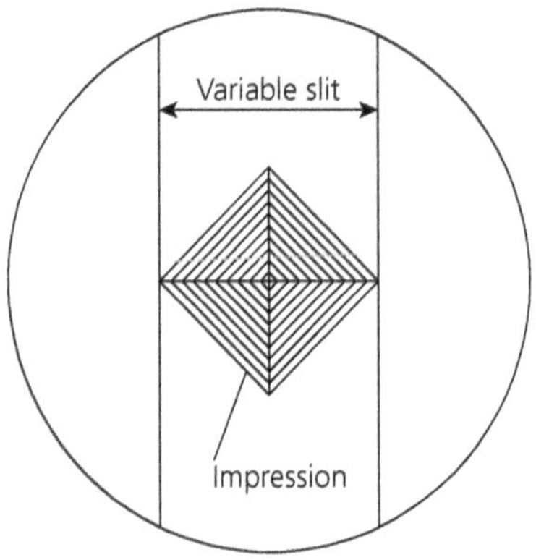

Figure 1.1.29 Vickers pyramid test.

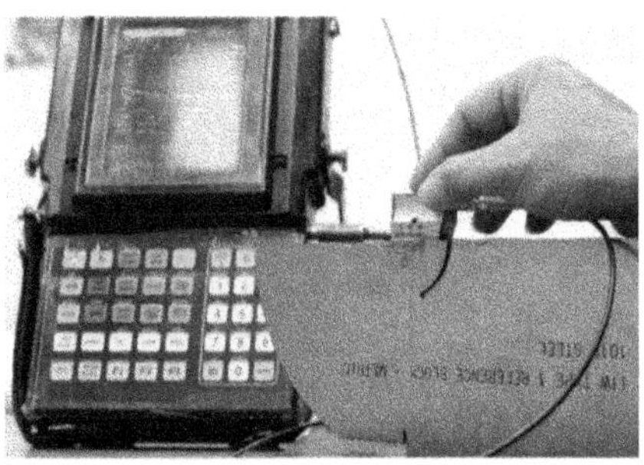

Figure 1.1.30 Ultrasonic testing for internal defects.

Ductility and malleability testing

A bend test is used to determine how well a material can withstand cracking during one continuous bend. A ductile material will have a lot of plastic deformation prior to failure.

The test piece is placed into a bending machine and held, supported at the ends. A mandrel or plunger loads the test piece at the centre and bends it to a predetermined angle, or until the test piece fractures. The material is then inspected for cracks or defects. Cracks on the outer bend indicate the level of ductility. Cracks on the inside of the bend indicate the level of malleability.

Non-destructive testing

Non-destructive testing (NDT) is usually carried out on products such as large castings where there is a likelihood of an internal defect or imperfection which would not be detected by other methods. There are two standard NDT methods: ultrasonic and x-ray.

Ultrasonic testing: a transducer generates sound waves which are pulsed into the material. The intensity of the reflected sound waves signal is recorded on a display unit. The sound waves travel through the material and if there are any defects such as cracks in the sound waves' path, the sound energy will be partially reflected and shown on the display unit.

Ultrasonic testing can be performed on all types of materials; it can be portable and has a high accuracy of flaw detection. It is widely used in the offshore and aerospace industries, and is commonly used to check the quality of welds in pipes for offshore oil applications.

X-ray testing: an x-ray beam passes through the material and an image is projected onto a display screen. X-rays and, in particular, micro-focus x-rays allow the observation of tiny details within the material. The magnified images enable minute flaws such as voids or hairline cracks to be reliably detected.

X-ray testing can be used to inspect integrated circuits (ICs) and printed circuit boards (PCBs), and to check for micro-voids in composite materials such as Formula 1 cars or track pursuit cycles where there is a monocoque (single shell) construction. It can also be used to detect faults in turbine blades or aeroplane engine blocks.

Electrical conductivity: a four-point probe method can be used to determine the electrical conductivity of materials. Four small-diameter wires are stretched, parallel to each other across a non-conductive polymer block, held in place and connected to copper terminal blocks. Two leads are attached to the inner wires and two to the outer wires. The outer leads are connected to a precise current and the two inner leads measure the voltage drop. Ohm's law (V=IR) can then be used to calculate the resistance of the material sample.

Thermal conductivity: using a heat flow meter, a square-shaped material test piece is placed between two temperature controlled plates. The temperature is increased at a controlled rate and the heat flow through the material is measured by heat flow sensors placed on the surface of the material. As the material is heated to a specific temperature, the sensors measure and record the rate of thermal conductivity.

ACTIVITY

Using notes and diagrams, describe a workshop test for material hardness. You should describe how to set up the test, the test procedure and how to determine the results of the test.

KEY TERMS

Manufactured board: a man-made, wood-based composite material.

Stock form: the set of standard sizes in which materials are available, as well as the format e.g. powders, granules, sheets etc.

Mechanical properties: associated with how a material reacts to an external force.

Physical properties: associated with the actual make-up or structure of the material.

Compressive strength: the ability to withstand being crushed or shortened by pushing forces.

Tensile strength: the ability to resist stretching or pulling forces.

Hardness: the ability to resist abrasive wear such as scratching, surface indentation or cutting.

Toughness: the ability to absorb impact force without fracture.

Composite: a material comprised of two or more different materials, resulting in a material with enhanced properties.

Smart material: a material whose physical properties change in response to an input or change in the environment, such as electricity, pressure, temperature or light.

Modern material: a material developed through the invention of new or improved processes, for example as a result of man-made materials or human intervention.

Ferrous metal: a metal containing mostly iron and carbon.

Non-ferrous metal: a metal which does not contain iron.

Alloy: a metal made of two or more metals, or combining two or more elements, one of which must be a metal.

Thermoplastic: a material that can be repeatedly reheated and reshaped.

Thermosetting polymer: a material which when heated undergoes a chemical change whereby the molecules form rigid cross links. Thermosetting polymers cannot be reheated and reshaped, even at very high temperatures.

Elastomer: materials which at room temperature can be deformed under pressure, and then upon release of the pressure will return to their original shape.

KEY POINTS

- Workshop tests can be carried out easily in a workshop using basic tools and equipment.
- Industrial tests are usually carried out in a lab at a materials testing facility with specific testing machinery, using standardised test pieces of materials.
- The Izod impact test is used to test the toughness of a material.
- The Rockwell, Brinell and Vickers pyramid tests are used to test the hardness of a material.

Check your knowledge and understanding

1 List the mechanical properties that would be required for a drill bit.

2 Describe the specific properties that make beech a suitable material for a workshop mallet.

3 Suggest a reason why industrial tests use standardised test pieces.

4 Using notes and diagrams, describe an industrial test to measure material hardness. You should describe how to set up the test, the test procedure and how to determine the results of the test.

Further reading

The New Science of Strong Materials or Why You Don't Fall through the Floor (2006) by J.E. Gordon (Author), Philip Ball (Introduction)

Explanations and examples of non-destructive testing: www.trainingndt.com

News, research, information and articles on materials science: www.azom.com

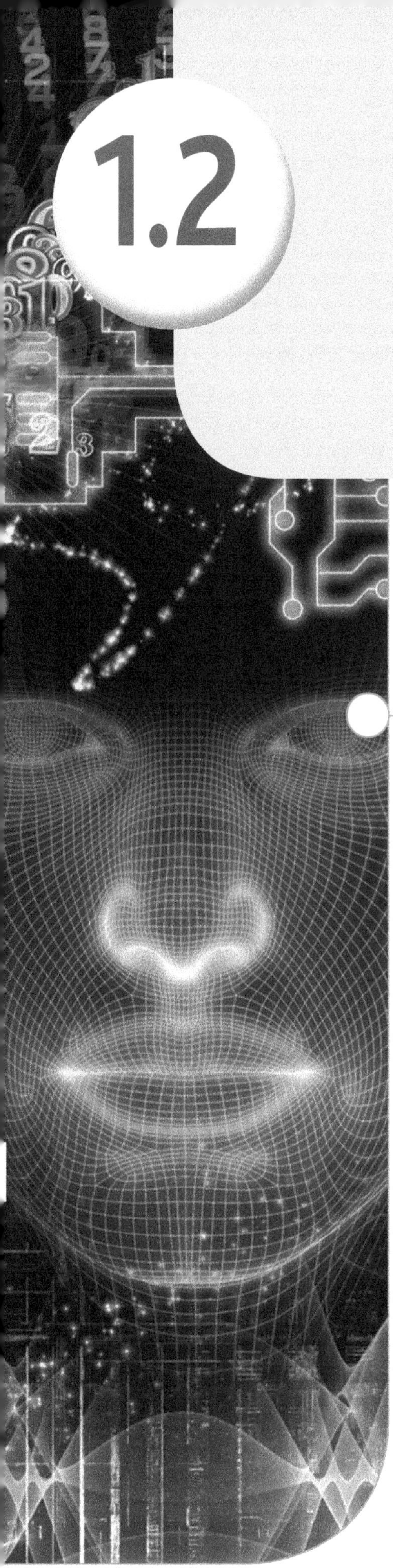

1.2 Performance characteristics of materials

LEARNING OUTCOMES

By the end of this section you should have developed a knowledge and understanding of:

- the specific names and classifications of materials
- the performance characteristics of different types of materials
- why different types of materials are suitable for different applications
- stock forms of materials.

Materials selection

Materials selection significantly impacts upon the potential success of a manufactured product. A great design may fail if the material properties are not suitable for the product function. A designer's knowledge of materials and how they could be used helps them to select suitable materials early in the design development process, therefore increasing the chances of a more successful product in terms of both function and commercial viability. When selecting materials, the designer must also consider the proposed manufacturing method, the material cost and any impact on the environment. Knowledge and understanding of the working properties of different materials is often the starting point when selecting a material for manufacture.

Paper and boards

Paper and boards are compliant materials, meaning that they are suitable for scoring, cutting and folding to produce items such as packaging. The original source of paper is wood pulp, which is grey in colour. During manufacture, bleaches are added to transform the grey pulp into whiter paper. The natural source makes paper and boards suitable for recycling, or if left in landfill they will naturally biodegrade.

Table 1.2.1 Different types of paper.

Name	Properties/performance characteristics	Uses
Layout paper	Thin translucent paper with a smooth surface.	Sketching, quick ink, technical drawing, tracing
Cartridge paper	Off-white paper usually with a slightly textured surface.	Sketching, rendering in pencil, ink and pastel Can be used for printing
Tracing paper	Translucent paper slightly thicker than layout paper.	For copying images when sketching
Bleed-proof paper	Similar to cartridge paper but has a bleed-proof layer on one side so colours do not run.	Spirit-based marker rendering
Treated paper	Plain paper with a clear binder or dye layer applied to help hold the image on the paper surface and brighten the image. Surface sheens such as high gloss or matt available.	Photographic printing
Watercolour paper	Available in absorbent, smooth, hot-pressed or the more textured cold-pressed and heavily textured rough.	Watercolour painting
Corrugated card	Usually with carton board outer layers and a corrugated middle layer, giving the material the ability to provide protection against impact.	Protective packaging, model making, prototyping ideas, food packaging such as take-away boxes
Bleached card	Chemically treated to brighten the surface to make it suitable for high-quality printing.	Greeting cards, high-quality packaging
Mount board	Made from compressed fine cotton fibres to produce a rigid board.	Presenting artwork, picture mounting, modelling
Duplex card	Made up of two layers of paper, with the exterior often coated to make it more water-resistant and to give it a glossy sheen and waxy feel.	Food packaging such as juice or milk cartons, disposable cups, plates
Foil backed and laminated card	Card with polymer film or foil applied to either one side or both sides to provide a water resistant and/or heat insulating layer. The foil or laminate layer must be removed before recycling the paper pulp.	Drinks packaging, milk cartons, take-away box lids
Metal effect card	High-quality card with a thin metal effect layer applied to the outer surface for enhanced aesthetics. Can be embossed.	Gift boxes and packaging, high-quality metal effect business cards
Moulded paper pulp	Recycled paper pulp moulded when wet and dried to a specific shape. Usually smooth on the visible inside surface and rough textured on the outer surface.	Moulded packaging products, eco-friendly packaging, egg boxes, fruit packaging

Stock forms

Papers are available in different standard sizes from A0 (largest size) to A8 (smallest), each 'A' size is half the size of the one previous, for example A4 is half the size of A3. Commercial printers may work with 'paper – untrimmed sizes' (raw format A – RA and supplementary raw format A – SRA) which are slightly larger than the standard A sizes. The larger sizes allow for any bleed from the printing process, and the papers are trimmed to size after they have been printed. Paper thickness is measured in grams per square metre (gsm), for example standard photocopying paper tends to be 80 gsm, mount board is 1,000 gsm. The lower the gsm, the more lightweight the paper, and the easier it is to bend, cut and score.

ACTIVITY

Find as many different types of paper product as you can. Make a list of the specific paper or board used for each product and why you think it was used for that particular product.

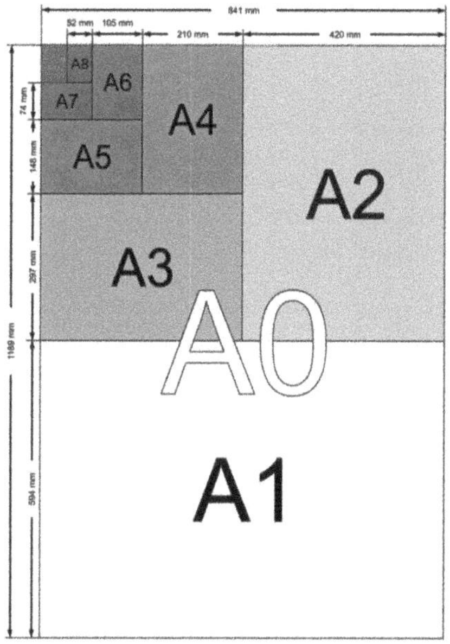

Figure 1.2.1 Paper sizes A0 to A8.

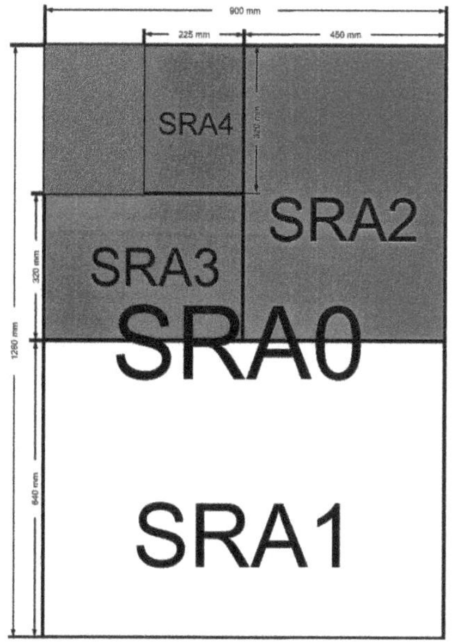

Figure 1.2.2 Paper sizes SRA.

ACTIVITY

Weigh a standard LDPE carrier bag and then put weights inside until the bag fails. Note how much weight the bag can hold before failing and how many times its own weight the bag can hold.

Polymer-based sheet and film

The materials in the following table are all manufactured from polymer materials, and are produced in sheet form of varying thicknesses for purposes from model making to packaging. In terms of sustainability, they have different features: some, such as cellulose acetate which comes from a natural source, will biodegrade; others, such as fluted polypropylene (PP), can be recycled. Although facilities are available to recycle foam board, it is more difficult due to the foam core and outer sheets of card; as such, foam board is more likely to be reused than recycled.

Table 1.2.2 Different types of polymer-based sheet and film.

Name	Properties/performance characteristics	Uses
Foam board	Two outer layers of card with a foam middle core, available in a range of thicknesses.	Mood board backing, presentation boards, modelling
Fluted PP sheet	Extruded sheet with integral 'flutes' or corrugations. Lightweight and easily bent along the flutes.	Signs (such as construction site signs, sale boards for estate agents), storage box construction, portfolio cases, small models
Translucent PP sheet	An extremely tough polymer sheet. Can be scored before bending to produce a plastic hinge, which can be folded many times. Virtually impossible to tear, water resistant.	Packaging products, folders, boxes
Styrofoam	Dense, closed cell foam, commonly blue in colour. Can be cut, shaped and sanded with standard workshop tools.	Product modelling, formers for moulding and laminating
Low density polyethylene (LDPE) sheet	Tough, available in thin sheet form, transparent, good chemical resistance, flexible.	Food wrapping, air pillow packaging, bubble wrap, carrier bags
Plastazote foam	Closed cell polyethylene foam, tough, flexible, good impact resistance, impermeable to liquids.	Protective packaging, swimming floats, gym and exercise mats, sound and pipe insulation, stage props
Cellulose acetate	Transparent, tough, naturally biodegrades.	Packaging film, membranes, photographic film
Polylactide	Transparent, tough, naturally biodegrades.	Biodegradable packaging film

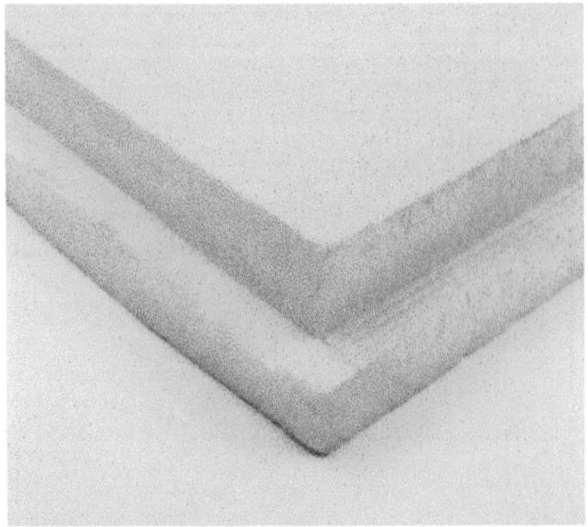

Figure 1.2.3 Styrofoam.

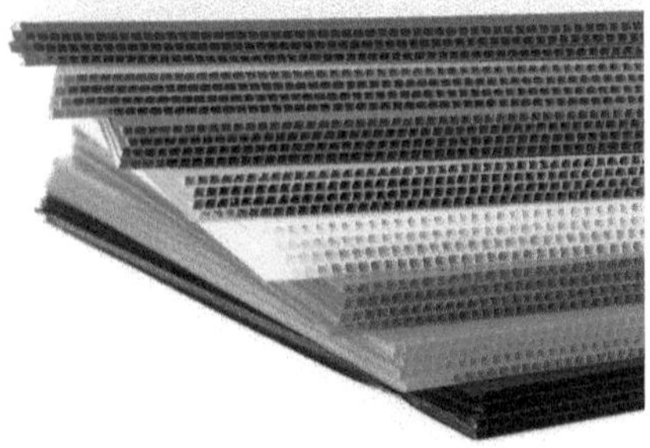

Figure 1.2.4 Fluted PP.

Woods

Wood is a natural material which is grown all over the world and has been used for building work, furniture and decorative items for thousands of years.

The wood fibres grow along the length of the tree or branch (grain) and these fibres consist of hollow cells (tracheids) supported by lignin resin. Timber is strong in the direction of the grain (along the grain) and weaker across the grain. This means that if you use a jack plane to smooth a piece of wood, the plane will cut well as it moves with the grain. If the plane is used across the grain, chipping will occur. In addition, when selecting wood for a table top, long planks running in the direction of the grain would be used for the table length. If the planks were placed lengthways with the grain running across the board, it would be very easy to split the table top in the middle, (a similar effect to when someone 'karate chops' across the grain of wood to break it).

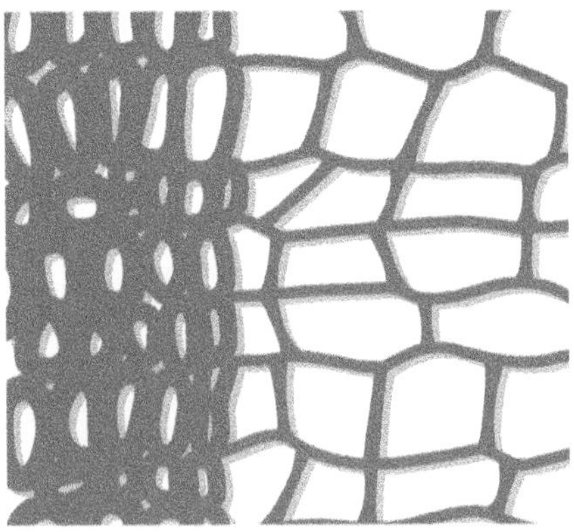

Figure 1.2.5 A magnified cross-section of wood cells. The large cells indicate rapid growth, the small cells slow growth. Slow growth produces denser wood.

Prior to most use, wood is converted (sawn into useable pieces) and then seasoned (removing the excess moisture, 80–90 per cent). There are two main forms of seasoning: **air seasoning** and **kiln seasoning**.

Air seasoning is a traditional, inexpensive method which involves stacking the wood under a shelter, protected from the rain. Air circulates between the planks to slowly remove the excess moisture. Air-seasoned wood is used for outdoor wooden products because it is seasoned to the same moisture content as its surroundings and therefore the wood will be less prone to defects.

Kiln seasoning is a more expensive but controlled method which is very quick and can take just a few weeks. Planks are stacked onto trolleys and placed in the kiln where both temperature and humidity are controlled. Initially, the kiln atmosphere is very steamy but this is gradually changed to become hotter and drier. Indoor products such as furniture will use kiln-seasoned wood because it has been seasoned to meet indoor conditions and will have a lower moisture content than air-dried wood.

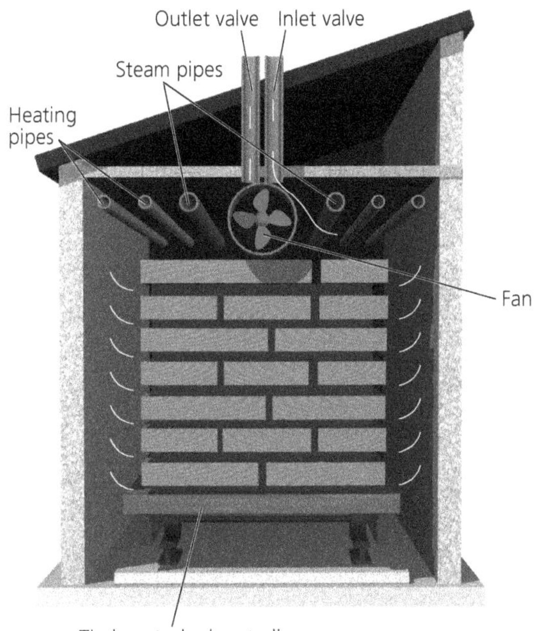

Figure 1.2.6 Kiln-seasoned wood.

If an unseasoned plank is placed in a heated room, it will twist, warp, cup or bow when drying out and cracks or gaps may appear. Seasoning the wood also gives the following advantages:

- greater immunity from decay and increased resistance to rot
- increased strength and stability
- helps preservatives to penetrate
- makes wood less corrosive to metals.

When ordering wood, the designer must ensure that the correct type is selected, for example:

- oak air-dried (A-D): suitable for exterior use
- oak kiln-dried (K-D): suitable for interior use.

As wood is a natural material, it expands and shrinks with changes in humidity (the amount of water present in the atmosphere). To reduce the time taken to season wood to the correct moisture content, trees should be felled in the winter when they grow less, have less sap and therefore contain least moisture. The largest amount of shrinkage occurs after the tree is cut down and while it is being seasoned.

Wood shrinks by different amounts in different directions, with almost no shrinkage in the direction of the wood's grain (lengthwise). Some shrinkage happens radially (perpendicular to the growth rings), with a greater amount of shrinkage occurring tangentially (along the curvature of the growth rings). The ends of the wood tend to lose moisture content more quickly than the core, and sometimes cracks can appear on the plank ends. Care must be taken when seasoning the wood so that the whole plank will shrink uniformly, and usually once the wood has been fully seasoned, any end cracks will close as the plank is now in equilibrium throughout.

Seasoned wood can also be susceptible to shrinkage due to seasonal changes in the ambient humidity, but this is less of a concern in modern homes where temperatures are fairly constant due to heating and air conditioning systems.

Classification of woods

Table 1.2.3 Classification of woods.

Classification	Examples	Properties/characteristics	Uses
Hardwoods	Oak	Hard, tough, attractive grain, good weather resistance. Contains tannic acid which will corrode steel screws or fixings.	Furniture, flooring, boat building, cladding, interior and exterior joinery
	Ash	Tough, attractive open grain pattern which makes it more flexible.	Tool handles, ladders, sports goods, laminating
	Mahogany	Can contain interlocking grain, making it more difficult to work. Rich dark red colour.	Indoor furniture, shop fittings and cabinets, veneers commonly used on manufactured boards
	Teak	Hard, tough, straight grain, natural oils resist moisture, acids and alkalis.	Outdoor furniture, laboratory benches, traditional boat decks
	Birch	Hard, straight close grained, resists warping.	Furniture, indoor panelling, veneers used for birch-faced plywood construction
	Beech	Tough, close grained, hard, available in steamed (white colour) and un-steamed (pink tinge), does not impart a taste to food.	Chairs, chopping boards, tools (mallets), steam-bent laminated furniture, turned bowls
Softwoods	Pine (European Redwood)	Straight grain, knotty, can contain resinous knots.	Construction work, roof beams and timber frame construction, interior joinery
	Spruce	Straight grain, resistant to splitting.	Indoor furniture
	Douglas fir	Straight or slightly wavy grain, few knots, stable, good resistance to corrosion.	Veneers, plywood construction, joinery and construction work
	Larch	Hard, tough, attractive grain pattern fades to silver upon exposure outside, good resistance to moisture.	Garden furniture, cladding, decking, fencing
	Cedar	Straight grain, can corrode ferrous metals due to acidic nature, low density, good sound damping, good resistance to moisture.	Exterior cladding, sheds, greenhouses, beehives, interior panelling
Manufactured boards	Plywood	Thin layers of wood are placed and glued at 90 degrees to each other and compressed to form the board. Good strength in all directions, no grain weakness, always has an odd number of layers.	Structural work, desk tops, indoor furniture, floorboards
	Marine plywood	Similar construction to plywood but is gap- and void-free and uses specialist water and boil proof (WBP) glue to provide resistance to moisture. Some higher quality marine plywood is also resistant to fungal attack.	Boat dashboards, boat lockers and panelling
	Aeroply	Plywood made from high-quality timber such as birch. Available in very thin sheets, lightweight, easy to bend around a support frame.	Gliders, laminated furniture, laser cut projects, jewellery items
	Flexible plywood	An odd number of layers glued together with the two outer layers made from open grained timber which allows the sheet to flex. Bent and glued around a former to achieve a solid shape.	Laminated furniture, curved panels
	Chipboard	Wood chips compressed with a resin such as urea formaldehyde.	Often veneered or covered with polymer laminate, kitchen worktops and units, shelving and flat-pack furniture
	MDF	MDF is compressed wood fibres, although sometimes urea formaldehyde is added as an additional resin. It has two smooth faces, available in either standard grade or veneered with a layer of timber.	Model making or mould making, furniture items such as bookcases, cabinets and desks
Veneers		Thin slices of wood less than 3 mm thick.	Decorative coverings for manufactured boards
MF laminates		Thin sheets of MF polymer, hard, tough, chemical resistant.	Decorative covering for chipboard for kitchen worktops, etc.

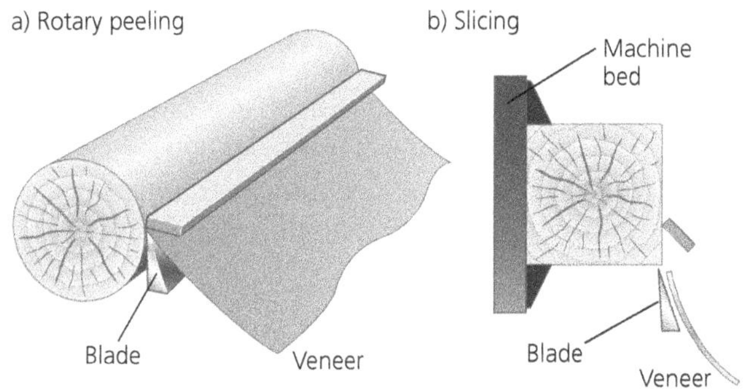

Figure 1.2.7 Cutting wood veneers.

Figure 1.2.8 MF laminates.

The toxicity of woods

The dust from woods as well as the sap can be a hazard and form a health risk to workers manufacturing with wood, as well as those in the vicinity. The toxicity of a wood depends on the species, so it is important that designers and manufacturers are aware of the different risks posed by different species.

Under the Control of Substances Hazardous to Health (COSHH) regulations 2002, both **hardwood** and **softwood** dusts have workplace exposure limits (WELs) of 5 mg per cubic metre. Both hardwood dust and softwood dust can cause respiratory irritation, and hardwood dusts are listed in COSHH as carcinogenic. As such, employers should ensure that there is adequate personal protective equipment (PPE), extraction and ventilation systems in place, and that exposure levels for workers are reduced so as not to exceed the WEL.

Stock forms

Wood is available in the following stock forms from wood suppliers. The use of standard stock sizes keeps material costs down; for example, two standard stock sizes for **rough sawn** wood are 32 mm thick and 50 mm thick, therefore the designer should choose a final processed size as close to the original size as possible. If the designer wanted a cabinet panel to be 25 mm thick, the rough sawn 32 mm should be chosen; using the 50 mm thick would mean excess material being wasted as the plank is planed down to the required size. The cost of the raw material increases with the amount of processing; for example, rough sawn is cheaper than planed all round.

Rough sawn

The wood comes directly from seasoning and has rough surfaces produced from initial conversion. Rough sawn wood will have nominal sizes rather than accurate finished sizes.

Planed square edge

Planed square edge (PSE) wood has only one edge that is planed accurately; the rest are rough sawn. Planing removes about 3 mm from the original nominal size.

Planed all round

Planed all round (PAR) wood has sides and edges that are all planed square, straight and level, leaving a smooth finish, and is ready to use. The PAR board is about 3 mm smaller all round than the original rough sawn nominal size.

Natural wood

Natural wood is only available at the maximum width of the tree. To produce a wide desk top, natural wood planks need to be joined together. Manufactured boards, however, can be produced in much larger sheets, typically 1,220 mm x 2,440 mm, meaning the wide surface can be produced from one sheet without the need for joining processes.

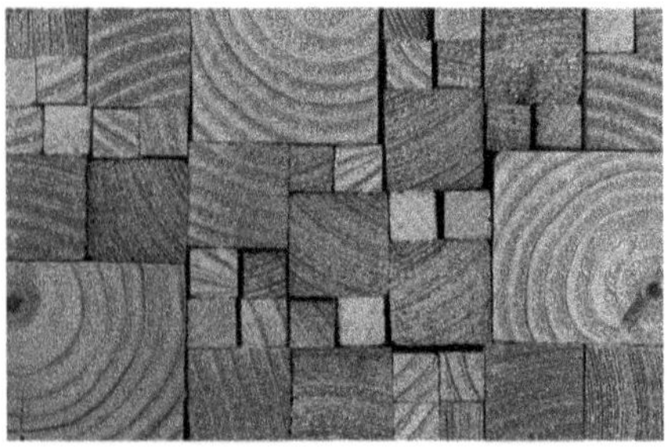

Figure 1.2.9 End grain on a wood pile.

When joining planks of wood together, the end grains must go in opposing directions to each other. This is to ensure that the internal forces of the wood do not pull the wood in one direction, and the top is joined in equilibrium.

Manufactured boards

Manufactured boards offer many advantages to the designer/manufacturer:

- available in long, wide boards of uniform thickness and quality
- very stable, not affected by temperature and humidity as much as solid wood
- uniform strength across the board, no grain problems
- difficult to split and available in flexible thin sheet form
- generally less expensive than solid wood of a similar size
- veneers can be applied to the surface for enhanced aesthetics.

Mouldings

Some wood is made into mouldings such as skirting boards or wooden trims and are readily available in standard lengths up to approximately 4 metres.

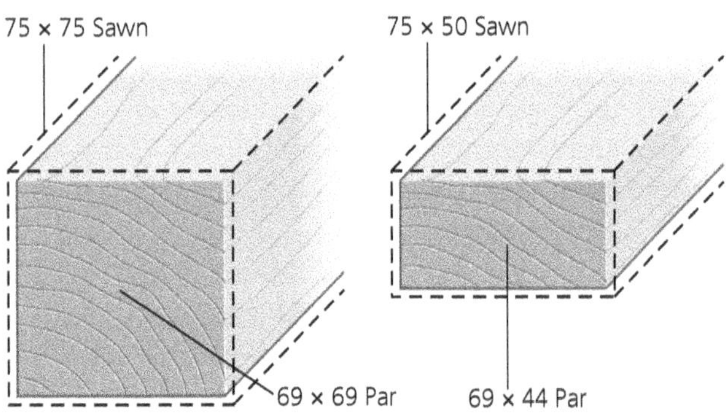

Figure 1.2.10 Typical planed wood sizes. Note the reduction due to planing.

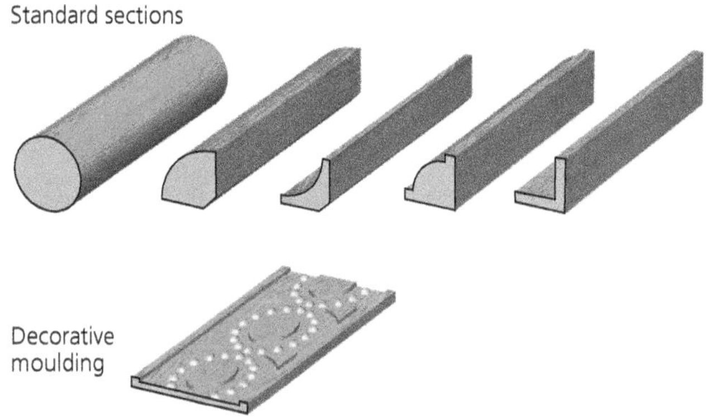

Figure 1.2.11 Wood mouldings.

Metals

Metal is a naturally occurring material and is mined from the ground in the form of ore. Approximately 25 per cent of the Earth's crust is made up of metal ores. Bauxite is the most common ore, followed by iron. The raw metal is extracted from the ore through a combination of crushing, smelting or heating, with the addition of chemicals and huge amounts of electrical energy. Most metals can be recycled, saving natural resources and limiting the amount of materials imported from abroad.

ACTIVITY

Using notes and sketches, describe how solid wood planks would be joined to make a large top for a dining table.

Table 1.2.4 Raw metal and the ore from which it is extracted.

Metal	Ore
Copper	Chalcopyrite
Aluminium	Bauxite
Tin	Cassiterite
Zinc	Zinc blende

Classification of metals

Table 1.2.5 Classification of metals.

Classification	Examples	Properties/characteristics	Uses
Non-ferrous	Aluminium	Lightweight, ductile, malleable, corrosion resistant, electrical and thermal conductor, can be metal inert gas (MIG) and tungsten inert gas (TIG) heat welded. Melting temperature 660 °C.	Drinks cans, aircraft bodies, bike frames, window frames, packaging, baking foil
	Copper	Ductile, malleable, tough, corrosion resistant, good electrical and thermal conductor, can be soldered and brazed. Melting temperature 1,083 °C.	Electrical wire, printed electrical circuits, water pipes, hot water tanks, central heating pipes
	Zinc	Low melting point, good corrosion resistance. Melting temperature 420 °C.	Galvanising steel as a protective coating for dustbins, buckets, farm gates; intricate die castings
	Silver	Malleable, ductile, corrosion resistant, can be soldered. Melting temperature 962 °C.	Jewellery, cutlery, used for plating other materials
	Gold	Malleable, ductile, corrosion resistant, can be soldered. Melting temperature 1,063 °C.	Jewellery, electronic components, connectors, switch components
	Titanium	Hard, similar strength to steels but more lightweight, high resistance to corrosion. Melting temperature 1,668 °C.	Joint replacements, tooth implants, spectacle frames, aircraft, spacecraft, golf clubs, bicycles, ship hulls
	Tin	Ductile, malleable, low melting point, corrosion resistant. Melting temperature 232 °C.	Soft solder, coatings for food cans, rarely used in pure form
Ferrous	Low carbon steel	Ductile, high tensile strength, tough, malleable, poor resistance to corrosion. 0.15–0.30 per cent carbon content. Melting temperature 1,427 °C.	Nuts, bolts, washers, screws, car bodies, outer panels for white goods
	Medium carbon steel	Harder than low carbon steel but less ductile, malleable and tough. 0.30–0.70 per cent carbon content.	Springs, gardening tools (trowels, forks)
	Cast iron	Hard outer skin but brittle core, good under compression. 3.5 per cent carbon. Melting temperature 1,149 °C.	Disc brakes, machine parts, engine blocks; sheet furniture, for example decorative bollards, bins and lighting
Ferrous alloys	Stainless steel	Tough, hard, corrosion resistant. 18 per cent chrome, 8 per cent nickel. Melting temperature 1,510 °C.	Sinks, kitchenware, cutlery
	High speed steel (HSS)	Hard, tough, high level of resistance to frictional heat. 18 per cent tungsten, 4 per cent chromium, 1 per cent vanadium, 0.5–0.8 per cent carbon.	Tool blades, drill bits, milling cutters, router bits
	Die steel (tool steel)	Hard, tough.	Blanking punches and dies, extruder dies, fine press tools

Classification	Examples	Properties/characteristics	Uses
Non-ferrous alloys	Bronze	Tough, corrosion resistant, can be cast. 90 per cent copper, 10 per cent tin. Melting temperature 850–1000 °C.	Statues, coins, bearings
	Brass	Corrosion resistant, good electrical and thermal conductor, low melting point, casts well. 65 per cent copper, 35 per cent zinc. Melting temperature 930 °C.	Boats fittings, door furniture, cast valves and taps, ornaments
	Duralumin	Equivalent strength properties to low carbon steel but more lightweight, ductile, becomes harder as it is worked. 4 per cent copper, 1 per cent magnesium and manganese.	Aircraft parts, vehicle parts
	Pewter	Malleable, low melting point, casts well. 85–99 per cent tin, with the remainder consisting of copper and antimony. Melting temperature 170–230 °C.	Tankards, flasks, goblets, photo frames, decorative items, candlesticks, sports trophies

Figure 1.2.12 Silver teapot body.

Figure 1.2.13 Brass door handle.

Stock forms

Generally, the more complex the stock form or the more material the stock form contains, the more expensive it is.

Sheet

Sheet is large flat sections of metal in varying thicknesses from less than 1 mm up to approximately 3 mm.

Plate

Plate is generally thicker than 3 mm and narrower than sheet metals.

Bar

Bar is solid form, flat bar, square bar, round bar or hexagonal bar.

Tube

Tube form is hollow in cross section and can be square, rectangular (also called box section), round or hexagonal.

Structural

Angular shapes such as H beams, L beams (angle), I beams, tee bar and channel.

ACTIVITY

Make a list of all the metal items you encounter during the day from when you wake up until you go to bed. Find out the classification for each metal and name another product made from the same metal.

Polymers

The majority of polymers are made via the fractional distillation of crude oil which produces fractions containing a mixture of hydrocarbons. The fraction called naphtha is the main one used in the production of polymers. The hydrocarbons undergo an additional process called cracking, which breaks the large hydrocarbons down into smaller, more useful forms. Polymers are made by either the polymerisation or polycondensation process, in which monomers are linked together to form long chains of monomers. 'Mer' is the single unit from which plastics are made, 'Poly' means many. Polymer is the name given to the many single units joined together.

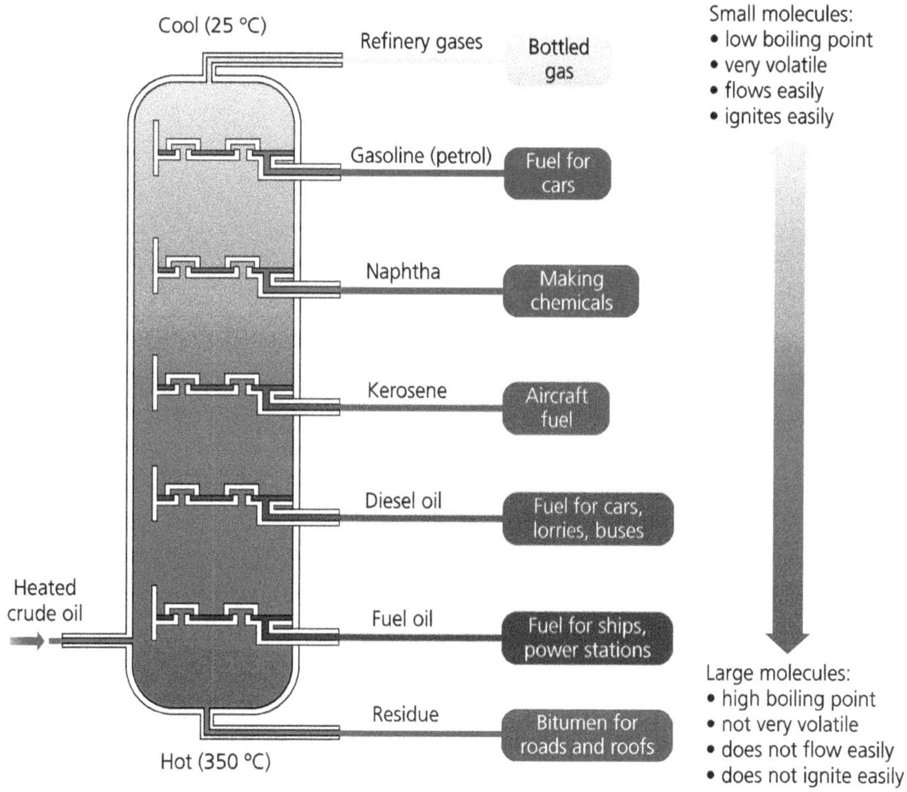

Figure 1.2.14 Polymer production process.

There are many different types of polymers, and they can be grouped into three main classifications:

- thermoplastics
- thermoset polymers
- elastomers.

In addition, there are also bio-based polymers which are designed to have a much shorter degradation time than the synthetic polymers made from crude oil. Synthetic polymers can take hundreds of years to degrade, whereas bio-based polymers can be designed to degrade in as little time as a few months to a couple of years.

Polymers are described as 'self finishing'. This is because they can be pigmented during the manufacture process to give colour and therefore require no additional secondary finishing process. Polymers are available in granules, sheet, film, rods and other extruded forms, tubes, bars, foam and powder stock forms.

When processing polymers, the **safe working temperature** should be noted. This is the temperature deemed safe for processing without possible degradation of the polymer properties.

Polymers are susceptible to damage from UV (Ultra Violet) light and unless the polymer is enhanced during processing, the polymer will disintegrate if exposed to sunlight for extended periods of time. Due to the differences in molecular make up, some polymers are more resistant to UV than others. Polymethyl-methacrylate (PMMA) is very UV stable and the polymers High Density Polyethylene (HDPE) and Polypropylene (PP) have more UV resistance than Acrylonitrile butadiene styrene (ABS). Polypropylene (PP) is often used for ropes and the outer fibres which are exposed to sunlight will become brittle and lose colour more quickly than the inner fibres of the rope. How to prevent degradation due to UV exposure is covered in Chapter 1.3.

Thermoplastics

Table 1.2.6 Classification of thermoplastics.

Name	Properties/characteristics	Uses
Low density polyethylene (LDPE)	Very tough, good chemical resistance, weatherproof, available in translucent, low level of rigidity. Safe working temperature 65 °C.	Squeezy detergent bottles, toys, carrier bags, bin liners, general packaging, food wrap film, food trays
High density polyethylene (HDPE)	Available in translucent form, weatherproof, tough, good chemical resistance. Safe working temperature 65 °C.	Chemical drums, jerry cans, toys, household and kitchenware, long life carrier bags, buckets, bowls
Polypropylene (PP)	Available in translucent, good chemical resistance, tough, good fatigue resistance (hinge property). Safe working temperature 100 °C.	Rope, folders, folio cases, food containers, medical equipment, hinged containers lids
High impact polystyrene (HIPS)	Hard, rigid, available in translucent, tough. Safe working temperature 70 °C.	Yoghurt pots, refrigerator linings, single use drink cups, toilet seats, instrument control knobs
Acrylonitrile butadiene styrene (ABS)	Extremely tough, hard, available in opaque. Safe working temperature 80 °C.	Telephone handsets, rigid luggage, domestic appliances (food mixers), handles, computer housings, remote control casings, calculator casings
Polymethyl-methacrylate (PMMA)	Tough, hard, good chemical resistance, available in translucent. Safe working temperature 95 °C.	Car light casings, computer numerically controlled (CNC) laser cut items, lighting units, lighting covers, baths
Nylon	Tough, corrosion resistant, good temperature resistance, low coefficient of friction. Safe working temperature 150 °C.	Bearings, gears, curtain rails, textiles, boil-in-the-bag food packaging, car engine manifolds, cable ties
Rigid polyvinyl chloride (uPVC)	Rigid, opaque, tough, hard, good weathering resistance, good chemical resistance, fire retardant. Safe working temperature 95 °C.	Window frames, external doors, guttering and downpipes for buildings, water service pipes, bank cards
Flexible polyvinyl chloride (PVC)	Available in translucent, tough, flexible, good weathering resistance, good chemical resistance. Safe working temperature 95 °C.	Hose pipes, cable insulation, medical grade tubing, inflatable products, imitation leather, seat coverings

Figure 1.2.15 PMMA car light casing.

Figure 1.2.16 Flexible PVC hosepipe.

Figure 1.2.17 uPVC guttering and downpipe.

Figure 1.2.18 MF laminates.

Thermoset polymers

Table 1.2.7 Classification of thermoset polymers.

Name	Properties/characteristics	Uses
Urea formaldehyde (UF)	Hard, heat resistant, good electrical insulator, brittle. Safe working temperature 80 °C.	Electrical fittings, adhesives
Melamine formaldehyde (MF)	Hard, opaque, tough, heat resistant food safe, chemical resistant. Safe working temperature 130 °C.	Decorative laminates, picnic ware, buttons
Polyester resin	Rigid, heat resistant, chemical resistant, brittle. Safe working temperature 95 °C.	Castings, used in lay-up process for glass reinforced plastic (GRP) for boat hulls, car parts, chair seats
Epoxy resin	Rigid, clear, hard, tough, chemical resistant. Safe working temperature 80–200 °C.	Adhesives, surface coatings, encapsulation of electrical components, cardiac pacemakers, aerospace applications, used in lay-up techniques with Carbon Fibre Reinforced Plastic

Stock forms

Elastomers

Rubber, the original elastomer, was first brought to Europe by Christopher Columbus in 1493. But it was not until hundreds of years later that it started to be used for products. Apparently, in 1770, Joseph Priestley noted how good it was for removing pencil marks from paper, and from then other uses developed such as elastic bands, flexible tubing, vehicle tyres and belts.

Further developments in materials technology have resulted in the creation of other elastomers such as those shown in the following table. All elastomers have the ability to be stretched to many times their original length and then, upon the removal of the tension, return to their original shape. As with other polymers, elastomers are self finishing, with pigments for colour added during the manufacture process.

Elastomers are commonly used for 'elastic' products such as wristbands. Thermoplastic elastomers (TPEs) such as polybutadiene (butadiene rubber BR),

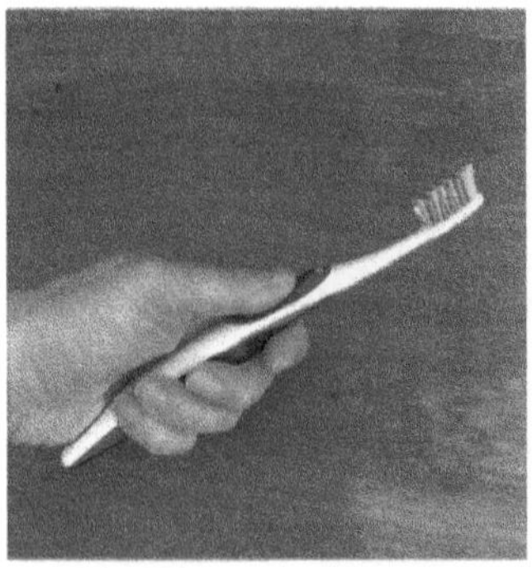

Figure 1.2.19 TPE overmoulding on a toothbrush. The TPE and texture provide additional grip.

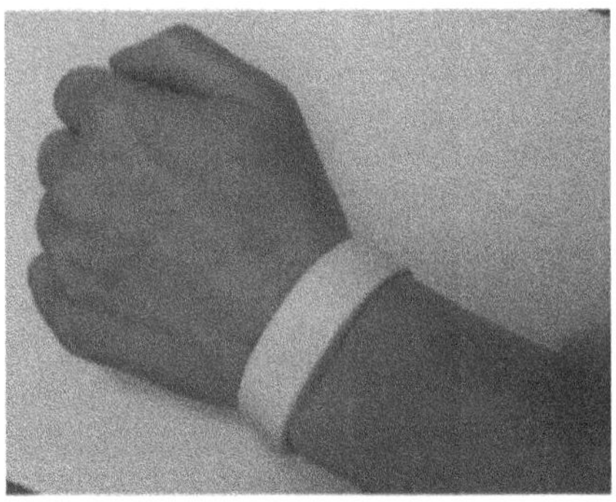

Figure 1.2.20 Silicone wristband.

neoprene (polychloroprene rubber) and silicone are often used as overmouldings to provide improved grip and textures on products such as toothbrushes, handheld electric drills and razors. They are also non-toxic to the user, which makes it ideal for gripped surfaces touched by the skin.

Table 1.2.8 Classification of elastomers.

Name	Properties/characteristics	Uses
Natural rubber (polyisoprene)	High tensile strength, low elongation, good hardness compared to other elastomers, tough, electrical insulator, good cold resistance.	Automotive industry such as tyres, tubes, hoses, gaskets, belts, balloons, toys, footwear
Butadiene rubber (BR)	Tough, excellent wear resistance against friction, good thermal resistance against friction, electrical insulator.	Vehicle tyres, shoe soles, toys, conveyor belts, water and pneumatic hoses
Neoprene (polychloroprene rubber)	Good thermal resistance, toughness, good oil and chemical resistance, excellent weather resistance, good abrasion resistance and electrical insulator.	Wetsuits, laptop cases, industrial wire insulation, automotive applications such as shock absorber seals, hose covers, transmission belts, gaskets and door seals
Silicone	Good flexibility at low temperatures, poor abrasion resistance, good thermal resistance and resistance to temperature extremes, good weather resistance, good lubricating qualities, electrical insulator.	Flexible ice cube trays, bakeware, cooking utensils, seals for refrigerators, machinery lubricant, sealants, mould making, medical uses such as lubricants for prosthetics, tubing for drug delivery systems

Biodegradable polymers

Most polymers are produced from finite, non-renewable resources such as crude oil, which do not break down easily or quickly. Discarded polymers, such as those contained in household waste, are a major cause of pollution and can kill marine and bird life, as well as cause environmental scarring due to landfill.

Polymers can however also be produced from natural and renewable resources such as wood (cellulose), vegetable oils, sugar and starch, therefore saving finite resources. As a result of technological advances as well as increased environmental awareness, manufacturers are making use of bio-polymers and biodegradable polymers. These can be used for many product applications such as shopping bags, food trays, disposable cutlery and plates, drinks bottles, plant pots, disposable gloves, disposable nappies and medical sutures such as 'dissolving stitches'.

Bio-polymers can be split into two main categories: natural and synthetic.

Natural bio-polymers: made from natural materials such as cellulose, starch and polysaccharides.

Synthetic bio-polymers: made from renewable resources but chemically engineered (synthesised) to break down more quickly. For example, corn kernels can be milled to extract dextrose from their starch; the dextrose is fermented into lactic acid, and lactic acid is then chemically processed to produce polylactide acid (PLA).

Polymer degradation occurs when the polymer undergoes a significant change in properties due to the influence of chemicals, heat or light.

There are two main descriptions for biodegradable polymers:

- The polymer is made from finite resources such as crude oil. The polymer contains additives that cause it to degrade more quickly than traditional polymers, for example in less than five years. Degradation occurs in the presence of light (photodegradable), oxygen (oxy-degradable) or water (hydro-degradable). Additives can be added to any thermoplastic and are often used in LDPE, HDPE, PP, PVC and PET.
- Degradation occurs because of the action of micro-organisms which convert the material into water, carbon dioxide (CO_2), biomass and possibly methane (CH_4). Some polymers degrade in a few weeks while others may take several months. The ability of a polymer to biodegrade is dependent on the structure of the polymer rather than the origin of the raw material.

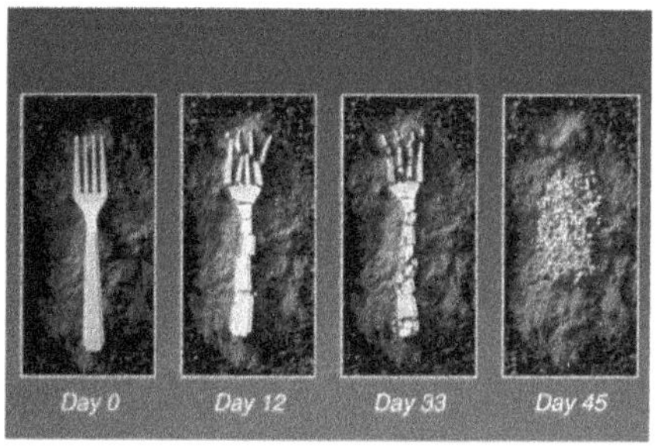

Figure 1.2.21 Biodegradable fork breaking down.

Oxy-degradable polymer

The polymer breaks down into a fine powder with exposure to oxygen and is subsequently degraded by the action of micro-organisms. The length of time for oxy-degradable polymers to degrade can be 'programmed' during manufacture and ranges from a few months to a few years. The oxy-degradable additive is commonly used in LDPE carrier bags which results in the carrier bag breaking down into small confetti shapes and then into a powder.

Photodegradable polymer

The polymer bonds are weakened and the polymer breaks down with exposure to ultraviolet (UV) light, such as UV rays from sunlight. Photodegradable polymers are often used in the agriculture industry where the ground is covered by the polymer sheet, which acts as a mulch, to prevent weed growth when growing crops. As the sheet slowly degrades, weed growth halts, therefore decreasing the need for herbicides. The sheet also helps hold water in the ground and extends the growing season by insulating the ground.

Hydro-degradable polymer

The polymer breaks down with exposure to water and subsequently micro-organisms. Hydro-degradable polymers tend to degrade more quickly than oxy-degradable polymers. The additive is often used in liquid detergent sachets for household use in washing machines or dishwashers. It can also be used for single use packaging such as chocolate box trays. Other uses include large-scale laundry bags, where the laundry is placed directly into a hydro-degradable bag, which will dissolve on contact with the water in the washing machine. This is particularly useful where there is a risk of cross contamination such as in chemical plants or hospitals.

Implications of the use of bio-polymers and biodegradable polymers:

- Biodegradable polymers can produce methane gas when they decompose in landfill. Methane gas is a greenhouse gas which contributes to global warming.
- Biodegradable polymers can take high temperatures to decompose and may leave behind toxic residues.

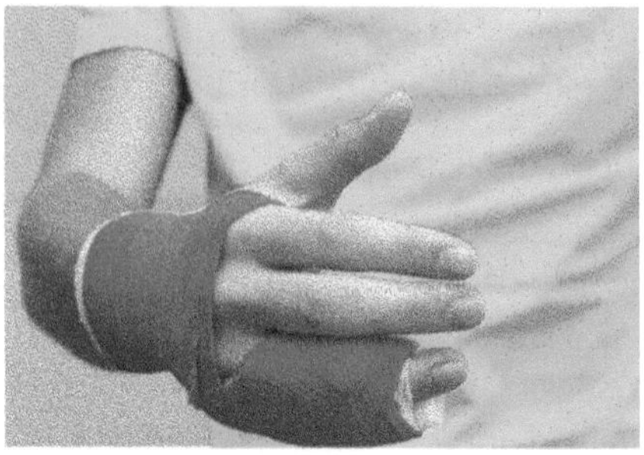

Figure 1.2.22 Degradable bio-polymers can be used for medical applications such as bone repair fixings or plates.

- Natural bio-polymers need land to grow the crops to make them. This could result in traditional agricultural land being lost to 'grow polymers'. The loss of land could lead to an increase in food costs for the consumer.
- Natural bio-polymers could be made from genetically modified (GM) crops.
- Bio-polymers and biodegradable polymers cannot be recycled. Their use may not encourage people to recycle and may contribute to a throw-away culture.
- Bio-polymers and biodegradable polymers can be processed in the same way as thermoplastics, for example injection moulded, blow moulded, calendered and vacuum formed. For more on these processes, see Chapter 1.4.
- The terms 'bio-polymer' and 'biodegradable polymer' have different meanings which may be confusing to the lay person, potentially making it harder for them to make a positive environmental choice when shopping.

Table 1.2.9 Classification of polymers.

Name	Type and source	Uses
Corn starch polymer	Bio-polymer (natural). Made from high starch vegetables such as corn, potatoes and maize.	Packaging products, straws, vending cups, disposable cutlery, bags, take-away food containers
Potatopak	Bio-polymer (natural). Made from potato starch.	Single use food items such as bowls, cutlery, food trays, serviettes, packaging beads or 'peanuts', bin bags
Biopol (bio-batch additive) Trade name for Polyhydroxybutyrate (PHB)	Bio-polymer (natural). Made from bacteria grown in cultures. Additive to promote degradation. Usually 1 per cent added to thermoplastics.	Packaging products such as film, carrier bags, vending cups, nappies, surgical stitches, pill coverings
Polylactide acid (PLA)	Bio-polymer (synthetic). Made from corn kernels or cane sugar, fermented to produce lactic acid, then synthesised to produce polylactic acid.	Packaging, single use bottles, carrier bags, plant pots, disposable nappies, medical sutures, 3D printing
Polyhydroxyalkanoate (PHA)	Bio-polymer (natural). Made from bacteria grown in cultures. Fully compostable.	Packaging, medical uses such as slow release medication patches, films, screws and bone plates
Lactide	Bio-polymer (synthetic). Fully compostable, water soluble. PLA and cellulose based.	Biomedical applications, slow release medication, bone repair fixings, detergent washing sachets
Glycolide (Lactel and Ecofilm)	Bio-polymer (synthetic). Fully compostable. PLA and cellulose based.	Food film, bags, packaging wrap, bin bags, agricultural ground sheet, flower wrap

ACTIVITY

Investigate a product made from a bio-polymer and/or biodegradable polymer. Note the specific polymer source and/or additive used, and explain the environmental implications of its use.

Composites

A **composite** is a material comprised of two or more different materials, resulting in a material with enhanced properties. For example, glass fibres are very brittle but when combined with polyester resin to produce GRP, the resultant material is both tough and hard.

Composites such as CFRP and GRP can be easily moulded into complex 3D forms using the lay-up method. This method allows monocoque designs to be produced which would be virtually impossible to create when using traditional materials such as woods or metals. Products made from CFRP and GRP tend to be lightweight, with good toughness and strength throughout the material.

CFRP is often used for high specification racing bicycles where the rider needs to transfer energy efficiently to move the bike. CFRP offers many advantages such as being lightweight, impact resistant with good resistance to the torque produced when the rider is initially pressing down on the pedals to initiate the movement.

Figure 1.2.23 Carbon fibre matting.

ACTIVITY

Find specific examples of structures made from concrete or reinforced concrete and find out how the structures are constructed from the ground up.

Table 1.2.10 Classification of composites.

Name and composition	Properties	Uses
CFRP Carbon fibres mixed with polyester resin or epoxy resin	Lightweight, corrosion resistant, tough, hard, good compressive strength, low thermal expansion.	Sports equipment such as bicycle frames, tennis racquets, fishing rods, racing car bodies and parts, helmets, prosthetics
GRP Glass fibres mixed with polyester resin	Lightweight, corrosion resistant, tough, hard, low thermal expansion, good compressive strength.	Boat hulls, pond liners, kayak shells, sports car bodies and parts, locomotive train cabs
Tungsten carbide A cermet (mixture of ceramic and metal particles) Ceramic tungsten and cobalt metal	Hard, tough, resistant to high temperatures, corrosion resistant.	Cutting tools such as drill bits, lathe tool tips, router bits, kitchen knives
Aluminium composite board Aluminium sheets with a polyethylene core	Lightweight, rigid, tough, malleable, good thermal and sound insulation, good vibration damping.	Sound-proofing panels in cars, buildings and boats, signage
Concrete Cement powder, sand, aggregate particles mixed with water	High compression strength, low tensile strength, few surface defects, easy to mould.	Pathways, beams, blocks, driveways, house foundations
Reinforced concrete Cement powder, sand, aggregate particles and low carbon steel rods (rebar) mixed with water	High compression and tensile strength, consistency across the structure, few surface defects, fire resistant.	Buildings, bridge piles and bridge spans, retaining walls, grid floors
Fibre cement Cement powder, sand, aggregate particles and polymer or steel fibres mixed with water	Lighter in weight than reinforced concrete, hard, tough, good at low temperatures and freeze/thaw situations.	Suspended floors, complex geometric shapes, pathways
Engineered wood (e.g. glulam – glued laminated timber) Timber laminates and MF adhesive	Good aesthetics, natural alternative to reinforced concretes, more lightweight than concrete alternatives, fire resistant, good structural stability, corrosion proof.	Beams, bridges, domes, arches, decking, roof beams, rafters

Smart materials

Smart materials are materials whose physical properties change in response to an input or change in the environment, such as electricity, pressure, temperature or light. Many products utilise smart materials, often bringing benefits such as increased safety and/or ease of use.

One of the most common smart materials is a shape memory alloy (SMA), called nitinol. When an SMA is heated to a certain temperature (the transition temperature), it will return to its original shape. Nitinol has a transition temperature of 70 °C, meaning it will return to its original shape if placed in hot water. The transition temperature can also be dictated by passing an electrical current through the material. For example, dental braces made from nitinol wire can be designed to change at body temperature; therefore when the braces are placed in the mouth, the wire maintains a constant tension.

Common uses for SMAs:

- Response in change in temperature, for example triggers for fire alarm systems, sprinkler systems, hot water valves in showers, industrial greenhouse windows.
- Repair of malformed or damaged products, for example spectacle frames which have been distorted can return to their original shape when heated to the transition temperature.
- A movement response from an electrical signal, for example artificial muscles in robot arms, electric door locks.

Table 1.2.11 Classification of smart materials.

Name	How does it work?	Uses
SMAs, e.g. nitinol	Changes shape in response to a change in temperature or electrical input. For example, nitinol wire in dental braces changes length in response to a change in temperature.	Dental braces, self-closing windows, aeroplane wing flaps, bioengineering such as stents, bone plates and screws
Thermochromic pigment	Changes colour in response to temperature change.	Room thermometers, medical thermometers for children, bath water thermometers, colour change mugs and kettles, food packaging to indicate food is hot to eat or cool for drinks, baby feeding spoons, battery charge indicator strips
Phosphorescent pigment	Absorbs light energy during the day and 're-emits' the light energy when it is dark.	Fire exit signs, 'glow in the dark' products such as adhesive stars, masks, night lights, watch hands
Photochromic pigment	Changes colour with light intensity.	Welding goggles, reactive spectacle lenses, security markers that show under UV light
Electroluminescent wire	Thin copper wire coated in a phosphorescent material which glows in response to an alternating current.	Glow bracelets, interweaving for clothing, home decoration, outdoor decorative lighting
Piezoelectric material	Gives off a small electrical charge when deformed. Increases in size (up to 4 per cent) when an electrical current is passed through it.	Airbag sensors in cars, lighters for barbecues, vibration damping in tennis racquets, musical greetings cards, pressure sensors

Modern materials

Modern materials will continue to evolve as technology advances. They are developed through the invention of new or improved processes, for example as a result of man-made materials or human intervention. Modern materials are not 'smart materials' because they do not react to external change.

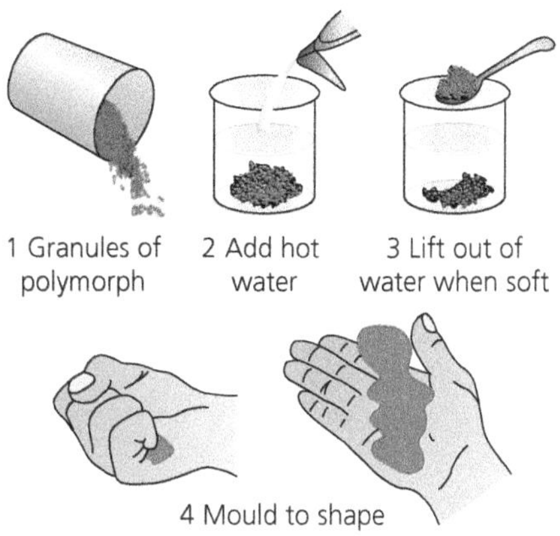
1 Granules of polymorph
2 Add hot water
3 Lift out of water when soft
4 Mould to shape

Figure 1.2.25 Polymorph.

Table 1.2.12 Classification of modern materials.

Name	Features	Uses
Kevlar	Aramid fibres with high cut and heat resistant properties.	Body armour, cut-proof gloves and aprons, aerospace applications, surfboard components, puncture-resistant bicycle tyres
Precious metal clay (PMC)	Clay consistency material made up of fine metal particles. Works like ceramic clay, easily mouldable, easy to shape and form, sets hard once fired with a kiln or butane torch. Inexpensive compared to solid metals such as silver.	Jewellery, decorative items, small sculptures, fobs
High density modelling foam	Polyurethane closed cell foam blocks or sheets. Lightweight, easy to work using traditional woodworking tools, CNC miller/router, sands easily to allow intricate shapes and forms to be created.	3D modelling, prototypes
Polymorph	Granules which become mouldable at about 60 °C. Can be heated in water or with a hair dryer. Solidifies at room temperature. Also available in liquid form. Liquid at room temperature and solidifies at 2 °C.	Modelling, shaping ergonomic handles, prototype mechanical parts

KEY TERMS

Air seasoning: a traditional, inexpensive method which involves stacking the wood under a shelter, protected from the rain. Air circulates between the planks to slowly remove the excess moisture.

Kiln seasoning: a more expensive but controlled method which is very quick and can take just a few weeks. Planks are stacked onto trolleys and placed in the kiln where both temperature and humidity are controlled.

Hardwood: a wood from a broad-leaved (deciduous) tree.

Softwood: a wood from a cone-bearing (coniferous) tree.

Rough sawn: wood that comes directly from seasoning and has rough surfaces produced from initial conversion. Rough sawn wood will have nominal sizes rather than accurate finished sizes.

Planed square edge (PSE): wood has only one edge that is planed accurately, the rest are rough sawn. PSE is about 3 mm smaller in width from the original nominal size.

Planed all round (PAR): wood has sides and edges that are planed square, straight and level, leaving a smooth finish, and is ready to use. The PAR board is about 3 mm smaller all round than the original rough sawn nominal size.

Safe working temperature (polymers): the temperature deemed safe for processing without possible degradation of the polymer properties.

Composite: a material comprised of two or more different materials, resulting in a material with enhanced properties.

Smart materials: materials whose physical properties change in response to an input or change in the environment, such as electricity, pressure, temperature or light.

Modern materials: materials that are developed through the invention of new or improved processes.

KEY POINTS

- A great design may fail if the material properties are not suitable for the product function. The designer must also consider the manufacturing method, the material cost and any impact on the environment.
- The original source of paper is wood pulp, and the natural source makes papers and boards suitable for recycling or, if left in landfill, they will naturally biodegrade.
- As wood is a natural material, it expands and shrinks with changes in humidity (the amount of water present in the atmosphere). The largest amount of shrinkage occurs after the tree is cut down and is seasoned.
- Metal is a naturally occurring material and is mined from ores in the ground. Approximately 25 per cent of the Earth's crust is made up of metal ores – Bauxite is the most common ore, followed by iron.
- Polymers are made via the fractional distillation of crude oil which produces fractions containing a mixture of hydrocarbons. Crude oil is a finite non-renewable resource and its disposal can cause a negative environmental impact. Some polymers are made from natural sources such as cellulose, sugar and starch, which naturally biodegrade.

Check your knowledge and understanding

1 Explain how the structure of plywood gives the material uniform strength across the board.

2 Describe the potential health risks posed to a worker using woods.

3 Name a composite material, the constituent parts and a product made from the named composite.

4 Explain the reasons why Kevlar is suitable for body armour.

5 For a specific product, describe the ways in which smart materials are used to enhance product performance.

Further reading

Stuff Matters: The Strange Stories of the Marvellous Materials that Shape Our Man-made World (2014) by Mark Miodownik (Author)

www.earthpac.co.nz/Earthpac is an interesting example of a company that produces starch-based products made from potato starch.

www.hse.gov.uk/pubns/wis30.pdf. This document provides information about common toxic woods, possible adverse health effects they can cause and precautions to take when using them.

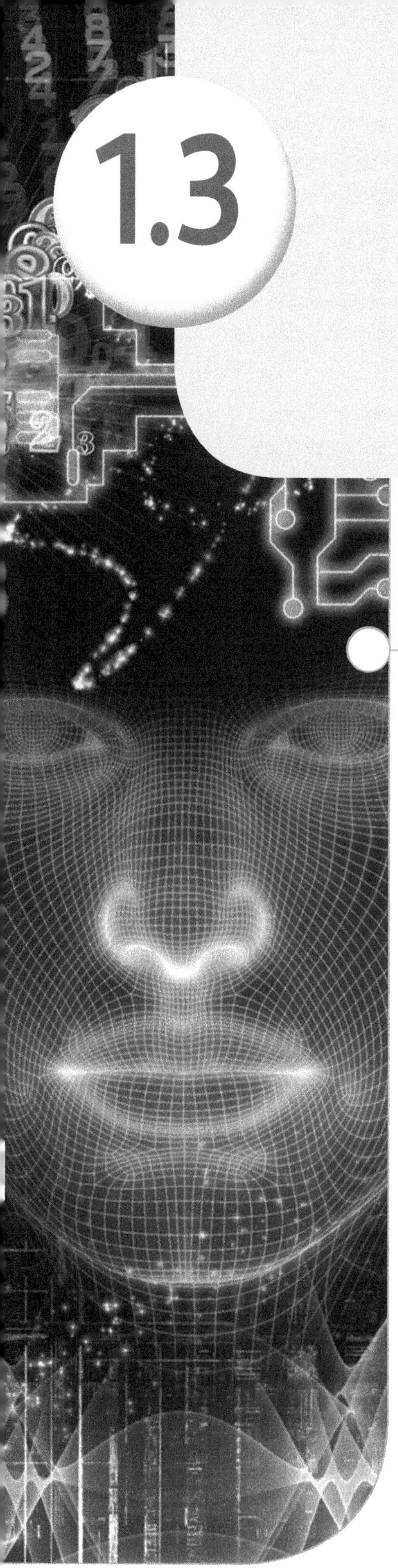

1.3 Enhancement of materials

LEARNING OUTCOME

By the end of this section you should have developed a knowledge and understanding of:

- enhancement methods for materials and their suitability for specific applications.

Material enhancement

Material enhancement is a way of improving a material's properties to better suit the requirements of the final product. Enhancements are commonly carried out on polymers, woods and metals. For example, some metals are very brittle, meaning they cannot be bent easily into shape without fracture; the use of heat treatments can alter the crystalline structure of the metal, therefore improving the toughness of the metal and making it less likely to fracture.

Polymer enhancement

Polymers are generally inert and will resist most chemical, insect and fungal attack. Certain polymer characteristics can be enhanced using additives which offer several advantages such as improved aesthetics and working properties.

Additives to make polymers easier and less expensive to process

There are many ways to process polymers, but most involve melting the polymer powder or granules and then forcing them into a specific shape. Each polymer has different working properties but may be processed in the same way, for example ABS and LDPE have different properties, but both can be processed via injection moulding.

Additives called process aids help to make the polymer flow more easily into the mould:

- **Lubricants:** the addition of wax or calcium stearate reduces the viscosity of the molten polymer, making it less 'sticky', allowing more intricate shapes to be formed. Lubricants can also allow the moulding temperature to be lowered, saving energy during the manufacturing process.
- **Thermal antioxidants:** these help to prevent the polymer oxidising or discolouring due to excessive heat during processing.

Additives to enhance aesthetics

- **Pigments:** these are tiny particles that are mixed into the polymer in its molten state to give colour to the final processed product.

Additives to improve product function

- **Antistatics:** due to their poor electrical conductivity, plastics are used for insulating products such as electric wire covers, plugs and sockets. However, the insulating property can lead to a build-up of static electrical charge which attracts dirt and dust. The addition of antistatics improves the surface conductivity by attracting moisture from the room surroundings, therefore reducing the static charge.
- **Flame retardants:** bromine, chlorine, phosphorous or metal, when added to polymers such as polyester, reduce the likelihood of combustion or the spread of fire. This is particularly important in products exposed to heat such as car engine components or potential electrical fires such as plug sockets.
- **Plasticisers:** these are added to allow plastics such as PVC (for hosepipes) to become less hard and brittle at normal temperature use. Plasticisers are added to LDPE in the manufacture of food wrap, allowing the wrap to be stretched over the food product. Plasticisers also help in processing because they allow polymers to be easily formed at higher temperatures.
- **Fillers:** these have a dual function – to provide bulk to the product, meaning that less polymer is required, and to improve the polymer properties. Commonly used fillers are sawdust and wood flour as well as mineral fillers such as chalk, clay and calcium carbonate. Sawdust and flour provide bulk, therefore reducing the amount of polymer required for the product. Mineral fillers can help increase the thermal conductivity of the polymer, meaning they will heat up and cool down more quickly, which gives shorter mould cycle times.

Additives to prolong life (prevent degradation)

- **Antioxidants:** help to reduce the environmental deterioration of the polymer from exposure to oxygen in the air. Polymer degradation can lead to increased brittleness, surface cracks and pigment discolouration.
- **UV light stabilisers:** prevent the polymer chains being broken down by sunlight. UV attack causes the polymer to lose colour and become more brittle This is evident in products such as polymer garden furniture or outdoor children's toys, which fade or lose colour when left out in the sun for a prolonged period. White pigmented items will turn yellow and darker colours such as green and blue take on a milky effect. PVC hosepipes can become more brittle and cracks may appear in the tubing as a result of extended UV exposure. UV stabilisers are also used in polymer items such as sports stadium seats and synthetic grass playing surfaces.

Additives to encourage degradation

- **Biodegradable plasticisers:** similar to plasticisers used to enhance processing, these make the polymer more flexible, softer and easier to break down, which means faster degradation time.

ACTIVITY

Investigate the way in which large supermarket chains have utilised biodegradable additives in their product packaging.

- **Bio-batch additives:** oxy-degradable (degrade in the presence of oxygen), photodegradable (degrade when exposed to UV light) and hydro-degradable (degrade in the presence of water) additives can be added to polymers to help reduce the degradation time from hundreds of years to a few years or even months. Oxy-degradable additives are commonly used in high volume single use items such as food packaging and carrier bags, whereby the bag will start to disintegrate after a pre-determined length of time. This means that these carrier bags do not contribute to landfill issues.

Wood enhancement

Natural wood can have defects such as splits or knots, which affect the overall stability. Natural wood can split due to the grain pattern, and this is one reason why manufactured boards such as MDF and plywood have been developed. Due to the lack of grain structure and consistent strength throughout the board, they are a popular choice for manufacturing purposes.

Natural wood as a manufacturing material has several disadvantages:

- The strength properties decrease when the wood is wet.
- Natural wood is highly combustible.
- Natural wood is susceptible to fungal attack, insect attack and rot.
- Natural wood is anisotropic, which means it has different properties in different directions, for example wood is easier to split along its grain than against the grain.

Despite these disadvantages, wood is still used as a building and manufacturing material, and wood can be enhanced to give improved quality such as improved resistance to weather, fungal or insect attack and fire.

Enhancement using preservatives

Wood for buildings is highly sought after due to the beautiful grain and colour, good insulating properties and, as a renewable material, it is an environmentally friendly source. To protect wood from fungal and insect attack, it can be pre-treated with preservatives which penetrate the wood to give the enhanced properties throughout the wood section. The use of preservatives is particularly important for outdoor timbers; these preservatives often use a copper based preservative because copper is a naturally occurring mineral which has excellent fungicidal properties. Pigments can also be added to the preservative to give different coloured shades to enhance the aesthetics of the wood product. Preservatives with pigments are often used for outdoor decking in garden projects. Preservatives with fire retardant properties are used to pressure treat wood for applications such as roof systems, cladding, joists and exhibition stands.

Wood can also be treated to make it harder and more resistant in high wear situations such as flooring. The wood is impregnated with a modified natural polysaccharide, similar to cellulose in timber. The polysaccharides effectively cure within the wood cell structure to produce wood that has increased hardness, toughness and stability. The addition of pigments means that inexpensive softwoods can be coloured to look like expensive hardwoods, as well as obtaining properties such as hardness and toughness. This enables manufacturers to provide more affordable and environmentally friendly products to consumers, because softwoods grow much faster and have a faster replanting time than hardwoods.

Figure 1.3.1 Larch wood grain for building cladding.

Figure 1.3.2 Wood grain and colour to enhance a building exterior.

Figure 1.3.3 Coloured preservative on wooden outdoor garden decking.

Resins and laminations

The wood industry makes extensive use of engineered wood to enhance the properties of the useable parts of trees. Engineered wood generally covers the classification manufactured boards, where long, wide sheet materials, typically 1,220 mm x 2,440 mm, are manufactured using wood parts such as sawdust, wood chips and fibres.

Figure 1.3.4 Structural composite lumber (SCL).

Manufactured boards such as chipboard are made from compressing wood chips with a resin such as urea formaldehyde. This produces a board which is very stable and is not affected by temperature and humidity as much as solid wood. Chipboard also has uniform strength across the board with none of the grain problems as seen in solid wood. The resins used for manufactured boards can also be impregnated with fire retardant additives, giving boards many more uses such as internal cladding, structural components and flooring. Most manufactured boards also have veneers laminated onto the surface for enhanced aesthetics.

Structural composite lumber (SCL) and Laminated veneer lumber (LVL) are engineered woods made by layering either veneers or strands of wood with resins such as urea formaldehyde, pressing and heat curing them to produce a stable wood billet. Both these timbers offer advantages over the original material such as being less prone to defects such as warping, splitting or shrinking along with greater load bearing properties.

Figure 1.3.5 Laminated veneer lumber (LVL) in wood joists.

SCL and LVL are used in a range of structural applications including, beams, joists, rafters and lintels.

ACTIVITY

Gather information on products made from glulam, an engineered wood.

Metal enhancement

Each individual metal has a range of properties such as toughness, hardness or ductility. Metals are made up of millions of tiny metal grains, and the structure of these grains influences the properties. The grain size and nature is dependent on the particular elements within the metal, as well as the temperature that the metal has been heated to and the subsequent rate of cooling during processing. Metal properties can be enhanced through the controlled application of heating and cooling, known as heat treatment.

Work hardening

When a metal is 'cold worked' such as by bending, rolling or hammering, the crystals within the metal are distorted and changed, leading to improved tensile strength and hardness in the worked area. This is known as **work hardening**. When the metal crystals are distorted, they are prevented from moving freely within the metal structure, which can result in less ductility, in cracking or damage in the worked area. The effects of work hardening can be removed by **annealing** the metal.

Annealing

Annealing is used to make the work hardened metal easier to work by making it less brittle and more ductile. The metal is heated and then very slowly cooled, allowing the metal crystals to grow and slowly move into place. In industry, this process is carried out in a specific temperature-controlled furnace. The process can also be carried out in a school or college workshop using a brazing hearth.

Case hardening

Case hardening is a process used for hardening the surface of steels with less than 0.4 per cent carbon content. Case hardening produces a steel with an outer casing of greater hardness and resistance to surface indentation, while the inner core of the metal retains the original 'softer' properties. Case hardening provides the metal with improved wear resistance and is used on components such as vehicle cam shafts, gears, bearings and valves.

Case hardening is a two-stage process:

- **Carburising**: this changes the chemical composition of the surface of low carbon steel so that it can absorb more carbon to increase surface hardness. In industry, the steel is placed in a ceramic box which is packed with carbon. The box is then heated to about 930–950 °C for a predetermined length of time. At this temperature, the carbon atoms can diffuse into the material's structure to build up the surface carbon content. The depth of carbon layer is determined by the length of time the material is exposed to carbon. The longer the time, the thicker the carbon layer. The product is then heated to about 760 °C and then quenched.
- **Quenching**: the hot metal is quenched in water. This is a fast cooling process which seals the hard surface case while not affecting the properties of the inner core.
  Case hardening can also be carried out in a workshop, via a similar carburising and quenching process. The steel is heated to red-hot, dipped in carbon powder, heated again and then quickly quenched in cold water.

Hardening and tempering

Hardening is the process of heating medium and high carbon steels to alter the crystalline structure, holding them at this temperature for a given time and then quenching them in water, oil or salt water baths. The hardening and subsequent quenching greatly increases the hardness property but it also increases the brittleness of the material.

After hardening, the steels have large internal stresses, and they are often much harder than required and often too brittle for many practical uses. **Tempering** is a heat treatment process for medium and high carbon steels that is carried out after hardening, to reduce some of the excess hardness and increase the toughness. The reduction in hardness leads to an increase in ductility and decreases the brittleness of the metal. To temper a metal, the metal is heated to below the **critical point** for a given time then slowly air cooled. The exact temperature will determine the amount of hardness removed.

The rise in tempering temperature results in:

- a reduction of hardness
- an increase in toughness.

ACTIVITY

Using notes and sketches, describe the industrial process for case hardening a set of low carbon steel gears.

KEY TERMS

Work hardening: improved tensile strength and hardness in the localised area when a metal is cold worked.

Annealing: heating work hardened metal and very slowly cooling it, making it easier to work by making it less brittle and more ductile.

Case hardening: a process for hardening the surface of steels with less than 0.4 per cent carbon content.

Carburising: changes the chemical composition of the surface of low carbon steel so it absorbs more carbon and increase surface hardness.

Quenching: rapid cooling of a heat-treated metal.

Hardening and tempering: heating medium or high carbon steels to a given temperature, rapidly cooling via quenching and then heating to a set temperature to remove excess hardness.

Tempering colour: the colour seen on metal that indicates the temperature at which brittleness is removed.

Critical point: the temperature at which the atoms of carbon and steel mix freely before bonding together to become a solid.

KEY POINTS

- Polymers can be enhanced by using additives which can make them easier and less expensive to process, improve aesthetics and product function as well as prevent and promote polymer degradation.
- Bio-batch additives can be added to polymers to help reduce the degradation time from hundreds of years to a few years or even months.
- Preservatives give wood enhanced protection against insect attack, rot and even fire if fire retardants are included in the preservative mix.
- Heat treatments for metal can improve properties by increasing hardness, improving toughness and reducing brittleness.

Check your knowledge and understanding

1 Explain the ways in which polymers degrade from UV attack.

2 Name two polymer additives that could be used to prevent degradation.

3 Describe the advantages that engineered woods offer over natural woods.

4 Explain what is meant by the term 'metal quenching'.

5 What colour in the **tempering colour** chart would steels be heated to when tempering a hammer head?

Further reading

www.bpf.co.uk. The British Plastics Federation has a 'Plastipedia' with lots of useful information on polymers, including an overview of the main additives and the role they play in polymer production.

www.iom3.org. The Institute of Materials, Minerals and Mining has useful answers to FAQs related to various metals and their applications.

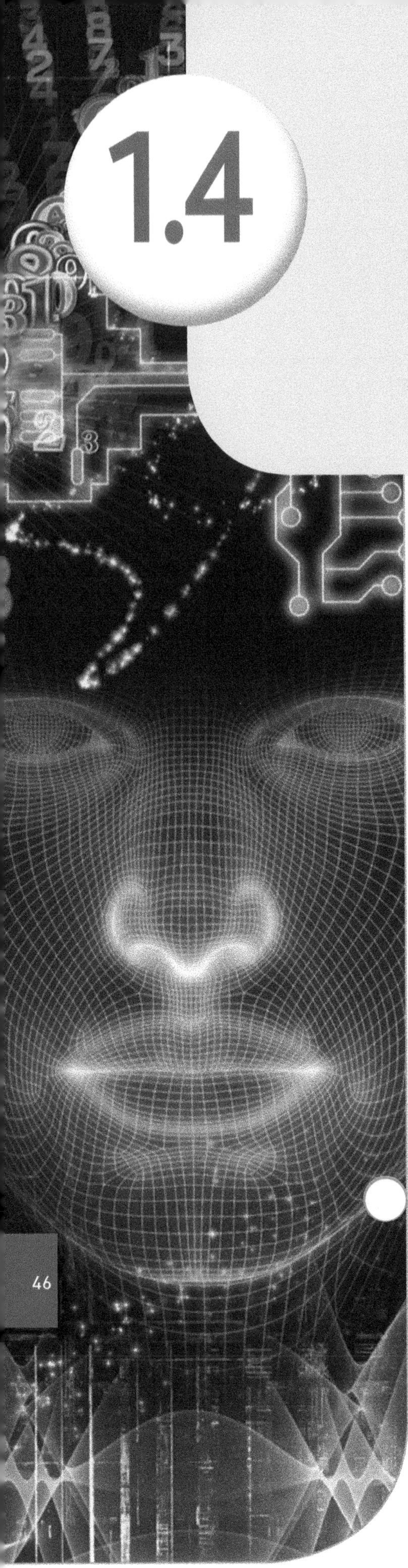

1.4 Forming, redistribution and addition processes

LEARNING OUTCOMES

By the end of this section you should have developed a knowledge and understanding of:

- the different ways that paper and board can be shaped into different products
- how polymers, metals and woods can be formed into 3D products for a range of specific products and different scales of production
- addition and fabrication processes for joining metals and woods
- 'wasting' processes for metals, and how they can be used to make specific products and components
- a range of adhesives and their application, including:
  - PVA
  - contact adhesives
  - UV hardening adhesive
  - solvent cements such as Tensol or acrylic cement
- how jigs and fixtures are used to aid the manufacture of products.

If you are studying at A-level you should also have developed a knowledge and understanding of the additional specific processes:

- calendering
- cupping and deep drawing
- drop forging
- investment casting
- welding
- temporary joining methods and fasteners.

Paper and board forming processes

Papers and boards are important materials in product design and manufacture. In an effort to reduce the effect on the environment, more and more products are being made from paper and board, replacing less sustainable materials such as polymers. You need to be aware of how paper and board is formed into 3D products using die cutting and creasing, laser cutting and bending.

Die cutting and creasing

Die cutters are used to cut out paper or card 'nets' or 'developments' which, when folded and assembled, will form a 3D shape such as a box. Die cutters use a plywood 'substrate' board into which steel cutting dies and creasing rules are inserted. The stock material is placed under the board, and a press pushes the board into the stock material, which then cuts out the net.

Where parts of the net are to be folded, the creasing rules score the paper or card. The creasing rules are not as high as the cutting die, and they are also blunt so they do not cut the stock material.

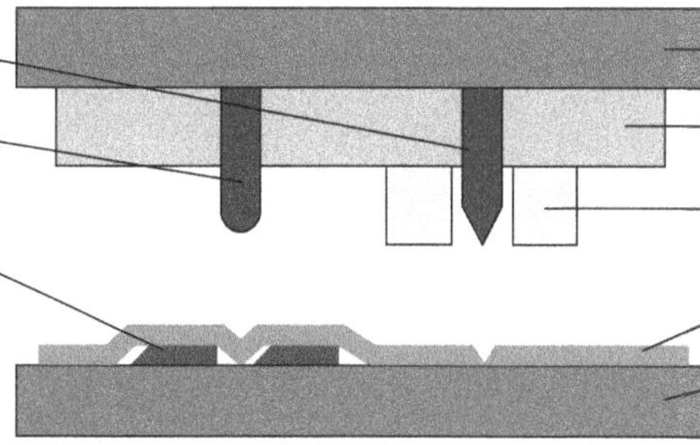

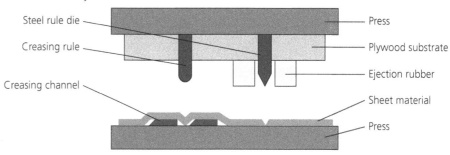

Figure 1.4.1 Die cutting machine.

The diagram above shows the different parts of a die cutting machine. Creasing channels are a raised part on the bed of the machine. These, together with the creasing rule, will make a line where the material can be bent. The cutting dies are sharp and will cut out the outline of the net. Rubber pads fixed to the press help to prevent the paper or card sticking as the press moves up or down.

ACTIVITY

Explain the die cutting process using diagrams, and explain how the process is used to cut and crease paper or card.

Bending

The stamped-out net is placed onto a folding table. Some die cutting presses might be equipped with moving parts that help to bend or fold parts of a net automatically. For example, when making boxes, the die cut net is held firmly while the sides of the box are bent up along the crease lines. Some of the trickier finishing might be done by hand.

In industry, larger machines might carry out die cutting, creasing and folding in one process.

Laser cutting

Figure 1.4.2 A laser cut box.

Laser cutters are perfect for prototype construction or small-scale production because they can be used to cut, engrave, perforate and carve. The use of lasers in packaging production is becoming quite common for prototypes and small batches because of their speed, accuracy and the high level of detail that is possible. Lasers are also very flexible, because they are not limited to making a single net shape or design. 2D drawings can be quickly downloaded to a laser to be cut, which is much quicker and more flexible than making a plywood die board.

Laser cutting can also be used to process paper, cardboard, corrugated card and polymer film.

Polymer processes

Polymers are used in a very wide range of products and are one of the most important groups of materials for you to research. Along with a wide range of polymers, there are a number of processes that are used to form polymers into 3D products. You need to be able to select the correct processes for products, describe the processes using notes and diagrams, and justify your selection.

Vacuum forming

Vacuum forming is used to make 3D products such as lightweight trays, box inserts and liners, coffee cups and yoghurt pots. Vacuum forming machines are often used in schools and colleges for a range of projects because the moulds are relatively easy to make. The technique is generally used for making larger numbers of items due to the time required to make the moulds which is not cast effective for one-off manufacture.

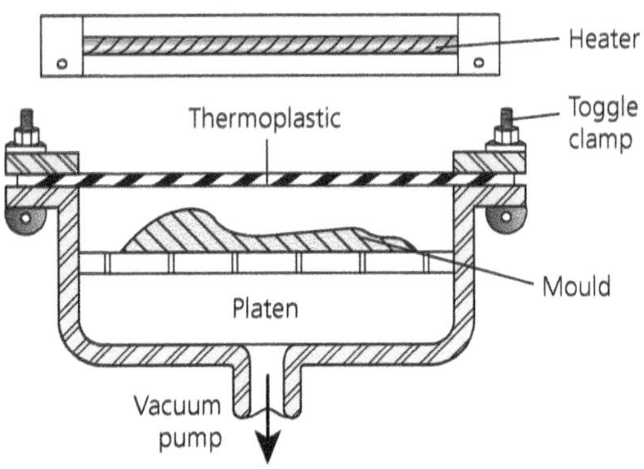

Figure 1.4.3 Vacuum forming.

The process is used with thermoplastic sheets such as HIPS. A mould in the shape of the product has to be made – typically from MDF. This has to have a small draft angle to allow the mould to be removed from the polymer sheet when formed into the moulding.

Because vacuum forming is quite a slow process when compared to other methods, it is normally limited to small-scale batch production. However, if a larger scale of production is required, moulds or formers can be made from cast and machined aluminium. Such moulds give consistent results with a high-quality surface finish. Vacuum forming is also limited to shaping polymer sheets of up to about 6 mm in thickness.

The vacuum forming process is described below.

1. The mould is placed on the bed of the machine, which is called the 'platen'. The platen is lowered to the bottom of the machine.
2. Polymer sheet is clamped over the mould and a heater is pulled over the polymer sheet.
3. When the polymer sheet has softened, the platen is raised into the polymer and the heat is removed.
4. The vacuum pump is switched on, which sucks the polymer onto the mould.
5. Once the polymer has cooled and returned to a solidified state, the platen is lowered and vacuum switched off.
6. The mould is removed from the moulding. Excess polymer is then trimmed off.

Thermoforming

Typically, thermoforming is used to mould additional detail into the surface of polymer or to mould thick polymer sheet such as 6 mm acrylic. A bath is a typical product that might be thermoformed. Alternatively, food packaging such as polymer fruit punnets, cake boxes and sandwich boxes can be thermoformed, especially where logos or product names might need to be moulded into the polymer.

Thermoforming is typically only used in industrial production. As with vacuum forming, it is quite a slow process, but fine detail such as embossed logos and lettering can be achieved.

Thermoforming is a very similar process to vacuum forming, except there is an additional mould that is pressed onto the surface of the polymer sheet at the same time as the vacuum is applied, sucking the polymer down on to the mould below. The two moulds trap the softened polymer in between them, giving extra detail to the moulding.

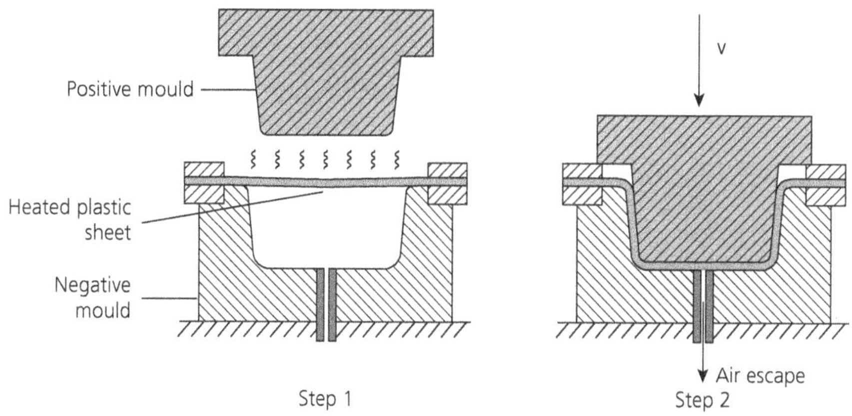

Figure 1.4.4 Thermoforming.

Calendering

Calendering is a smoothing and rolling process used towards the end of manufacturing paper. It is also used in making thin polymer sheet and film as stock material for further processing into products later in the production cycle. Calendering works by heating pellets of thermoplastic so that it melts into a dough-like consistency. It is then extruded between a series of heated rollers so that it becomes squashed and stretched to make it thinner. It is then finished on cooling rollers, before being chopped into standard stock sheet sizes, or in the case of polymer film, it is rolled up for storage and distribution. Calendering is purely an industrial production method, carried out by specialist manufacturers. Typically, this process is used for continuous production.

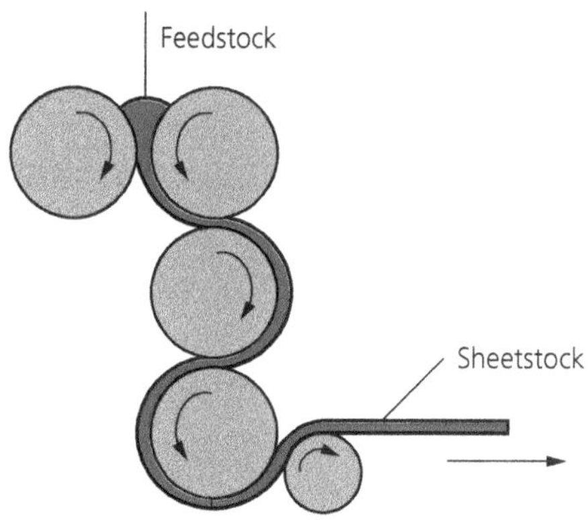

Figure 1.4.5 A typical roll configuration in calendering.

Line bending

Line bending is the process used to produce bends in sheet thermoplastic such as acrylic. The line bending process uses an electrically heated element that provides heat along a line. Typically, line bending might be used to make products such as acrylic boxes, shelves or point of sale displays.

This process is suitable for one-off or limited batch production because it is quite a slow and labour intensive process. School or college workshops might be equipped with a strip heater for line bending because it is ideal for making polymer prototypes.

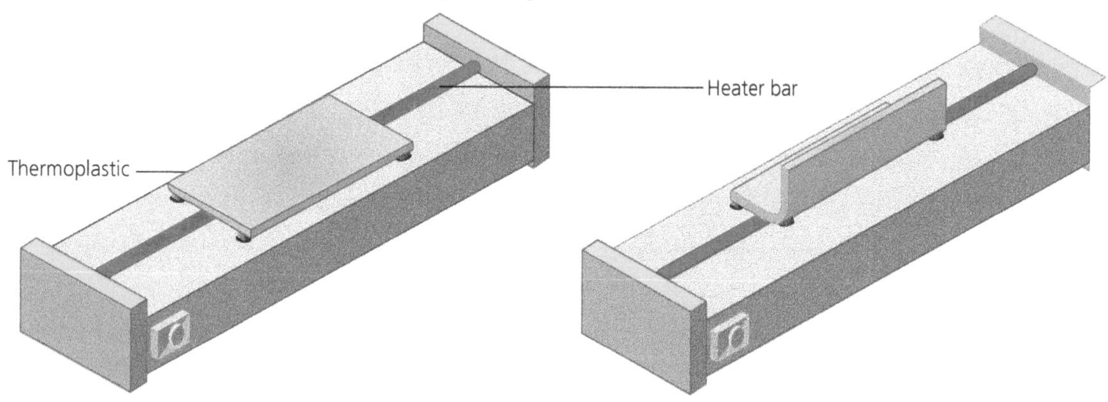

Figure 1.4.6 Line bending process.

Lamination (lay-up)

The lay-up lamination process is used to make 3D products from fibre-based composites such as GRP or CFRP. Examples of products that are made with the lay-up process include: boats and yachts, kayaks, trains, scenery or props for theatre and films, as well as cars for theme park rides.

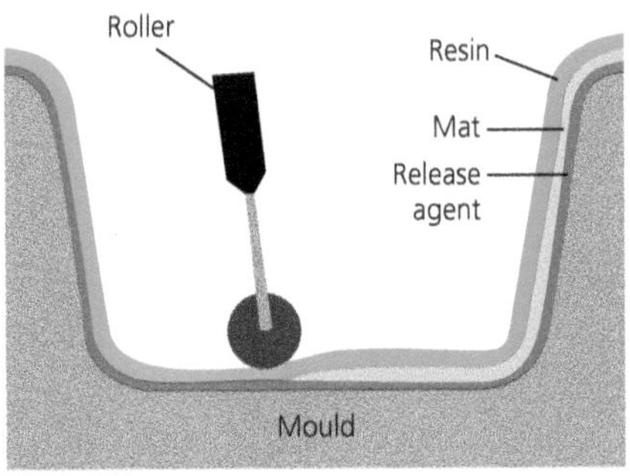

Figure 1.4.7 The lay-up lamination process.

The lay-up process for moulding GRP is described below.

1 A mould or former in the shape of the product or component is prepared. This might be made from timber, manufactured boards such as plywood or high density foam.
2 The mould or former is coated with a release agent such as wax or PVA, or is covered with parcel tape.
3 A top layer of gel coat is applied. The gel coat is a coating of polyester resin often mixed with a pigment to give a colour to the moulding. The resin might also have additives included to prevent degradation from UV and hydrolysis. It using CFRP then curing takes place in a specialist oven; know as an autoclave.
4 Fibreglass matting is cut to size and laid over the former. Polyester resin is brushed onto the matting and a small roller is used to push out any air bubbles that may form and to smooth out the matting (this process is called stippling). The matting is available in a range of stock forms including chopped strand, woven and 'tissue', which is a fine grade matting for covering the surface.
5 Step 3 is repeated until the desired thickness is achieved, and fine tissue matting is used on the top layers. A vacuum bag may be used to compress the layers of GRP before it sets.
6 The GRP is allowed to set.

Injection moulding

Injection moulding is an industrial process used to manufacture complex items made from thermoplastics. For example, the casing on electrical products such as computers, TVs and vacuum cleaners would be injection moulded. Such products have very complex parts that may include integral clip fastenings, screw posts, battery housing and circuit board holders. Such complex parts can only be made by melting the polymer and forcing it into a mould using high pressure. This process is usually used for large-scale mass production or continuous production, due to the high cost of the equipment and the moulds.

The injection moulding process is described below.

1 Thermoplastic granules are poured into the hopper.
2 A screw thread is rotated by a motor. This pulls the granules through the chamber and past electric heaters.
3 The heaters melt the polymer.
4 When a sufficient charge of polymer has melted and formed at the end of the screw, a hydraulic ram forces the screw thread forward. This injects the polymer into the mould.
5 The mould is water cooled, which enables the molten polymer to harden quickly.
6 The mould opens and ejector pins push the moulding out.
7 Any excess polymer is trimmed off the moulding. Formers or jigs may be used to maintain the dimensional accuracy of the moulding while it cools and hardens completely.

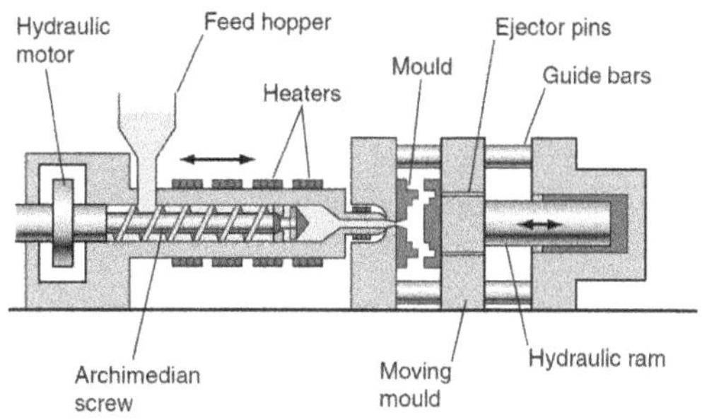

Figure 1.4.8 The injection moulding process.

Blow moulding

Blow moulding is the common process used to produce bottles and a wide range of other hollow products. The most common polymers are PET, LPDE, HDPE and PP. Blow moulding is normally used for continuous production due to the high setup costs associated with the equipment and moulds.

The extrusion blow moulding process is described below.

1 The polymer is fed into the hopper.
2 An Archimedean screw pulls the polymer through a heated section, melting the polymer.
3 The melted polymer is extruded as a tube, which is called a 'parison'.
4 The mould sides close around the parison and air is injected into the mould, forcing the polymer to the sides.
5 The polymer is allowed to cool for a few seconds, the mould opens and the finished bottle is ejected.

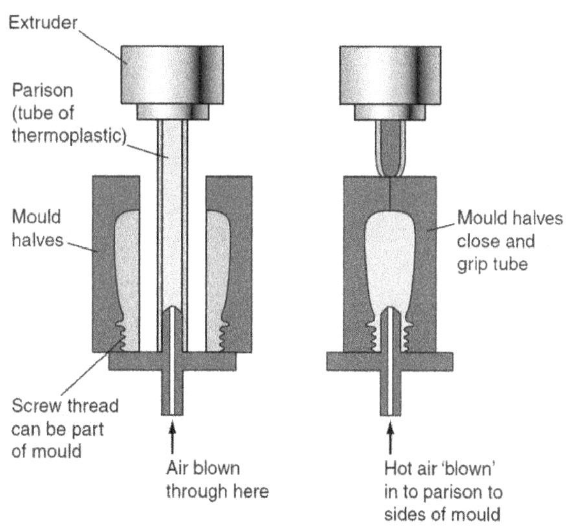

Figure 1.4.9 The blow moulding process.

Many products are now produced by injection blow moulding using an injection moulded preform. This is especially used for products that require a greater level of tolerance (for example, for threads).

Rotational moulding

Rotational moulding is used to produce heavy duty, seamless hollow objects that need a large wall thickness; for example, traffic cones, kayaks, water tanks and children's play equipment. Blow moulding is not suitable for these products because the sides or walls of the product would be stretched too thin in the moulding process. Typical polymers that are used in rotational moulding include HDPE and PP. The moulds can be changed relatively quickly so that different products can be made on the same machine, however the setup costs are high so it is normally used for large-scale batch or mass production.

The rotational moulding process is described below.

1. Polymer powder or granules are loaded into a mould, which is clamped and sealed.
2. The mould is transferred to an oven where it is heated to 260 – 370 °C, depending upon the polymer used. The mould is rotated slowly (fewer than 20 rotations per minute) around two axes, and as the polymer is heated, it coats the inside of the mould.
3. Once the polymer has achieved the correct thickness, the mould is cooled. Usually a fan and/or water is used to cool the polymer.
4. When the polymer has solidified, the part will shrink slightly, allowing it to be removed.

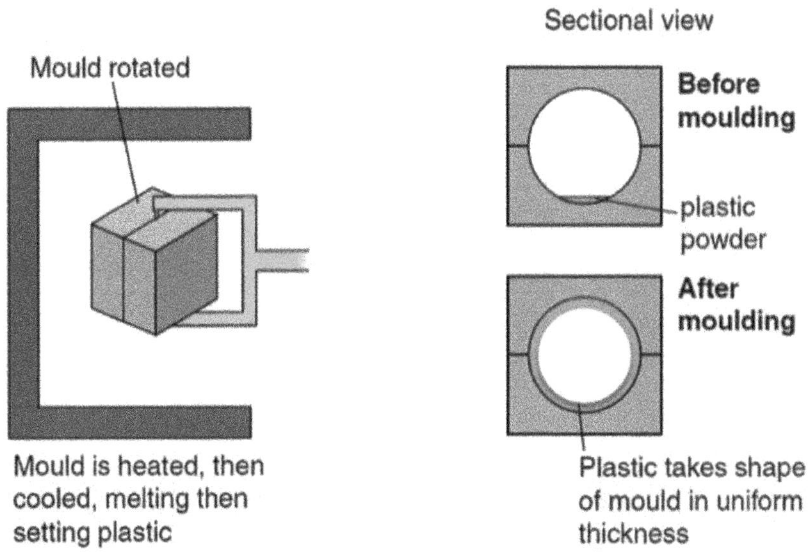

Figure 1.4.10 The rotational moulding process.

Extrusion

Extrusion is used to produce either solid rod, hollow tube, angle, sections such as 'I' and 'T' sections or channels in polymers. It can also be used to coat electrical wire with a polymer such as PVC for insulation. The extrusion process is similar to injection moulding, except that the polymer is forced through a die to form the shape of the extrusion. Again, this process is used for continuous production of stock material, and is therefore only used by specialist manufacturers.

The extrusion process is described below.

1. Polymer granules are loaded into the hopper.
2. The Archimedean screw moves the polymer granules past heaters.
3. The heaters soften the polymer.

4 When sufficient polymer has melted, the hydraulic ram pushes the Archimedean screw, forcing the polymer through a steel die. The shape of the die determines the shape of the extrusion.
5 The extrusion may be supported by rollers as it leaves the die and is cooled by water or air.
6 The extrusion is then cut to the desired length.

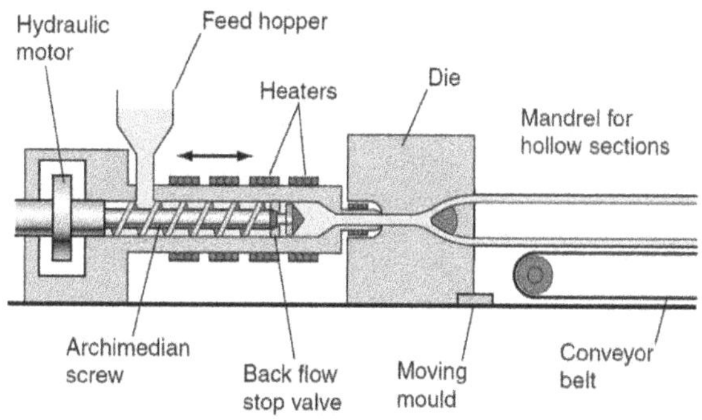

Figure 1.4.11 Extrusion.

Compression moulding

Compression moulding is a common process used to mould thermoset polymers such as urea formaldehyde and melamine formaldehyde. Typical products that are made from urea formaldehyde using this process include electrical light fittings, switches and electrical plugs. MF products might include polymer plates, mugs and bowls for picnic sets or a children's table set.

The compression moulding process is described below.

1 A 'slug' of pre-weighed thermoset polymer is inserted into pre-heated moulds.
2 The moulds are closed and hydraulic pressure is applied. The pressure ensures that the polymer takes the shape of the mould.
3 The moulds remain closed while cross-linking takes place and the thermoset 'cures'.
4 When the moulding has cured, the machine opens and the product is removed.
5 Excess polymer known as 'flash' is removed.

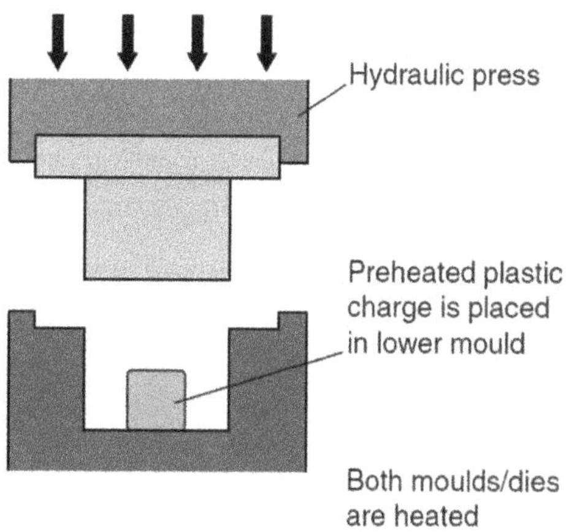

Figure 1.4.12 Compression moulding.

ACTIVITY

For each of the products shown, name the manufacturing process used to make it and explain why this process is suitable.

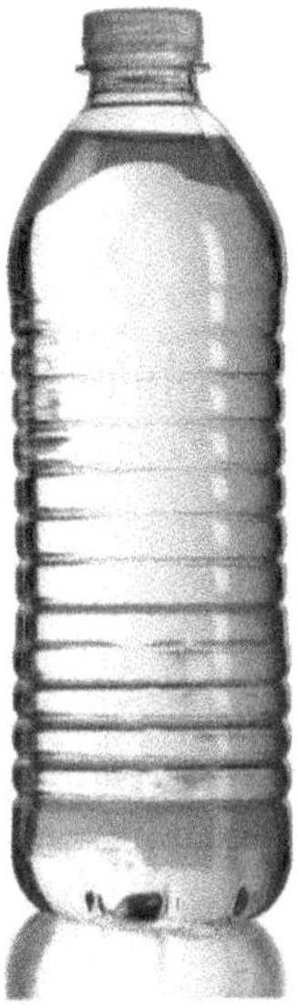

Figure 1.4.13 Water bottle.

Figure 1.4.14 Plug casing.

Figure 1.4.15 Guttering.

Figure 1.4.16 Games console.

Figure 1.4.17 Carrier bag.

Figure 1.4.18 Speed boat.

Figure 1.4.19 Storage barrel.

Metal processes

You need to be able to correctly match manufacturing processes for a range of products made from metals. You also need to explain why each of the manufacturing processes is suitable for the products given. Finally, you need to be able to describe the processes, using step-by-step notes and diagrams.

Metal processes can be grouped into **forming** processes (processes in which no material is removed, but that deform materials to produce required shapes), **redistribution** processes and **wastage** processes (processes that cut away material to leave the desired shape).

Forming processes

Press forming

Press forming is used to shape sheet metal into 3D forms, for example metal seats, car body panels, boxes and containers. The press forming process is often combined with punching (a wastage process) to remove parts of the sheet or to trim excess metal off the sheet. Press formers can be used in conjunction with robots; the sheet metal is lifted and placed into the machine, and finished pressings are transferred to the next process.

Metals such as medium carbon steel and aluminium are typically used in press forming because of their malleability and ductility.

Press forming is only usually used in mass production or large-scale batch production due to the cost and complexity of making the dies. Press forming dies are made from hardened steel, often machined using spark erosion techniques (where a material is worn away using sparks) and hand burnished to give a high-quality surface. Making dies is a highly skilled, specialist job, and is therefore costly. Only when manufacturers can recoup the costs by producing and selling many thousands of the same item would such expenditure be justified.

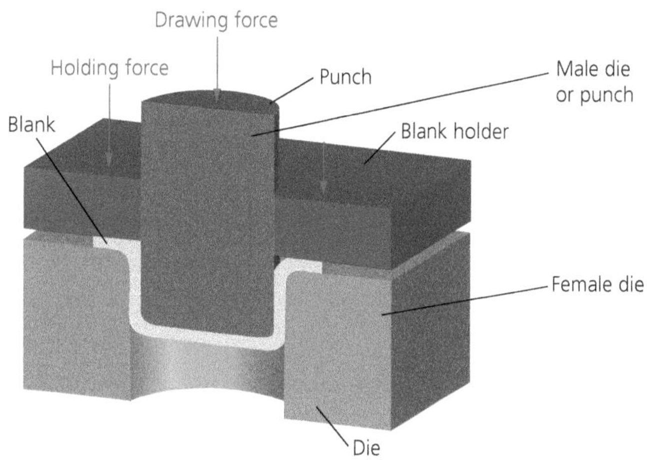

Figure 1.4.20 Press forming.

The press forming process is described below.

1. Sheet metal is clamped over a die of the product that will determine the final shape of the pressing.
2. A hydraulic press pushes the die into the sheet metal. Cutting blades may be included to punch holes into the sheet and trim the excess from the edges.
3. The hydraulic die is lowered and the pressed sheet component is removed.
4. The sheet may be placed into further press forming machines for additional pressing, where the shape is complex.

Spinning

Spinning is an alternative process to press forming for shaping objects such as the body of stainless steel kettles, saucepans or other products with radial symmetry. The process involves spinning the sheet of metal at high speed as it is pushed over a former or mandrel. Spun products can be identified by parallel lines that are formed on the surface of the metal during the process.

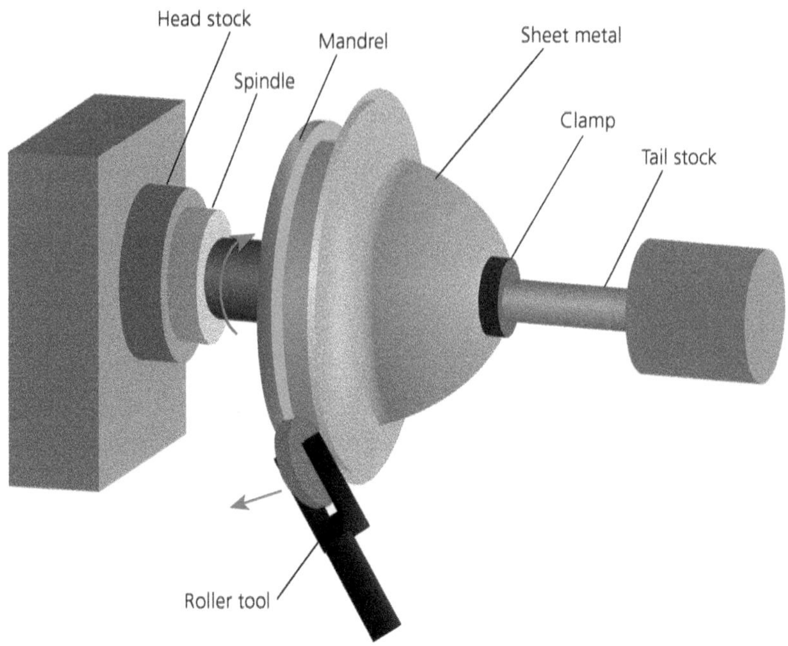

Figure 1.4.21 Spinning.

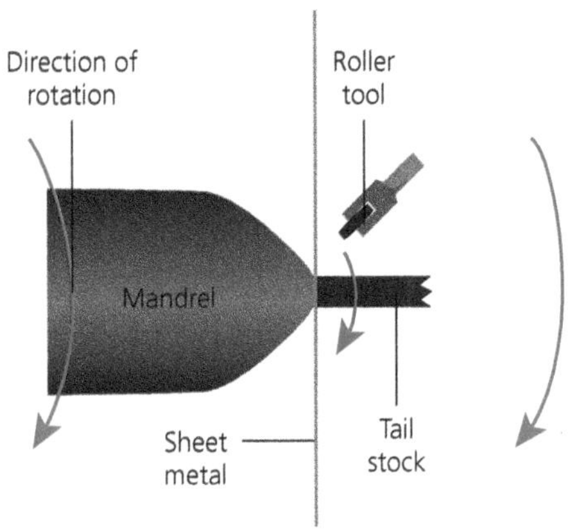

Figure 1.4.22 Stage 1.

Spinning is typically used in mass production, and can be set up using computer controlled machines so that thousands of identical products can be made. However, as the formers are relatively simple, it can be used in batch production, particularly where dome-shaped items need to be made but the quantity required does not justify the costs associated with press forming.

The spinning process is described below.

1. A former called a 'mandrel' is put into the chuck. The sheet metal blank is held in place between the mandrel and the tail stock.
2. The roller tool is moved into the blank and is rotated with the mandrel. This starts to stretch the metal over the mandrel.
3. The roller tool is moved along the mandrel as pressure is maintained against the rotating blank.

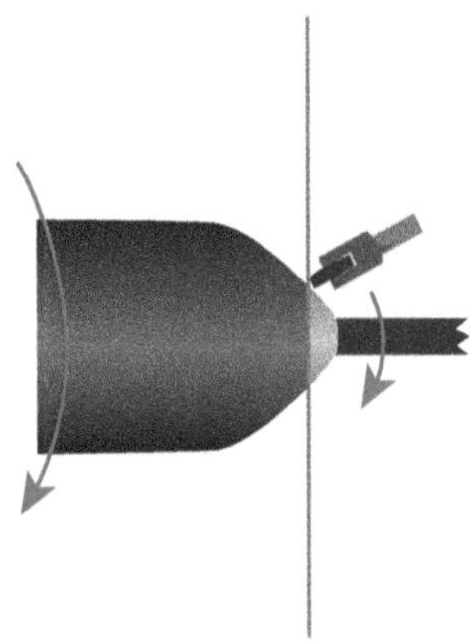

Figure 1.4.23 Stage 2.

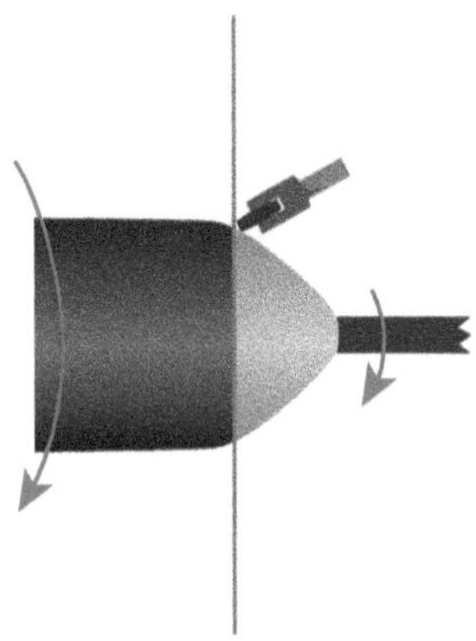

Figure 1.4.24 Stage 3.

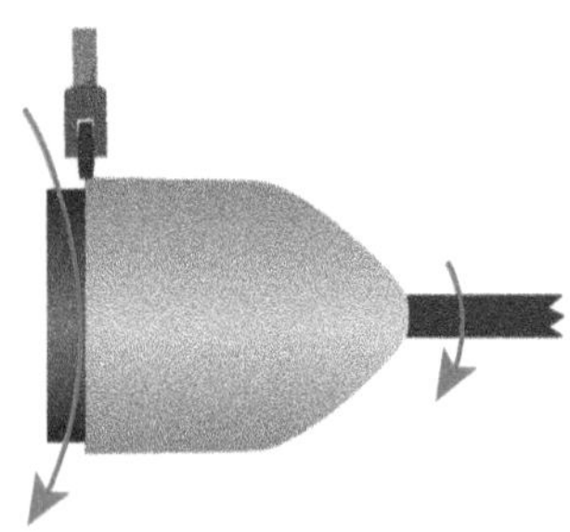

Figure 1.4.25 Stage 4.

4 The roller tool is moved to the end of the mandrel, while still maintaining contact with the blank. This finishes the shape of the product.

5 The finished product is removed from the mandrel.

6 Excess material is trimmed off following the spirring process.

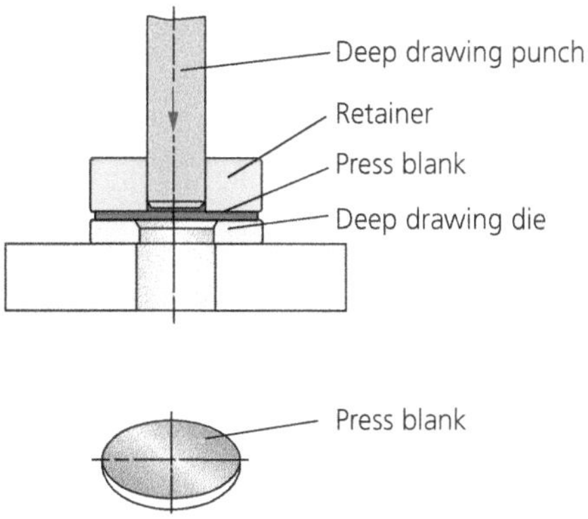

Figure 1.4.26 Stage 1.

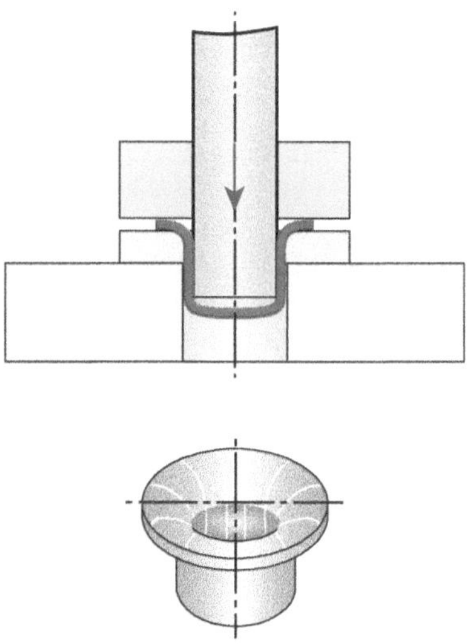

Figure 1.4.27 Stage 2.

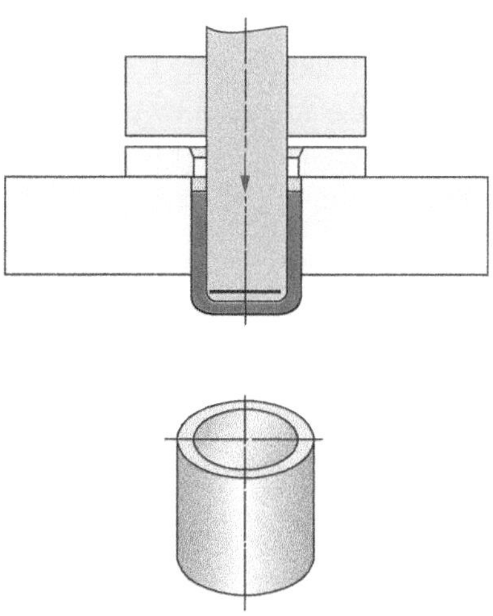

Figure 1.4.28 Stage 3.

Cupping and deep drawing

A level only

The cupping and deep drawing process is used to form tube-like shapes such as fire extinguishers and high pressure aerosol cans. It is considered 'deep drawing' when the depth of the pressing exceeds the diameter.

Cupping and deep drawing is a multi-operation process. The process is similar to piercing and blanking metal with a punch except that in deep drawing, the corners of the punch are rounded. This causes the metal to stretch rather than being sheared.

The high setup costs of the hydraulic presses and dies used in cupping and deep drawing mean that this process is only suitable for mass or continuous production of identical items.

The process is described below.

1 The pressing blank is clamped over a deep drawing die using a pressure pad or clamping ring known as a retainer.

2 A hydraulic press moves the deep drawing punch to be in contact with the blank. It then pushes the blank into the die cavity to make a cup shape.

3 The 'cup' is then pressed further down through the deep drawing die to make the desired tube shape.

Drop forging

Drop forging is used to shape hot metal into finished products. Manufacturers use this process when the finished product needs to be tough (impact resistant) and hard. Typically, products such as spanners, pliers and screwdriver shafts are made using this process. Drop forging allows hot metals to be formed and maintains the internal grain structure and thus the strength required.

Drop forging is used in mass production of identical items due to the fact that the dies are dedicated to making one specific item. However, the dies can be changed quickly to make other items using the same hydraulic ram.

The drop forging process is described below.

1 A die is made from cast tool steel (which resembles a mould) and this is secured to the top of an anvil.
2 A ram is also equipped with a die that resembles a mould.
3 The metal 'billet' to be forged is heated to above its recrystallisation temperature (the temperature below the melting point shape of the metal at which point it is possible to change the size and shape of the grains that make up the metal). This stops the product from work hardening as it cools, which would make it brittle.

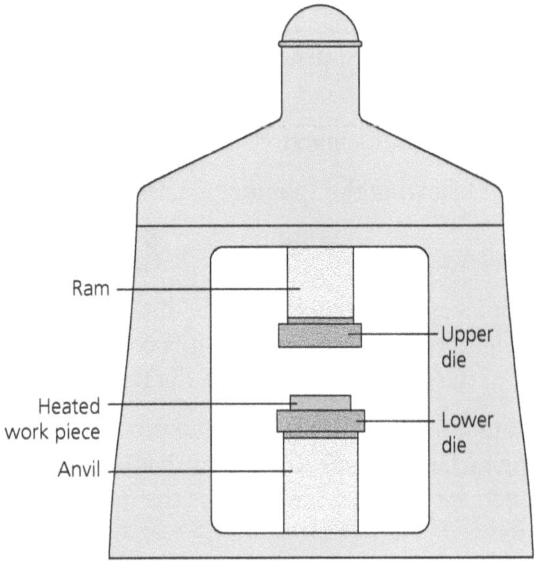

Figure 1.4.29 Drop forging.

4 Using tongs, the heated billet is placed by an operator into the anvil die, and the hydraulic ram is brought down with force. This makes the hot billet spread around the shape of the die.
5 The ram is lifted and the completed product removed for cooling and finishing.

Wrought iron forging

Wrought iron is a form of iron that is suitable for forging, rolling and bending rather than casting. It has a very low carbon content of less than 0.08 per cent, which makes it malleable and suitable for hammering into shape.

In addition to forging using hydraulically or mechanically powered rams and an anvil, wrought iron can be shaped using hand tools.

Typically, the wrought iron is heated in a gas or coke-fired forge. It is then shaped by holding it with tongs, hammering it over an anvil or using other tools such as scroll formers or twisting bars.

Wrought iron forging is suitable for one-off or limited batch production because there is no requirement to make formers or dies to shape the product. The process in its simplest terms uses heat, an anvil and a hammer.

Figure 1.4.30 Heated iron being shaped into a scroll.

Figure 1.4.31. Heated iron being hammered over an anvil.

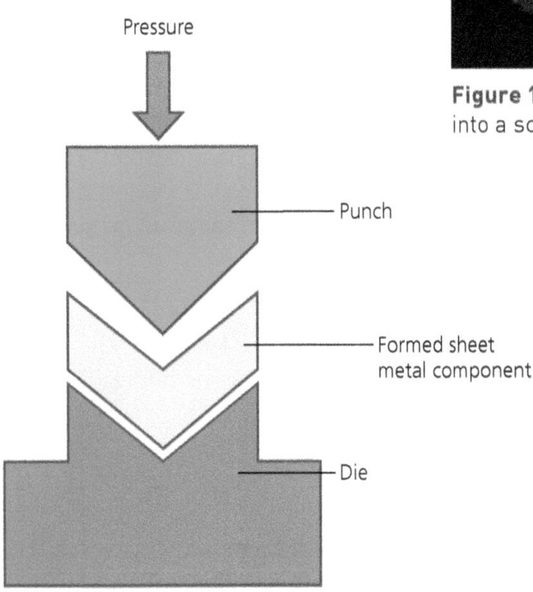

Figure 1.4.32 Bending sheet or plate metal in industry is done using a machine called a press brake.

Bending

Bending sheet or plate metal in industry is done using a machine called a press brake. The desired bends are achieved by clamping the stock metal between a matching punch and die. A hydraulic, pneumatic or mechanical brake holds the metal sheet or plate, and lowers the punch to bend the material to shape.

Modern press brakes use a device known as a back gauge to accurately position the piece of metal, so that the brake bends the metal in the correct place. In addition to this, the back gauge can be controlled by computer so that it can bend the metal repeatedly for bending complex shapes.

The most common dies used in press brakes are 'V' shaped, 90 degree and radius. Press brakes and dies can also be used to make seams between two sheets of metal and hemming sheet to make it safe to handle by the edges. Unlike press forming, bending does not usually combine any punching or trimming.

Bending is a process that can be done as a one-off production method. However, in industry where press brakes are used, it is more typically used in large-scale batch production.

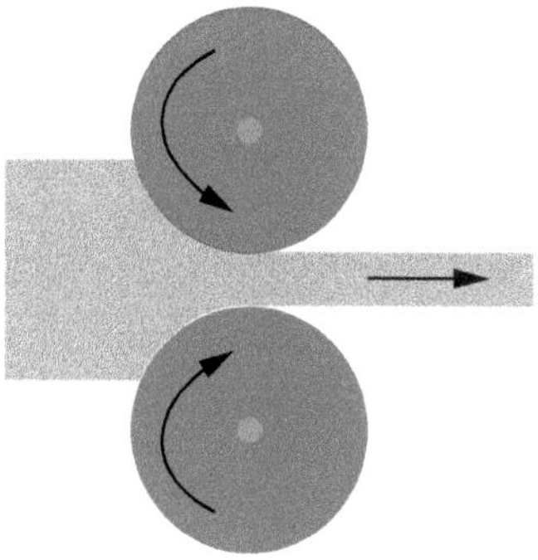
Figure 1.4.33 Metal forming by rolling.

Rolling

Rolling is a metal forming process in which the stock metal material is passed through sets of rollers to reduce the thickness of the material. This process is usually carried out with hot metal that has been heated to above its recrystallisation temperature (known as hot rolling), but can be rolled below its recrystallisation temperature (cold rolling).

Rolling is typically used to make structural steel members such as I beams for the construction industry, and metal stock forms such as angle, bar, plate and channel. Other examples might include rails for railway tracks.

Hot rolling metal results in material with mechanical properties that are uniform throughout the sample. Rolling the metal while it is hot means that it will not have any deformation or stresses which could result in a fault in the material. The disadvantage of hot rolling is that the surface is usually coated with carbon deposits, which have to be removed using acid pickling. Hot rolled materials have a more generous tolerance applied to their dimensions because of the carbon deposits on the surface.

Cold rolling metal (usually at room temperature) results in a material that has a tighter tolerance because carbon deposits are not formed during the rolling process. The surface finish of cold rolled metal is therefore much better. Cold rolled metals are typically used in products such as home appliances, filing cabinets, chairs, steel drums, electrical cabinets, containers and saucepans. However, many schools and colleges might have a set of bench rollers which can be used in making tubes from sheet metal, or hoops from small diameter bars.

Metal redistribution processes

Sand casting

Sand casting is used to form high melting point metals into components and products.

The sand casting process is often carried out in specialist factories called foundries, using moulds that are made from sand. Because the process is quite slow and labour intensive, and the moulds are only single use, it is typically used for one-off and batch production of products. Examples of products made by sand casting include railway carriage wheels, wood working clamps and vices, motor and pump housings as well as bollards, heavy duty park benches, drain covers and post boxes.

Sand casting does not give a very high-quality surface finish because the molten metal will pick up the grainy texture of the sand.

The sand casting process is described on page 60.

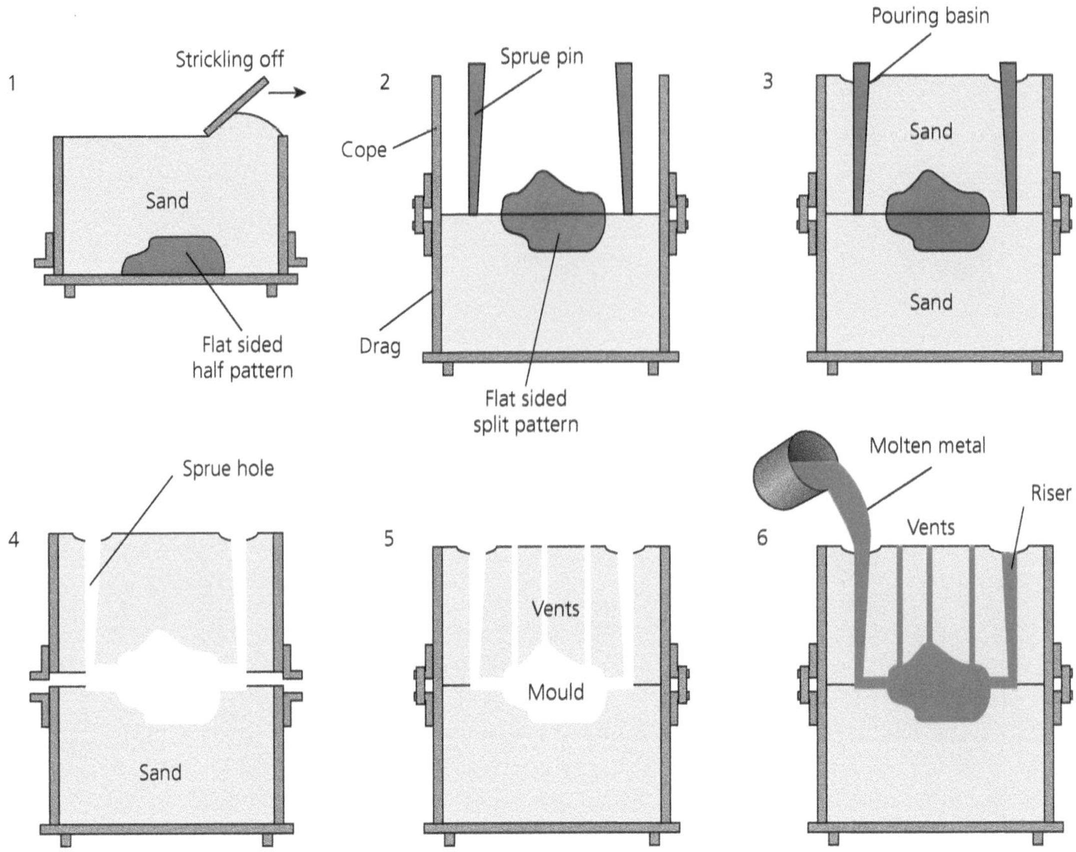

Figure 1.4.34 The sand casting process.

1 A 'pattern' is made, usually from wood. This is a replica (sometimes divided into two halves) of the item that will be cast, and it is placed in the bottom of a steel box called a 'drag'. The drag is then filled with sand which is packed or rammed in tight around the pattern and levelled.
2 The drag is turned over and a second box called the 'cope' is clamped into position over the top of the drag. The top half of the pattern is placed into this to mate with the bottom half of the pattern. Wooden stakes are positioned in the cope. These will form the sprue or runner, and riser later in the process.
3 Sand is packed into the cope around the runner, riser and pattern. A small depression is made on the surface around the sprue to make a pouring basin.
4 The cope and drag are separated, and the stakes and patterns are carefully removed. Connecting channels are cut to join the sprue to the pattern cavity, and from this to the riser. The cope and drag are then re-assembled and the mould is ready for pouring to begin.
5 Small metal spikes may be inserted and removed to make vent holes. These will allow gases from the casting process to escape.
6 The molten metal is poured into the pouring basin. It flows down the runner, into the cavity. When the cavity is full, the molten metal flows up the riser, indicating to the worker that the cavity is full. Once cool, the sand is removed to reveal the casting. The runner, channels and riser are cut off with a hack saw and the casting is ready for machining.

Die casting

Die casting is generally used to mould lower melting point metals such as aluminium, alloys of aluminium and zinc-based alloys. The die casting process uses tool steel moulds that are reusable. This, along with the complexity and cost of making the dies, means it is only used in very large batch and mass production. Die casting produces castings with a very high-quality surface finish. Typical examples of products that are die cast are alloy wheels, engine components, toy cars, collectible figures, door knobs and handles.

There are two main types of die casting: gravity and pressure die casting.

Gravity die casting is the simplest form of die casting because it involves melting the metal and then pouring it into the mould. This process relies on gravity to help the metal to flow through the mould. As in sand casting, there is a runner and riser. The runner is used to pour the molten metal into the mould, while the riser will indicate when the mould is full. Once the metal is cool, the mould is opened and the casting can be removed.

Gravity die casting is generally used to make parts that have a thicker or heavier section than pressure die casting, but thinner sections than sand casting.

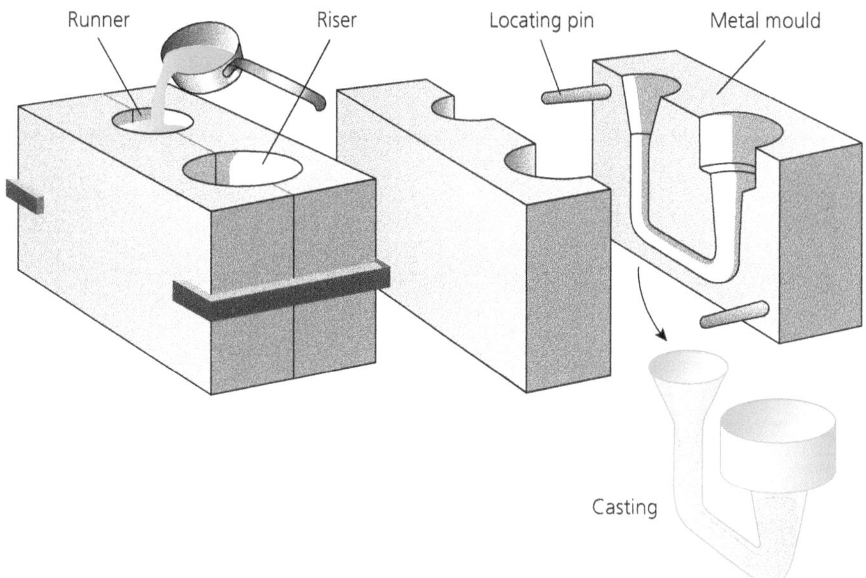

Figure 1.4.35 Gravity die casting.

Pressure die casting is used to produce cast items quickly and in high volumes.

There are two forms of pressure die casting: hot chamber and cold chamber.

With hot chamber pressure die casting, the molten metal is stored in a chamber which is part of the high pressure die casting machine. A pneumatic or hydraulic plunger forces a 'shot' of molten metal through the 'goose neck' into the die. Because this process uses high pressure, all of the mould is filled and it allows fine detail to be moulded. The process is also very fast because the molten metal is not stored separately and then transported to the casting machine. Aluminium is not cast using this process because it picks up iron from the steel chamber. Instead, cold chamber pressure die casting is used.

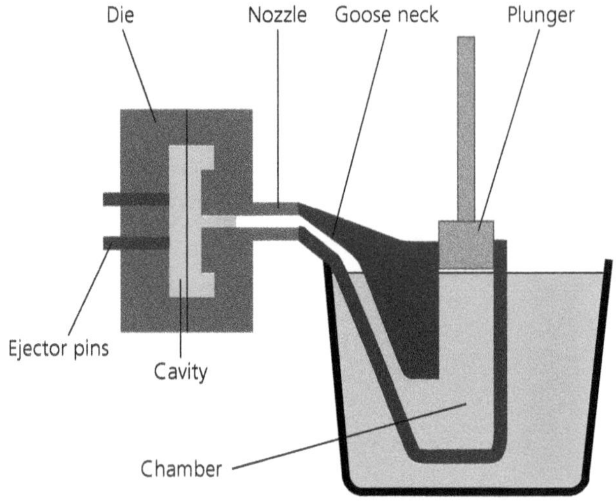

Figure 1.4.36 Hot chamber pressure die casting.

With cold chamber high pressure die casting, the molten metal is kept separately in a melting crucible. The molten metal is then ladled into the shot chamber, and a hydraulic ram forces the molten metal into the mould cavity. When the metal has hardened, the mould opens and ejector pins push the finished casting out.

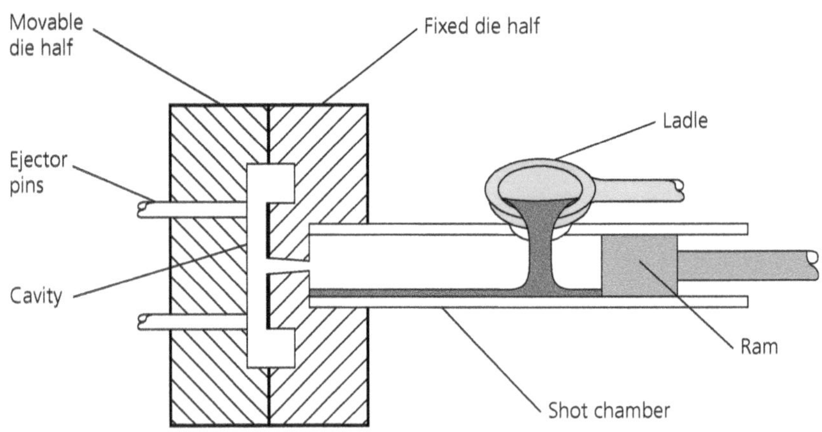

Figure 1.4.37 Cold chamber high pressure die casting.

Investment casting

Investment casting (also known as lost wax casting) is a process used to cast items that are intricate or awkward shapes which would be difficult or impossible to mould using any other casting process. Examples might include cast jewellery or collectable figures, as well as medical applications such a hip replacement joints.

A wide range of metals are suitable for investment casting, and include stainless steel, brass, aluminium and carbon steels. The process allows the manufacture of high-quality products with an excellent finish, and allows for repeatability because the wax patterns are cast from a master mould.

The investment casting process is described below.

1 An exact replica or pattern of the product to be cast is made using wax. (This might be made using a master mould machined in steel or aluminium if the product is to be batch produced.) Where several items are to be cast, further wax patterns might be joined together in a 'tree', including a replica of the runner that will be used to pour the molten metal in.
2 The wax pattern is dip coated with a refractory clay. It is then fired in a kiln to bake the clay hard. The wax is burned away, leaving a hollow clay mould.
3 Molten metal is poured into the clay mould.
4 Once the metal has filled the mould, it is allowed to cool.
5 The clay mould is then broken away, revealing the casting.
6 The runner and any connecting channels are machined off.

Low temperature pewter casting

Pewter is an alloy with a low melting point, which is often used in school or college projects and in commercial crafts. Pewter is ideal for making small-scale items such as jewellery, key fobs and decorative components for larger projects. Due to the fact that it is easy to make the moulds for pewter casting from a range of materials, it is very suitable for one-off production. However, if a higher scale of production is required, moulds can be machined from aluminium or steel, which of course have a higher melting point than pewter.

The pewter casting process is described below.

1 A mould is made from MDF, plywood or high density modelling foam. If made from MDF or plywood, the mould might be laser cut or cut with a fret saw. The mould will include a sprue or runner which will be used to pour the pewter into.

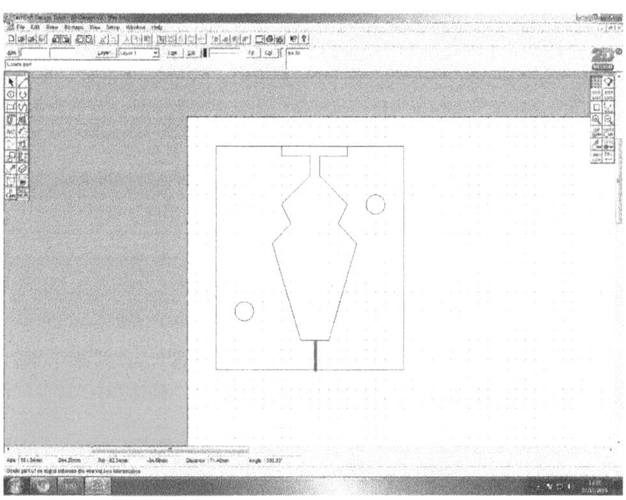

Figure 1.4.38 CAD image to be pewter cast.

Figure 1.4.39 Pewter cast mould in MDF.

Figure 1.4.40 The finished pewter product.

2 The mould is sandwiched between two pieces of MDF and clamped together. The top of the mould will be level with the top of the side pieces.
3 The pewter is melted in a ladle and then ladled into the sprue.

Figure 1.4.41 Heating the pewter.

Figure 1.4.42 Pouring the pewter.

4 Once the casting is cooled, it is removed from the mould.
5 The sprue or riser is removed with a junior hacksaw.
6 The casting is then filed, and cleaned up using abrasive wet and dry paper.
7 The casting would then be polished.

Addition/fabrication processes

Addition is the term used to describe how components and products are made by adding pieces (fabricating them) together using either permanent or temporary joining methods. This next section will describe some of the most common fabrication methods.

MIG, TIG, spot and oxy-acetylene welding

A level only

MIG welding

MIG (metal inert gas) welding is a fabrication process used to weld thin gauge metal (usually medium carbon steel) or aluminium (where aluminium electrode wire is used), particularly tube and box section. MIG welding is suitable for thin gauge metals because the heat generated by the electric arc is localised to a small area, and if done correctly, the process will not burn through the metal or distort the surrounding metal. MIG welding is often the preferred fabrication process for joining tubular steel in products such as climbing frames, bike frames and vehicle exhausts. It is a process that is suitable for one-off fabrication, but it can be used as a process stage in the mass production of a product. For example, in making cars on a production line, MIG welding might be used to join structural parts of the chassis where spot welding is not strong enough.

MIG welding uses an electric arc to create heat which melts the joint area. A wire electrode (made from the same metal as the material being joined) also melts in the arc and fills the gap between the two pieces being joined. The operator swirls the welding gun as they move it over the joint, to form a continuous bead of weld.

The electrode wire is stored on a reel and advances through the welding 'gun' as the trigger is pressed.

MIG welding uses an inert gas such as CO_2 or argon to form a 'flux' shield over the area that is being joined. The gas shield replaces the oxygen at the joint area, which helps to prevent oxidisation that would prevent the weld from forming properly.

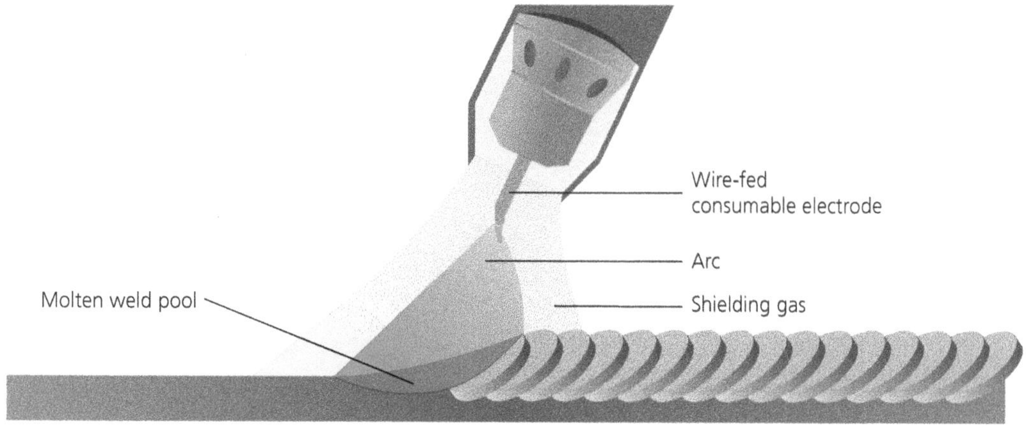

Figure 1.4.43 MIG welding.

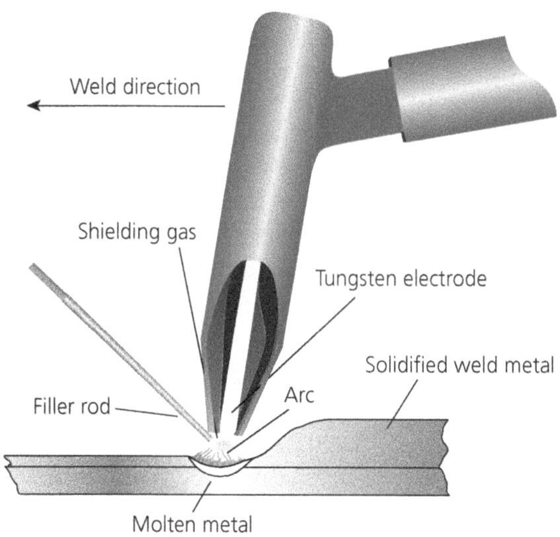

Figure 1.4.44 TIG welding.

TIG welding

TIG (tungsten inert gas) welding is a process used to weld metals such as stainless steel as well as non-ferrous metals such as aluminium and copper or magnesium alloys. It is an electric arc welding process that is similar to MIG welding, except that the electrode made from tungsten does not melt in the process. Instead, a separate filler rod is used. A gas shield of argon or helium is used to protect the weld area from oxidisation.

The TIG process allows for greater control by the operator and more accurate, stronger welds can be created. However, the process requires higher levels of skill and it is quite a slow method.

Typical examples of where TIG welding might be used is in fabricating stainless steel ladders for boats or swimming pools and welding stainless steel car exhausts.

Oxy-acetylene welding

Oxy-acetylene welding is used to weld low carbon steel sheet, tube or plate where arc welding processes are not available. Processes such as MIG welding, or electric arc welding, have made the use of oxy-acetylene almost obsolete, but it is useful for quick repair jobs or in remote locations where there is no electric power supply.

Oxy-acetylene welding uses a mixture of two high pressure gasses – oxygen and acetylene – to form an intense flame that can burn at temperatures of around 3,500 °C. The two gases are stored in separate bottles and mixed in a blow torch. The intensity and temperature of the flame can be adjusted by changing the gas and oxygen mixture through valves on the bottles and torch, to allow for either flame cutting, welding or brazing.

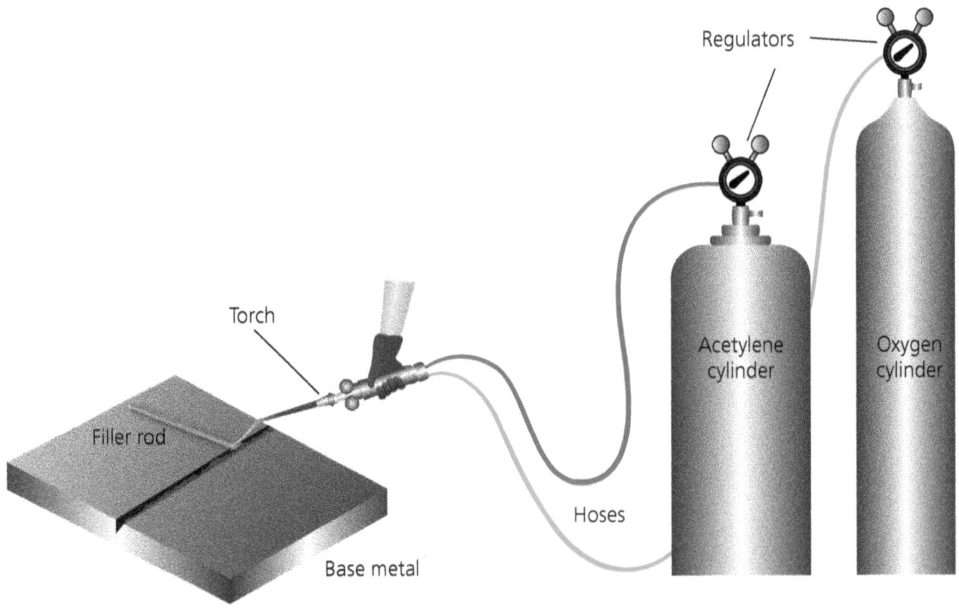

Figure 1.4.45 Oxy-acetylene welding setup.

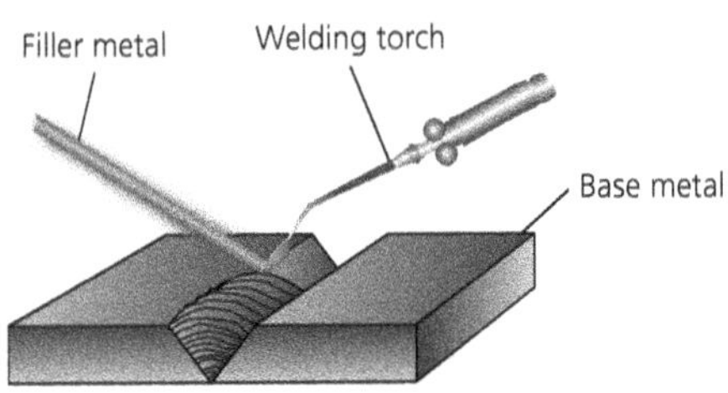

Figure 1.4.46 Oxy-acetylene welding.

The oxy-acetylene welding process is described below.

1 The metal is prepared by grinding an angle on the edges of the two pieces to be joined to form a 'v' shape. This is done to ensure that the weld runs through the entire thickness of the metal.
2 The joint area is heated to form a melt pool and at the same time, a steel filler rod is introduced to the joint area. The melt pool is extended to form a continuous bead along the length of the joint. The molten metal will flow to the hottest part of the metal, therefore by moving the torch along the joint line, a continuous seam is formed.

Brazing

Brazing (also known as hard soldering) can be carried out using either oxy-acetylene or a gas and compressed air brazing hearth. The process uses a lower temperature than welding, so it is suitable for joining thinner gauge low carbon steel tube or bar. The resulting joint is not as strong as welding but it is ideal for general fabrication. Brazing uses a filler rod made from brass which melts at 850 °C. Brazing is ideal for use in school or college projects for fabricating prototypes from low carbon steel bar or tube. An example might be the steel framework for a small item of furniture, a plant stand or ornamental light. Brazing can be used to join dissimilar metals such low carbon steel sheet to aluminium, copper and nickel.

The brazing process is described below.

1 The material to be joined is cleaned and degreased.
2 The two pieces are clamped together.
3 A flux is applied (this helps to prevent the joint from oxidising).
4 The joint is heated using an oxy-acetylene or gas/air torch to a temperature of approximately 850 °C.
5 The brazing rod is applied to the joint area. The brazing 'spelter' will flow along the joint by capillary action to the hottest part, so it can be made to follow the joint line by manipulating the torch.

Figure 1.4.47 A soldering iron being used on a circuit board.

Soldering

Soldering is a similar process to brazing, but it is only used for lightweight applications and thin gauge metals. It is commonly used for joining precious metals such as gold and silver, for example joining the seam on a ring or joining copper pipe in plumbing. In school or college projects, students might use soldering when making prototype jewellery.

Soldering requires the metal to be very clean, and it is important that there are no gaps between the areas being joined.

Soldering uses a filler material of a lower melting point than the metal being joined. Typically, solders are an alloy of tin and copper. Such solders are known as soft solders. If soldering silver, solders containing a high percentage of silver are used. As lower melting point metals are used, a gas/air torch or simple gas powered blow torch is used to create the heat required.

Soldering is also used in joining components to printed circuit boards. Electrical solder is an alloy of tin and lead (60/40) and it contains flux. (Lead solders are gradually being phased out due to their toxicity). The components are pushed through pre-drilled holes in the circuit board. The joint is made by heating the leg of the component with the circular pad where the leg is pushed through. The heat is provided by a soldering iron which melts the solder quickly.

The soldering process is described below.

1. The metal is cleaned and degreased.
2. The joint area is wired up or clamped.
3. The metal is heated up to the melting point of the solder.
4. The solder is added to the metal. The solder will flow along the joint using capillary action.
5. The metal is cleaned to remove any flux residue.

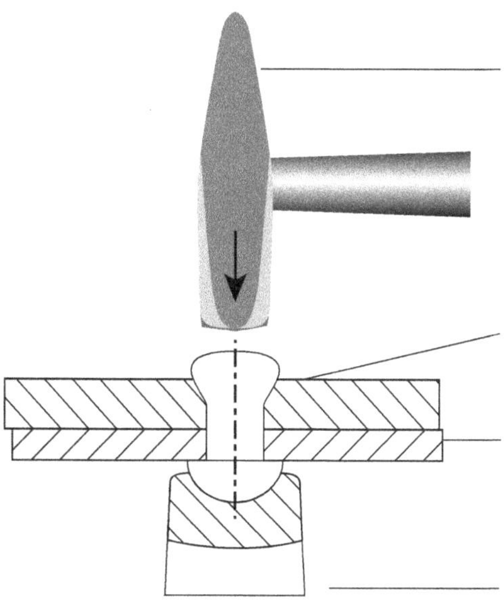

Figure 1.4.48 Riveting.

Riveting

Riveting is a permanent joining method usually used to join sheet metal or plate. Rivets are metal fasteners that have a 'head' at one end, and a shaft or tail at the other end.

In traditional cold riveting, the two pieces to be joined are overlapped and drilled. The rivet shaft is inserted into the hole. The head of the rivet is dome shaped and a 'set' tool is placed over this (also known as a snap). The end of the shaft is then hammered over to squeeze the two pieces together. An example of where you might use traditional riveting is in joining the handle of a trowel onto the blade.

Pop riveting

Pop riveting is a more modern form of riveting, and is used for joining thin sheet metal. The rivet used in pop riveting has a rivet and a pin. The rivet head is pushed through a hole drilled through the two pieces of metal being joined. Riveting pliers grip and pull the pin and, as this happens, the head of the rivet squashes and pulls the two pieces of metal together. The pin breaks off and is disposed of.

Pop riveting is ideal for use where the underside of the joint is inaccessible, for example when making sheet metal ducting for ventilation. A form of pop riveting is also used in aircraft production to join sheet aluminium to structural parts.

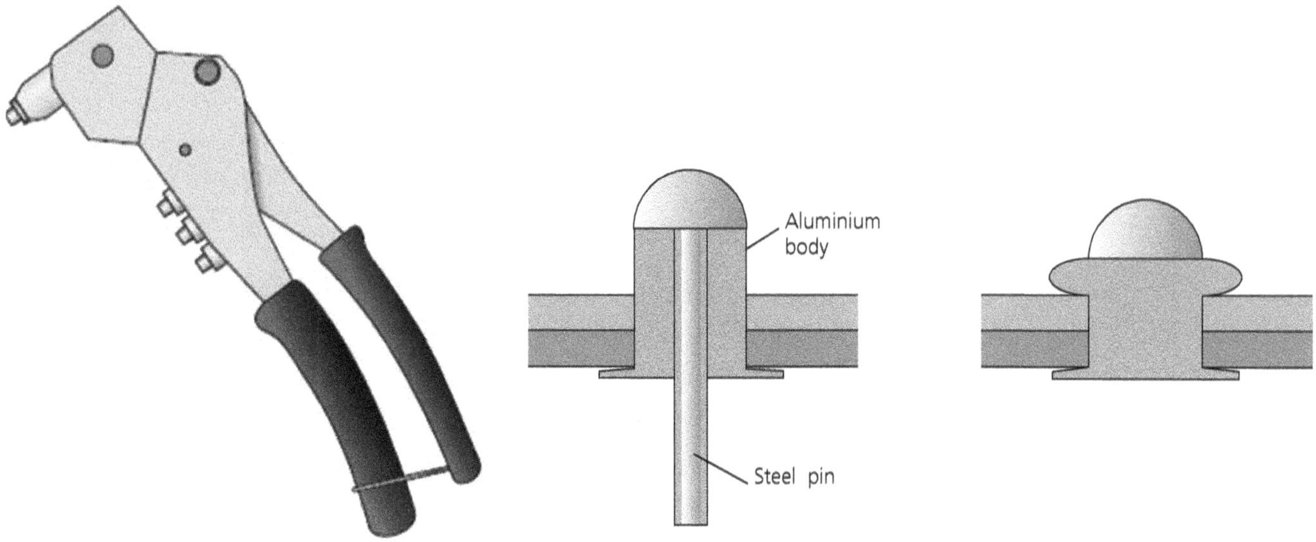

Figure 1.4.49 Pop riveting.

Temporary fasteners and joining methods

Temporary fasteners are used where a permanent joining method is not required, or in situations where either a product needs to be able to be disassembled or parts of the product need to be accessible, for example for maintenance.

There are generally three types of fasteners used for metals: self-tapping screws, machine screws and nut and bolt.

Self-tapping screws

Figure 1.4.50 Self-tapping screw.

Self-tapping screws are used for joining thin sheet metal. A pilot hole is drilled through the metal parts to be joined. The screw has a coarse thread which is made from hardened steel. When it is screwed into the pilot hole, it cuts its own thread, which holds the screw in place. Self-tapping screws are commonly used on products where panels need to be removed for maintenance, for example on computers and in white goods such as washing machines and driers. Self-tapping screws are also used on polymer products such as electronic children's toys where battery compartments must be secured with a screw for safety. Forming a screw thread within an injection moulded component would also vastly increase the cost of the mould.

Machine screws

A level only

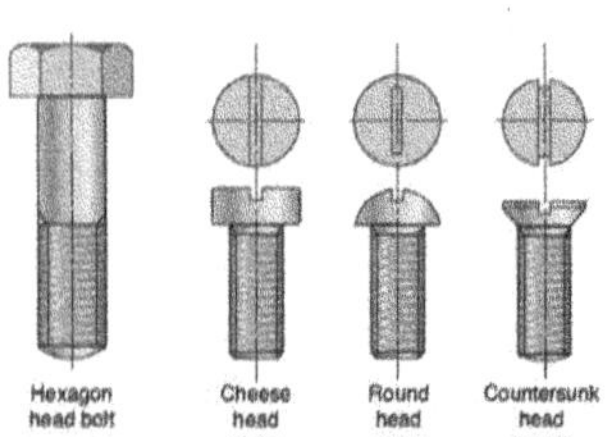

Figure 1.4.51 Machine screws.

Machine screws (usually with a metric thread) are a type of bolt that is used to join thicker pieces of metal together – typically machine parts such as inspection covers on motor or gear housings, or parts that need to be removed for maintenance. They have a thread the entire length of the shaft.

Figure 1.4.52 Nut and bolt.

Figure 1.4.53 Nyloc nuts.

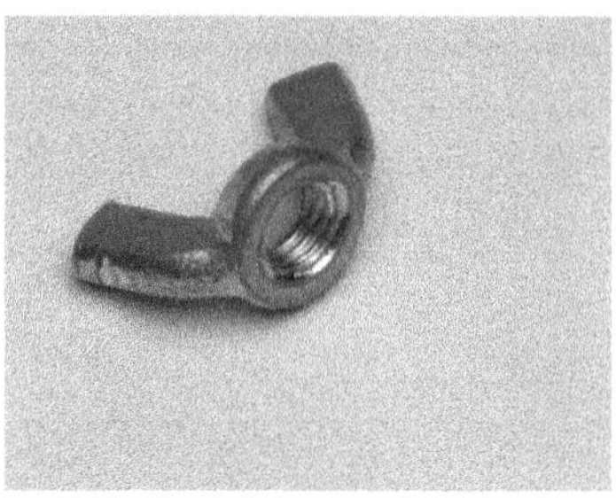
Figure 1.4.54 Wing nut.

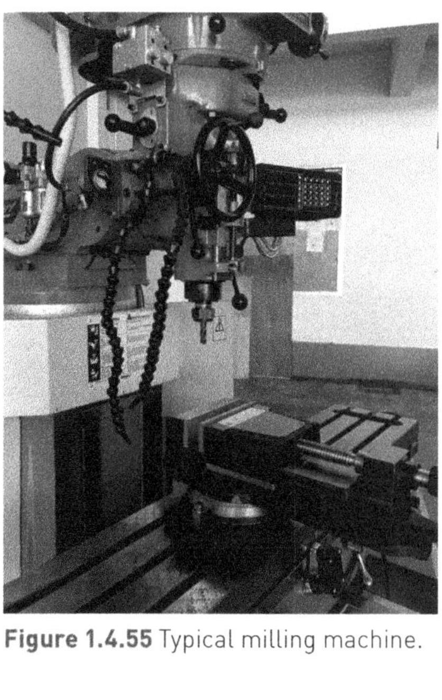
Figure 1.4.55 Typical milling machine.

Where there are two pieces to be joined, the top piece will have a clearance hole larger than the thread on the bolt. The bolt goes through this, and is then screwed into a threaded hole in the second piece.

Machine screws are often tightened with a spanner or allen key.

Nut and bolt

Bolts are similar to machine screws, but instead of being inserted into a threaded hole, they are put all the way through both pieces of metal, and then a nut is tightened on to the end of the bolt to keep the pieces together. Spacers known as 'washers' might be put under the head of the bolt and under the nut, in order to spread the tightening force over the surface of the material and to stop the bolt head and nut from digging into the surface. Spring washers may be used to keep tension on the joint, to prevent it from coming undone if the item may be subject movement or vibration.

Bolts are available in a wide range of types and in different materials such as alloy steels, nylon, brass and aluminium. Nuts are also available in a wide range of materials and there are different types. Nyloc nuts have a nylon collar insert at the end of the nut. As the nut is tightened over the bolt, the nylon elastically deforms over the bolt thread. The nylon insert helps to stop the nut coming undone through vibration, so they are often used on machines or equipment that are subject to movement.

Wasting processes

Wasting processes are making techniques which remove material in order to shape it. Wasting processes cut away parts of the material, using processes such as milling, turning and cutting.

Milling

Milling uses a machine that looks similar to a drill. The work to be machined is clamped onto the table of the machine. The table can run in the x direction (left and right horizontally), y direction (forwards and backwards horizontally) and z direction (up and down vertically); therefore the workpiece can be machined in a range of directions.

Depending upon the type of cutter used, milling can be used to cut slots, to shape the edge or the surface of the work piece or to drill and even thread holes. Milling can be done manually or by using computer controlled machines.

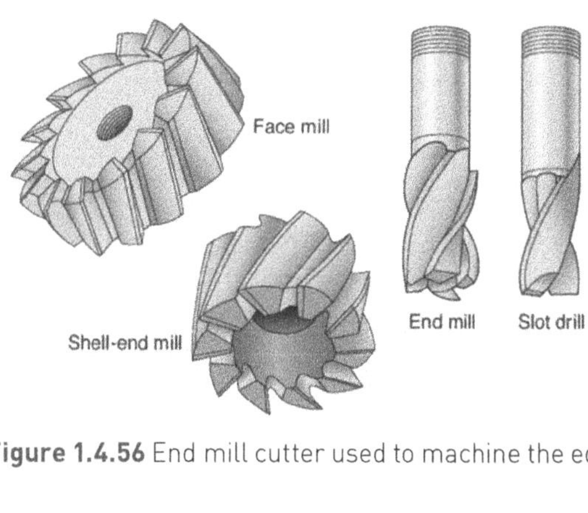

Figure 1.4.56 End mill cutter used to machine the edges of a piece of material.

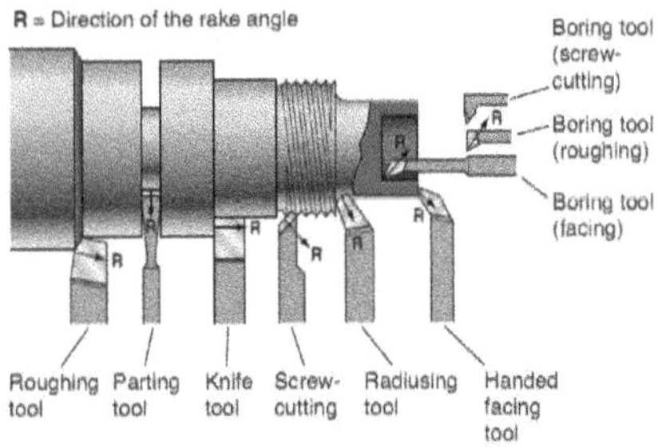

Figure 1.4.57 range of lathe tools for different turning processes.

Turning

Turning is a process carried out on a centre lathe. Work such as a bar can be held in a rotating chuck and machined to reduce the diameter, and to square or 'face off' the end, thread and drill. Where the piece being machined is long, it can be supported in the centre by a 'tail stock' on the lathe. A variety of shaped bars can be turned including round, hexagonal and square bars. Tubes can also be machined on lathes.

Turning can be done manually or using computer numerically controlled (CNC) lathes. The material is machined by cutting tools held in a tool post which can be moved in and out, and along the bar. Most lathes have a rotating tool post which can hold a variety of tools for different machining operations. The cutting tools are usually made from either high-speed steel or tungsten carbide. For many machining operations, a liquid coolant is flooded over the cutting tool and piece being machined, to prevent the tool from blunting and to maintain a good finish on the component being machined.

Flame cutting

A level only

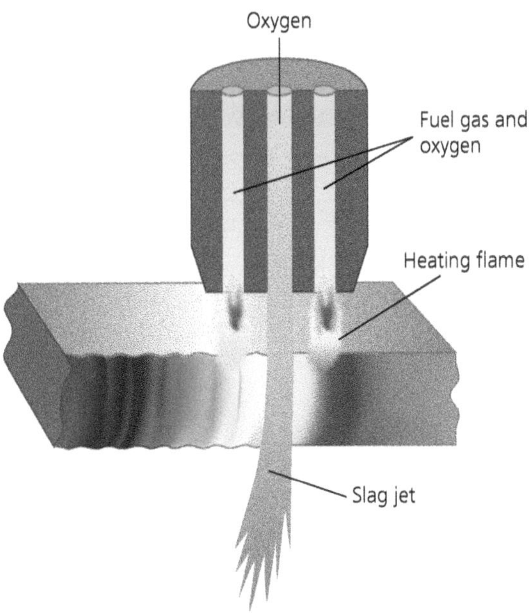

Figure 1.4.58 Flame cutting.

Flame cutting is a wasting process that uses oxy-acetylene gas and a special flame-cutting torch to deliver a very intense and focused flame above 3,500 °C. It is used to cut low carbon and alloy steel plate.

The flame cutting process resembles welding – the metal is heated and a melt pool begins to form. At this point, an additional jet of oxygen is introduced. This intensifies the flame and pierces the metal, forcing a jet of melted metal and carbon (known as slag) with it. The flame is then moved along the cutting path to continue the cut.

Manual flame cutting is an economical process to set up because the equipment is readily available. It can be carried out in almost any location because electricity is not required. Flame cutting can be set up with a CNC system where there is a need for repetitive cutting. The disadvantage of flame cutting is that it is difficult to maintain a parallel line with high levels of tolerance. There may also be deformation, structural changes and tempering on the cut edge.

Plasma cutting

A level only

Plasma is a super-heated ionised gas that is electrically conductive. A plasma cutter will use this conductive gas to transfer the energy from the power supply to a conductive material such as steel plate. The resulting cut is faster and cleaner than using oxy-acetylene.

The plasma arc is directed out of a torch where a gas such as oxygen, nitrogen, argon or compressed air is forced through a tiny nozzle. An electric arc is generated from a transformer and, combined with the gas, forms a jet of plasma. The heat generated by the plasma can be as high as 28,000 °C, which quickly burns through the material and blows it way.

Plasma cutting can be used for one-off production, for example cutting silhouette shapes from heavy steel plate to make a sculptural sign.

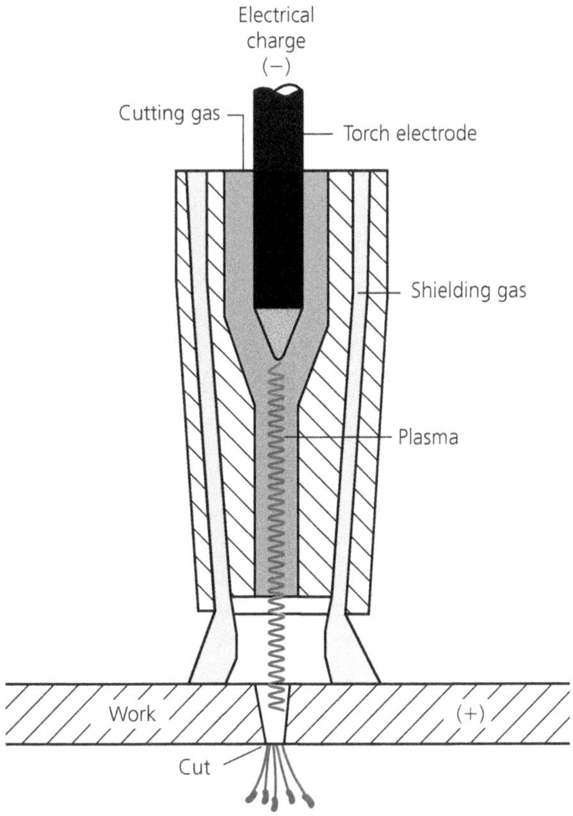

Figure 1.4.59 The typical setup of a plasma cutter.

The typical setup of a plasma cutter system includes:

- a power supply which converts AC mains to DC – this is usually varied between 200 VDC to 400 VDC depending upon the material thickness being cut
- an arc starting console – this provides a spark inside the torch to start the plasma arc
- a plasma torch – this contains an electrode and nozzle which are consumable parts. The torch can be used manually or controlled by CNC systems for accuracy and repeatability.

Laser cutting

A level only

Laser cutting is a wasting process that uses a computer aided design (CAD) file to direct a laser to cut materials. Laser cutting has been used in industry for a number of years for precision cutting metals, but lower-powered lasers are now readily available in schools and colleges, and are commonly used to cut manufactured boards and acrylic sheet.

Laser cutting uses the power from a high-powered laser that is directed through optics. The laser melts the material and a high-pressure gas or compressed air blows the melted material through the sheet.

The laser beam is emitted from a 'laser tube' where it is reflected through a series of mirrors in a similar way to a periscope into a 'laser head'. The head contains a lens which focuses the laser into a fine beam for cutting and engraving. Laser beams usually have a very fine tolerance, and the amount of material removed in the cutting process can be less than 1 mm.

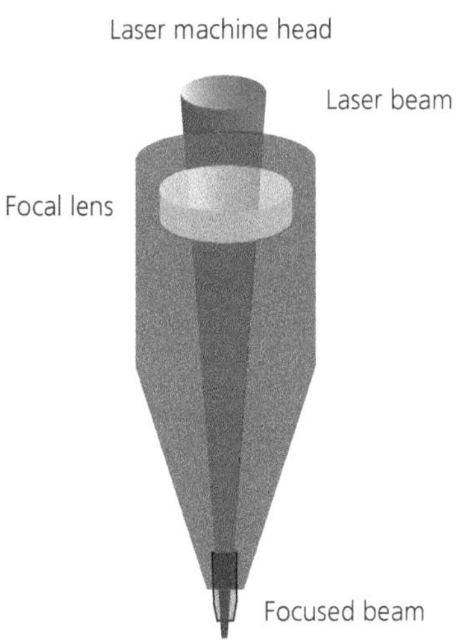

Figure 1.4.60 Laser cutter.

Laser cutting is a fast process, and processing software used in laser cutters is relatively easy to operate. This means that laser cutting can be used for one-off production as well as larger-scale batches. Laser cutting produces a fine cut with a high-quality surface finish. With laser cutting, there is less warping or distortion because the heat zone is much smaller than in other processes such as plasma cutting. Laser cutting is also more accurate and uses less energy than plasma cutting. In industry, laser cutting is typically used to cut flat sheet material, but it is not able to cut to the same thickness as plasma cutting.

There are modern industrial lasers that are approaching the power of plasma cutting, but are currently much more expensive.

Punching/stamping

Punching is a wastage process that uses computer controlled machines which stamp out sections of sheet material using hardened punches following a CNC program. The program moves the table of the machine in the x and y direction (x moves left and right, y moves forwards and backwards) under the punch.

Figure 1.4.61 Punching machine.

Modern CNC punching machines can be programmed using a graphic user interface so that specialist programming skills are not needed. Information is taken from a 2D CAD drawing to select the correct tooling to make the desired part. Software is also used to establish the most efficient layout of parts from a given sheet, known as 'nesting'.

The process uses the shearing action on a sheet of metal placed between an upper tool (punch) and a lower tool (die). The punch pushes through the sheet material, producing a punching slug that drops through a hole in the die. The pieces are collected via a chute for further work or recycling if the punched sheet is the desired part.

Alternatively, the punched parts can be tabbed into the sheet with micro joints. This enables the sheet to be taken out as one piece for later use or subsequent machining.

Punching machines use either a single tool head or a multi-tool turret to punch out shapes from the stock material. The process is suitable for small- and medium-size production runs, and it is normally used for processing metals from 0.5 mm to 6 mm thickness.

Wood processes

Manufacturing processes associated with woods can also be grouped into addition and fabrication processes, forming processes and wasting processes. You need to be able to identify which processes are used with specific products, and recognise how the production method changes with different scales of production.

Addition and fabrication processes

One of the most common ways of making products from wood is to join it together with other pieces of wood. This is known as addition. Products are 'fabricated' by using traditional wood joining techniques or by using knock-down (KD) fittings.

Traditional wood jointing

There are a wide range of different wood joints, and each one has a specific use. Where there are alternatives that have the same function, each will have its own merits. This section will describe the most common wood joints that you should be aware of.

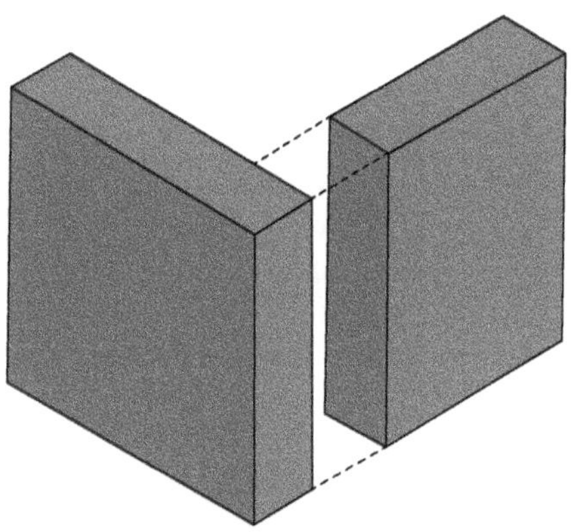

Figure 1.4.62 A butt joint.

Butt joint

A butt joint is the simplest method of joining two pieces of timber together. It relies upon an **adhesive** such as PVA to make the bond. It is only suitable for very lightweight applications such as modelling. A butt joint is very easy to complete – the pieces to be joined are cut square, PVA is applied and then the joint is clamped until dry.

Dowel joint

Dowels are small round pegs made from a hardwood. They come in a range of diameters for timbers of different thicknesses. Some dowels are machined with grooves that allow wood glue to flow up the dowel when it is hammered into place.

A dowel joint is made by measuring and marking the position for the dowels. Holes for the dowels are then drilled. Wood glue is put into the holes, then the dowels are hammered into position and the two pieces are clamped together.

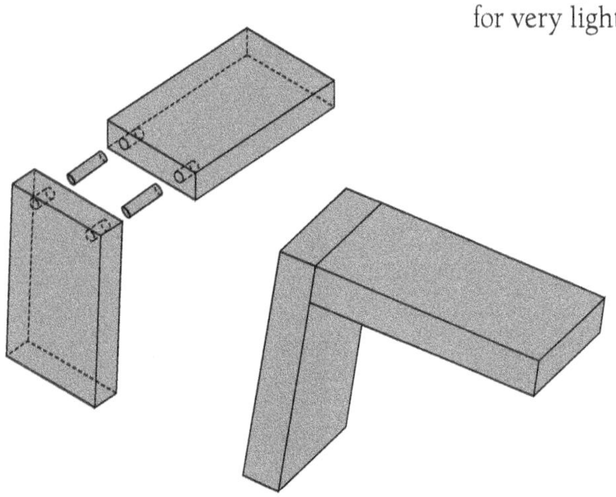

Figure 1.4.63 A dowel joint.

Dowelling is a simple joining method, and stronger than using just a butt joint because the dowels interconnect the two pieces.

Dowel joints are typically used in flat-pack furniture such as a bookcase or wardrobe, where dowels will be used to secure the top and bottom to the sides. They are easy to use and do not require any specialist skills or complex tools. Dowels tend to be used to help hold panels or pieces together, and the joining is usually done by an additional knock down fitting.

Mitre joint

Mitre joints are similar to butt joints in that they rely upon simply gluing and clamping the pieces together. The difference is that the ends of the pieces being joined to make a 90-degree corner are each cut to 45 degrees. They are then glued and clamped, often using a mitre jig to keep the corners straight. Mitre joints are used to make picture frames.

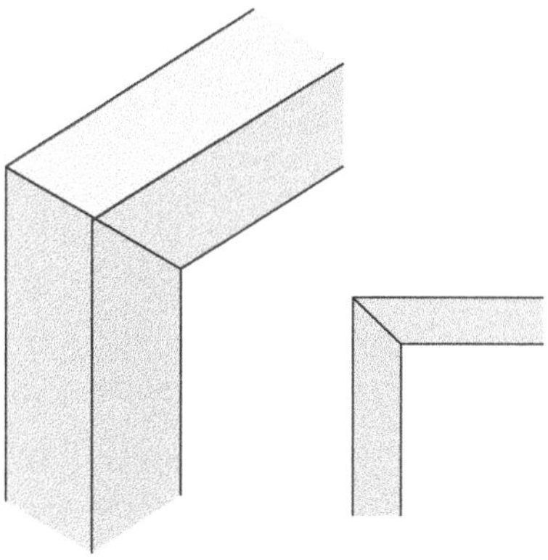

Figure 1.4.64 A mitre joint.

Comb joint

A comb joint is a common joint used to make boxes. It is perfect for this application because the two pieces being joined interlock. The combs also create an increased gluing surface area, which makes a very strong joint. Comb joints can be cut relatively easily using a band saw or laser cut, or with more skill using a tenon saw and wood chisel.

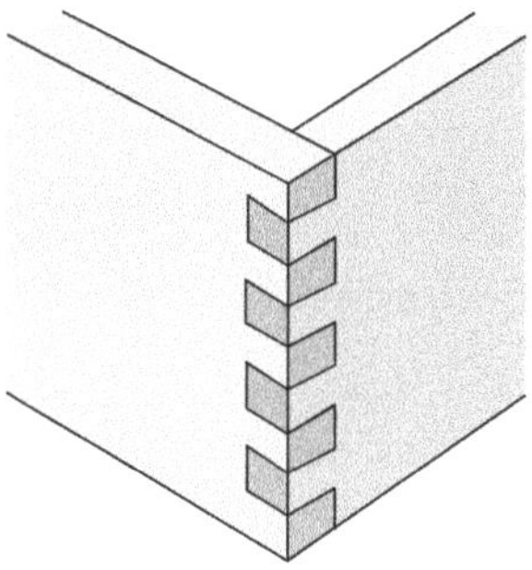

Figure 1.4.65 A comb or finger joint.

Dovetail joint

Dovetail joints are used in making drawers. They are perfect for this application because they have directional strength. Once glued together, they are impossible to pull apart, so they are excellent for joining the front and back of a drawer to the sides. Dovetail joints can be cut by hand using a dovetail saw or, more commonly, machined with a router and jig that requires much less skill.

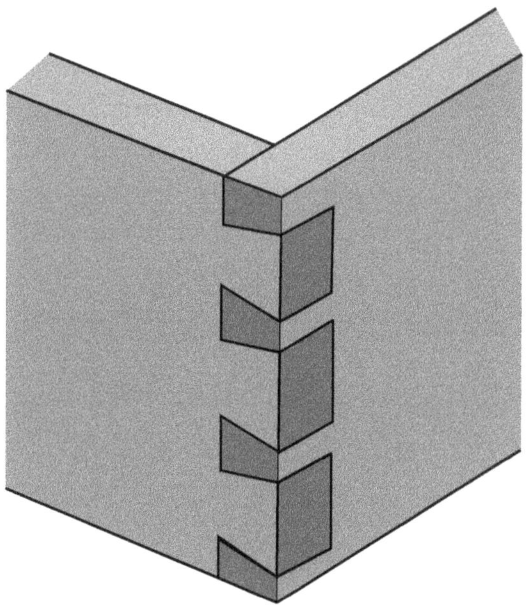

Figure 1.4.66 A dovetail joint.

Mortise and tenon joint

Mortise and tenon joints are used to make heavy-duty frame constructions for furniture or similar applications.

This particular joint involves making a square or rectangular hole – known as a mortise – in one of the parts to be joined. This might be done with a mortise machine or router, or by simply drilling and cutting using a chisel. The second piece of timber is then cut using a tenon saw or band saw to make the tenon. PVA glue is spread into the mortise and the two parts are clamped together.

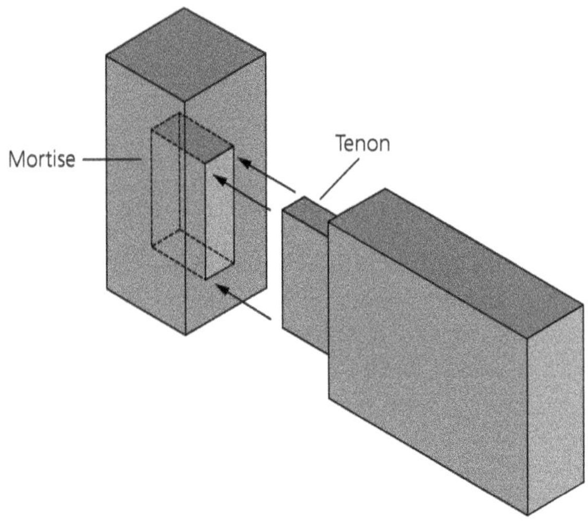

Figure 1.4.67 A mortise and tenon joint.

Housing joint

Housing joints are useful for making framework construction, cabinets and shelving. A groove is cut across one piece, and the end of the second piece is inserted into it. This could be glued in position to make the joint permanent, or left unglued so that it can be taken apart or adjusted. This type of joint is structurally strong, and can be made using a tenon saw and chisel or by using a router.

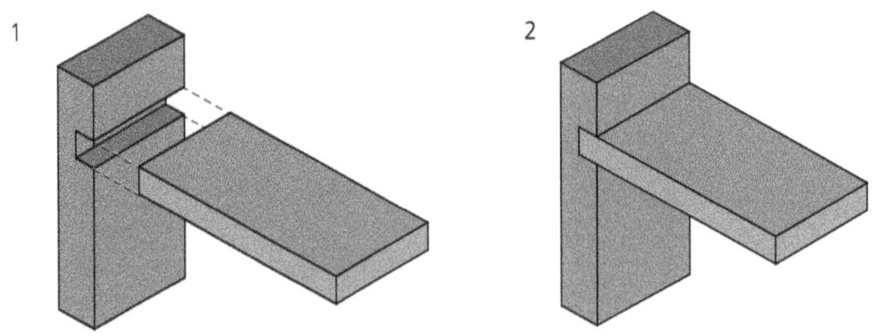

Figure 1.4.68 Housing joints.

Half lap joint

A half lap joint is used to make simple frames or boxes. It is made by cutting a 'step' in the end of each piece. The step is simple to make but it has a larger gluing surface area than a butt joint, making it stronger. Half lap joints are easy to mark out and cut with a tenon saw or band saw.

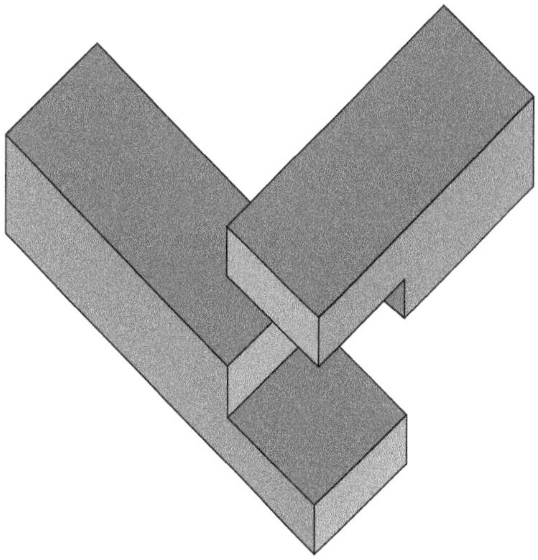

Figure 1.4.69 A half lap joint.

Knock-down fittings

Knock-down (KD) fittings are used to manufacture flat-pack furniture. They enable manufacturers to supply products that are not assembled, therefore reducing the cost for both the manufacturer and customer. This reduces the making time, and makes it much easier to store the products and transport them. For the customer, KD fittings are relatively easy to use with a limited number of simple tools (sometimes supplied with the product). Flat-pack furniture is often available for customers to buy in a store and take it home with them straight away. It is also easier to carry through doorways and upstairs than fully assembled items.

There are a wide range of KD fittings that are made by specialist component manufacturers and supplied to the manufacturers of the furniture. KD fittings are standardised and interchangeable, so that they can be used on a very wide range of products.

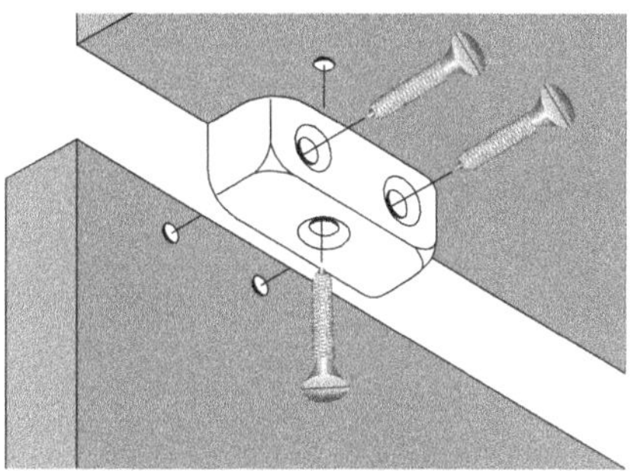

Figure 1.4.70 A modesty block.

Modesty blocks

Modesty blocks are small, rigid, polymer blocks. They have moulded holes that take screws which are used to join the block to panels. Typically, this type of block might be used on cupboards and storage units. It is simple to use but it does not make a particularly strong joint and it is unattractive, so is becoming outdated in mainstream furniture manufacture.

Figure 1.4.71 Barrel nut and bolt.

Barrel nut and bolt

A common KD fitting is the barrel nut and bolt. This uses a cross dowel that is fitted into one of the pieces to be joined. The bolt is inserted through the other piece of timber and tightened into the cross dowel (often using an allen key).

Barrel nut and bolt fittings are used to assemble parts of flat-pack furniture such as the frame of a bed or table.

Figure 1.4.72 A barrel nut and bolt is tightened with an allen key.

Cam-lock connector

Cam-lock connectors consist of a metal dowel that is screwed into one of the pieces by inserting a screwdriver into the slot on the side. The cam is a disk that fits into a pre-drilled hole in the other piece. When the disk is rotated with a screwdriver, the collar on the dowel locks into the cam and pulls both pieces tightly together.

Cam-lock connectors are typically used in flat-pack furniture such as bookcases, where the horizontal shelves are attached to the sides.

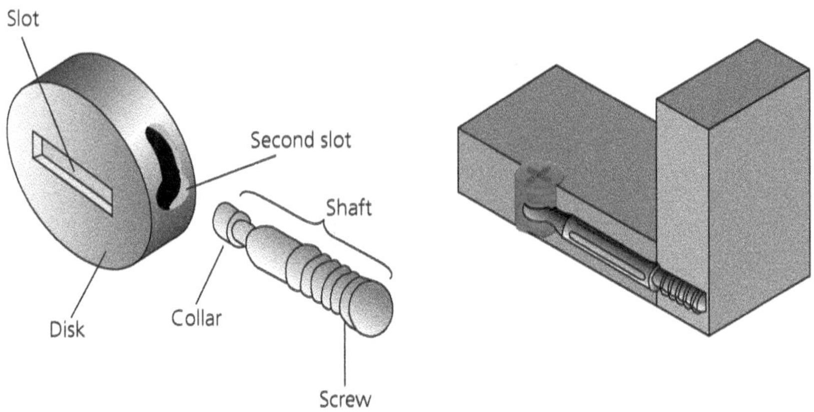

Figure 1.4.73 Cam-lock connector.

Wood screws, nuts and bolts and coach bolts

Wood screws

Figure 1.4.74 A wood screw.

Wood screws have a coarser pitch (fewer threads per inch), compared to screws used for sheet metal. Often, part of the screw – known as the shank – does not have a thread. This makes it easier for screwing two pieces of timber together where the thread is only needed at the bottom.

When joining two pieces of wood together, the top part is drilled with a clearance hole that the screw simply pushes through. The bottom piece of timber has a pilot hole (thinner than the thread) so that the coarse pitch of the thread bites into the timber.

Wood screws can have a countersink (conical head). This allows the screw to fit flush with the surface of the wood providing that a countersunk hole is drilled.

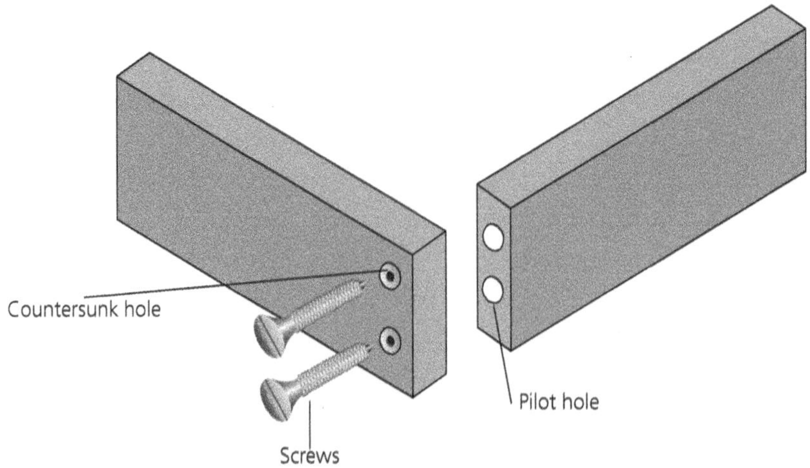

Figure 1.4.75 Wood screws can have a countersink (conical head).

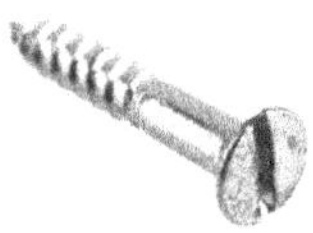

Figure 1.4.76 Slot head screw.

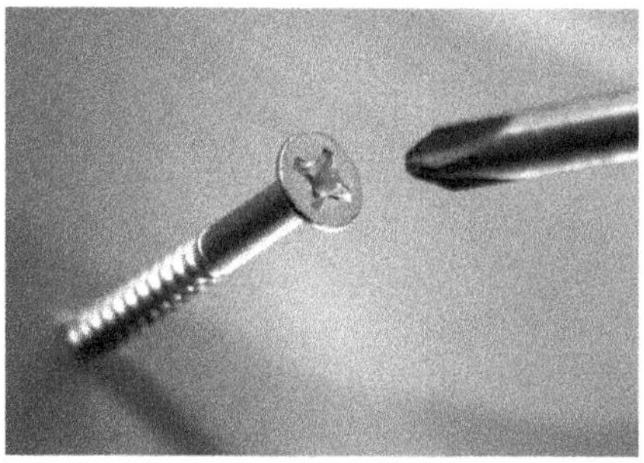

Figure 1.4.77 Phillips screw.

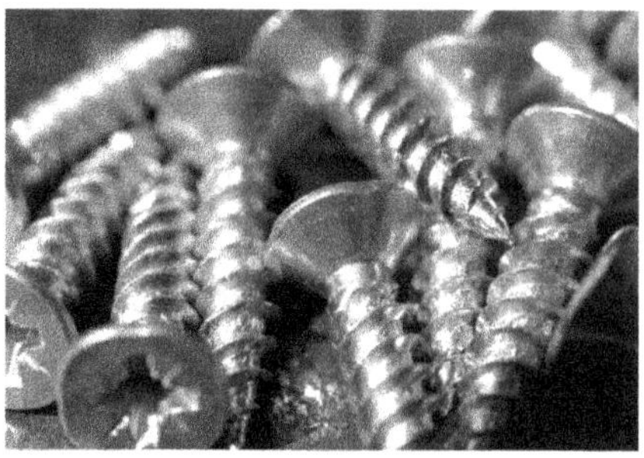

Figure 1.4.78 The Pozidriv screw was developed to make it easier to use with power tools.

Screws come in three different forms:

- slot head
- Phillips
- Pozidriv.

Slot head screws are the simplest and most economical screw. They are tightened using a slot head screwdriver, but they are not really suitable for use with a power screwdriver. Slot head screws are often supplied with products to be fitted to a wall, such as mirrors or light fittings.

The Phillips screw was developed to improve contact with the screwdriver and increase the torque that can be applied. The design of the screw is such that it accepts the tip of a screwdriver with an angle of 57 degrees. If too much torque is applied, the screwdriver will slip. This avoids stripping the thread on the screw.

Pozidriv screws are a further refinement of the Phillips screw. They improve contact with the screwdriver and allow greater torque. Pozidriv screws are particularly good for using with power screwdrivers. Pozidriv screws are the preferred type of screw for woodworking jobs where multiple joints need to be fastened.

The addition of further diagonal slots on the screw make a better contact with the driver bit.

Bolts

A level only

Bolts that are used to join wood pieces together – called 'coach bolts' – have a thread that runs about two-thirds of the way along the bolt. Under the domed head of the bolt is a square piece that digs into the wood when tightened, preventing the bolt from rotating. This means that when the bolt is tightened, the domed head cannot be undone with a spanner. This is useful in fitting bolts or locks to wooden doors because it makes them secure if the domed part is on the outside.

Wasting processes

There are several methods of shaping wood using wasting processes, and you need to be able to select the most appropriate process for given products. The most common wasting processes for woods are turning, routering and milling.

Turning

Turning involves machining wood on a wood lathe. There are three methods for doing this: turning between centres, turning on a faceplate or turning in a chuck.

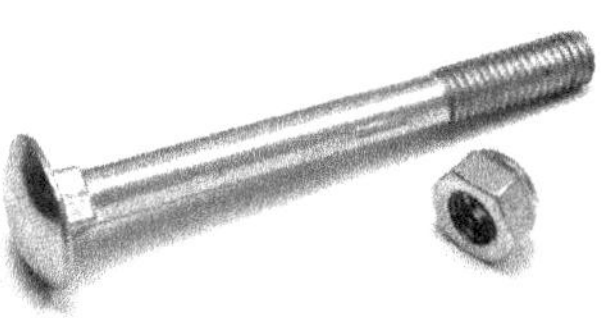

Figure 1.4.79 Coach bolt.

Turning between centres is used to machine a spindle such as chair legs and table legs. Turning between centres enables the spindle to be machined to a reduced diameter. Turning is normally done by hand, using tools that resemble large chisels known as gouges and scrapers. These are held on a tool rest, and moved in, out and along the length of the spindle being turned.

Figure 1.4.80 Turning wood between centres.

Figure 1.4.81 Turning a bowl on a lathe.

Turning on a faceplate is used to machine items such as domes or bowls. This involves screwing a thick piece of timber to a faceplate, where it can be machined to turn the outside circumference and to remove the inside in order to make a bowl.

Sometimes it is necessary to turn wooden items on a lathe using a chuck to grip the item while it is machined. An example is where the end of a spindle needs to be drilled. Chucks can also be used to machine the sides and inside of wooden bowls and vases. The rim on the base of a bowl or vase would be gripped by the jaws of the chuck, allowing access to the sides and inside for machining.

Routing timber

Routers can be used to machine timber in order to make slots and holes, or to make decorative 'mouldings' on the edges, for example on a table top. The most common type of router is a portable electric plunge router. These are handheld tools that can take a variety of routing bits, depending on the desired effect. Such routers are adjustable so that the depth of a slot can be set. They are normally supplied with a range of accessories such as guides or 'fences', which make it easy to produce continuous grooves either along or across a piece of timber.

Routers can also be computer numerically controlled. CNC routers can be used to machine a range of natural timbers and manufactured boards, in addition to high density modelling foam. CAD drawings are downloaded to CNC routers, and the drawing is converted into a control program which will move the router over the material. 2D CAD drawings can be downloaded to machine sheet material, for example when making flat-pack items or to create wall plaques and signs. 3D CAD drawings might be used to manufacture 3D shapes such as moulds for vacuum forming. CNC routers are usually fully enclosed and extracted so that dust is safely removed.

Figure 1.4.82 Plunge-type routers are used to cut holes in manufactured board.

Figure 1.4.83 A decorative panel being CNC routered.

Milling timber

In the same way as metals can be shaped on a milling machine, timbers can also be machined. Generally, milling machines run at too slow a speed for machining timber accurately, however they can be used for small sized, basic jobs such as a rough prototypes or roughing out a small hole or channel. Milling machines do not have the same range of movement or 'work area' as CNC routers, so they are generally only used to machine small items. They can be operated manually or they can be computer numerically controlled.

Forming processes

Forming processes concern changing the shape of timber or manufactured boards. There are two main methods used to form timber: laminating and steam bending.

Lamination

Lamination is the process of bonding materials together. Typically, wood veneers (thin slices of natural timber) or thin manufactured boards such as 3 mm plywood can be glued together and bent over a former, so that when dry, they form a thicker board in the shape of the former. While drying, the laminates may be held in place with clamps or, more commonly, using a vacuum bag.

With vacuum bagging, the former and veneers are placed under a polythene sheet. The edges of the sheet are taped down to the table. A valve is fitted to the polymer sheet and a vacuum pump sucks out air, and the resulting pressure pulls the laminates hard together. This ensures that there are no gaps in the laminating.

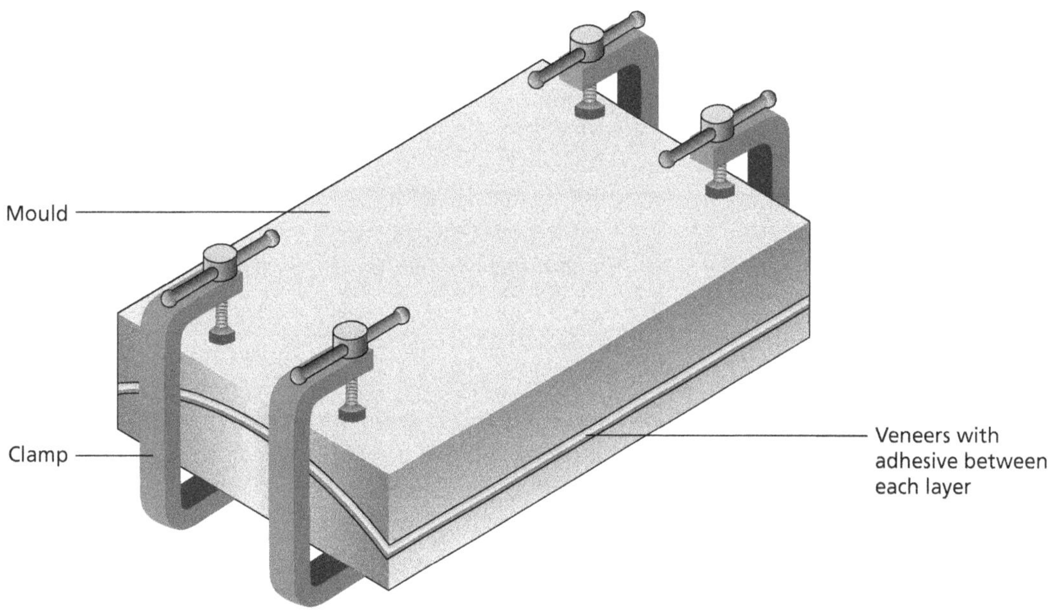

Figure 1.4.84 Laminating wood.

Steam bending

Combined heat and steam enable strips of timber to be made pliable so that they can be shaped over a former.

The process involves putting the timber into a steam box where the timber will absorb the steam. The timber is then usually bent over a former and clamped to it until it dries.

Steam bending has advantages over laminating in that it is quicker than waiting for glued layers to dry. It is also less wasteful because laminated parts usually require trimming to final size.

A basic steam box can be made from plywood. A wallpaper stripper steamer provides the steam. The steam box is usually angled slightly and a drain fitted, so that condensed steam can run away.

ACTIVITY

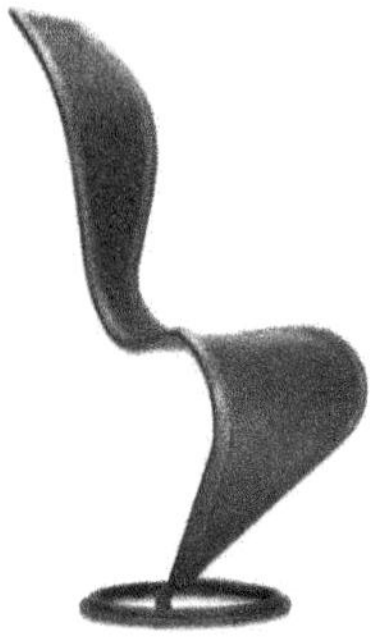

Figure 1.4.87 Curved chair.

Study the photograph of the chair shown above and answer the following questions:

1 Name a suitable material that could be used to make the chair.

2 Use notes and diagrams to describe how the curved parts of the chair could be made.

3 Describe a suitable method of attaching the seat to the frame.

Figure 1.4.88 PVA is a water-based adhesive used to bond timbers together.

Figure 1.4.89 Contact adhesive is used to join large areas such as sheet material together.

Figure 1.4.85 A basic steam box.

Figure 1.4.86 Steam bending jig used to bend wood into a curve.

The use of adhesives and fixings

Adhesives are substances used to stick materials together. Different types of adhesives are used for different materials and applications.

Polyvinyl acetate

Polyvinyl acetate (PVA) is commonly used to bond most woods and wood-based materials together. It is a water-based adhesive which is white in colour. PVA soaks into the surface of the wood and sets once the wood has absorbed the water content. PVA is not usually a waterproof adhesive, although some PVA adhesives have been developed that are water resistant. PVA can be used to bond wood joints such as dowel joints together, and is widely used to bond components when making indoor furniture products and other wooden products in the home.

Contact adhesives

Contact adhesive is used for large areas such as sheet material. It can be used to join the same or different materials together such as wood sheet to wood sheet, metal sheet to wood sheet, metal sheet to polymer sheet, etc. The two surface areas to be joined are coated in contact adhesive and left for approximately 10 minutes, or until the adhesive feels 'tacky'. On contact with the other surface, adhesion is instant, which means that clamping is not required. Contact adhesive is used when there is a large surface area to be bonded, such as an MF sheet bonded to chipboard for a kitchen worktop.

UV hardening adhesives

UV hardening adhesive is a clear liquid which 'cures' to form a bond when exposed to UV (ultraviolet) light. UV hardening adhesive contains a photo initiator, which means that as it absorbs the UV light wavelength, it begins to cure and set to a solid bond. It has numerous applications and can be used to join metal, glass and polymers. It is often used for workshop projects made from polymers such as lighting, as well as desktop stationery holders fabricated from parts. Many glass furniture products such as tables constructed from toughened sheet glass will make use of UV hardening adhesive to form the joints. The advantage of UV hardening adhesive is that any excess adhesive can be wiped away with a cloth prior to being exposed to UV light, thus giving a solid and clean joint. It has a fast curing time and, unlike adhesives such as epoxy resin, there is no need to mix the adhesive and hardener prior to application – the hardening aspect is an

Figure 1.4.90 Table made from toughened sheet glass joined with UV hardening adhesive.

integral part of the adhesive. UV adhesive is used where transparency or clarity is important; for example in joining glass to metals, or glass to glass.

Solvent cement

There are different types of solvent cement such as Tensol 12, which is an acrylic cement. Solvent cement is most commonly a clear liquid called dichloromethane, and is used to join polymers such as acrylic, ABS and PVC. Solvent cement works by softening the surface of the polymers to be joined, allowing them to fuse together. Solvent cement can be used in the plumbing industry to bond non-pressure pipes made from ABS or PVC together. It is also commonly used in school and college workshops for joining acrylic parts of project work.

ACTIVITY

Make a list of all of the adhesives available in your school or college workshop. Note what material(s) each adhesive is used for and how it is applied. Make a note of any health and safety or COSHH requirements for each adhesive that you find.

Jigs and fixtures

A fixture is something that holds work in a given position while a manufacturing process takes place. A jig both holds the work and guides a tool.

Jigs and fixtures are used for a range of purposes. They are often used to ensure that parts or components can be made repeatedly and accurately. Jigs can also be used to speed up manufacture.

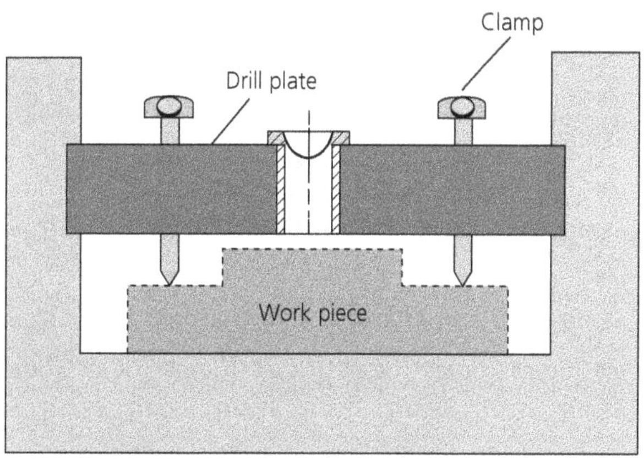

Figure 1.4.91 Open type jig. In this type of jig, the work piece is inserted into the jig and clamped. The work piece is then drilled through a bush made from hardened steel.

A typical example of how a jig might be used is for drilling parts in the same spot consistently and accurately. By using a jig, there is no need to measure and mark out each time.

Mitre blocks are a type of jig used to cut 45 degree angles in wood pieces. The mitre block is placed in the vice or clamped on to a bench. The wood is then held in the mitre block and cut at 45 degrees with a saw such as a tenon saw.

Figure 1.4.92 A mitre block.

Sanding jigs

Sanding jigs are used to hold and guide timber as it is sanded on disk or belt sanders. Typically, the angles on the end or edges of timber are sanded, and using a jig means that the angle will always be accurate and consistent.

Router jigs

Jigs can be used in conjunction with routers to shape wood accurately and consistently. Typically, a jig might be used to angle the router in a fixed position, while the edge of a piece of timber is run against the router bit.

Another example of a router jig is a circle-cutter jig, which is used to make circles in timber or manufactured boards. These work like a beam compass or trammel, where the router is fixed to a long piece of timber which is adjustable, to allow for cutting pieces of different diameters.

Figure 1.4.93 Sanding jig.

ACTIVITY

Match each product in the table to the correct forming, redistribution and addition processes.

Process	Product
Blow moulding	
Injection moulding	
Vacuum forming	
MIG welding	
Laminating	
Die casting	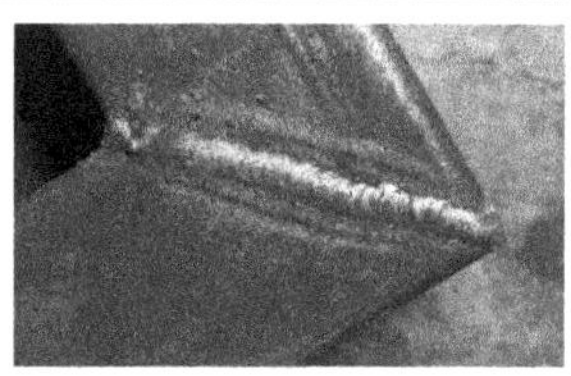

KEY TERMS

Die cutter: a machine using a hydraulic press and a plywood substrate with cutting dies and creasing rules to cut a net from paper, card or polymer film.

Forming: shaping stock form material through processes such as bending, rolling, press forming and steam bending.

Redistribution: shaping materials by melting them and reforming them in a mould that resembles the finished product or component.

Wastage: shaping stock material by machining it using processes such as milling, turning and routing.

Addition: shaping materials by adding additional pieces to them, either using the same or a different material.

Adhesive: a substance used to stick materials together.

KEY POINTS

- Redistribution processes such as injection moulding or casting are normally used when the product is too complex to be manufactured by any other method.
- Fabrication methods are generally much slower than manufacturing by forming and redistribution. Where fabrication is used in batch or mass production, manufacturers look for ways in which it can be simplified by using jigs and fixtures, or automated, for example by using robots.
- Wasting processes can be done manually, but where accuracy and repeatability is required, they may be done using CNC machines.
- Most polymer redistribution processes use thermoplastics as these can be heated and re-shaped.
- Redistribution processes are usually only suitable for large scale production due to the complexity and cost of moulds or dies.
- Knock down fittings are useful where it is important to supply products flat-pack.
- Temporary fixings are used where it is necessary to remove parts such as covers.
- Fabrication techniques such as welding or traditional wood jointing require specialist manufacturing skills.

Check your knowledge and understanding

1 Explain why injection moulding is used to make products such as a computer mouse, the casing for a portable electric drill or the casing for a vacuum cleaner.

2 Compare and contrast die casting with sand casting. Explain why manufacturers would select die casting or sand casting.

3 Use notes and diagrams to describe the blow moulding process.

4 Why would a manufacturer choose to use rotational moulding instead of blow moulding when making hollow polymer products?

5 What are the benefits for the manufacturer and the consumer in using KD fittings to assemble products?

6 Name a suitable adhesive for the following applications:

(a) MF sheet to chipboard sheet

(b) veneered oak MDF to a section of solid oak

(c) acrylic sheet to acrylic round tubing.

7 Give a specific application for UV hardening adhesive and give two advantages offered by the use of UV hardening adhesive.

Further reading

Manufacturing Processes for Design Professionals (2007) by Rob Thompson. This book has over 1200 photographs and technical illustrations covering established, new and emerging production techniques.

How forces make things stick, how adhesive products work and why adhesives do not stick to the container: www.explainthatstuff.com/adhesives.html

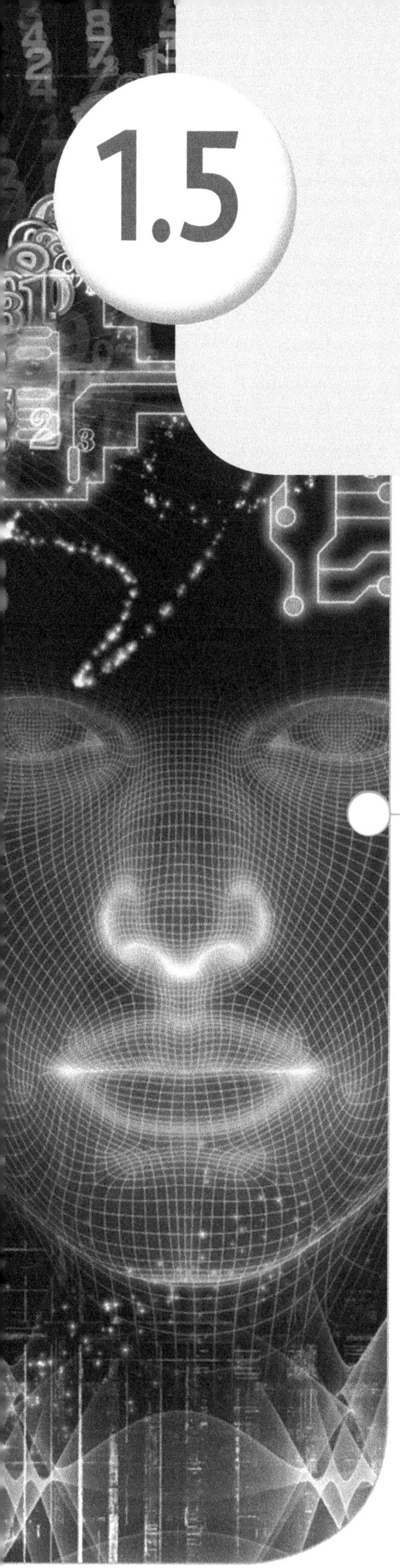

1.5 The use of finishes

LEARNING OUTCOMES

By the end of this section you should have developed a knowledge and understanding of:

- how different materials, including paper and boards, polymers, metals and woods, are finished to enhance their appearance and improve their function
- different types of printing processes and their suitability for various scales of production and for different products.

If you are studying at A-level you should also have developed a knowledge and understanding of:

- the use of acrylic spray paints and TPE
- methods of preventing corrosion and decay of metals, including sealants, preservatives, anodising, plating, coating and cathodic protection.

Materials and applications

Materials are finished to enhance the aesthetics of the final product as well as provide protection against aspects such as weathering, moisture, decay and insect attack.

There are three main terms associated with finishing:

- self finished: a material that has an acceptable finish after processing
- self coloured: a material that has an acceptable colour after processing
- applied finish: a substance applied to the material after processing.

The general aims of an applied finish are to prevent the material from absorbing moisture, protect against decay or corrosion, protect against insect attack and enhance the appearance of the final product.

Paper and board finishing

Finishes for papers and boards should not be overlooked in graphic design. Not only will a finish provide protection for the material, for example from liquid spills or dirt, but it will also allow for material enhancement, which is especially important if you wish to attract the consumer to a specific product feature.

Laminating

There are two main lamination methods to consider in relation to papers and boards.

- Lamination via encapsulation: This is carried out via a desk top laminator and the sheet of paper is encapsulated by a polymer pouch or film roll. The paper is fed through the desk top laminator and heat sealed, leaving a polymer sheet cover on both sides of the paper, along with a border around the paper edges. The laminating polymer is usually a mix of PET and ethylene-vinyl acetate (EVA).
- Lamination via a surface coating: The surface coating lamination of papers and boards can happen in two ways: liquid lamination or film lamination. The purpose of both is to coat the paper with a liquid to seal the paper and protect it.
  - Liquid lamination can be applied with a roller or spray and is used for applications such as signage to protect from the effects of moisture and dirt or UV light, for example causing colour fading. Film lamination can be done via a hot or cold process and is used for products such as menu cards and business cards.
  - Film lamination is usually made of PP with an adhesive to make the laminate stick to the paper. The laminate is fed from a roller and pressure is applied as the paper is fed through.

Figure 1.5.1 Film lamination being applied to the product.

Embossing

Embossing creates a raised design on the surface of the paper or card to give a visual and tactile effect. It is a popular decorative technique used on greetings cards, invitation cards and packaging products such as chocolate boxes, or packaging where the trade name or special feature is on a raised area.

The embossing process uses two dies – a male and female die – made from stainless steel or brass. The dies can be cold or heated. The **substrate** sheet is placed between the two dies and held under pressure until the embossed area is formed. The raised area can have ink or foil applied to it, or it can be left natural. An embossed area without any ink or foil is called a blind emboss.

Figure 1.5.2 Imprinted letters sitting above the surface of the card: embossing.

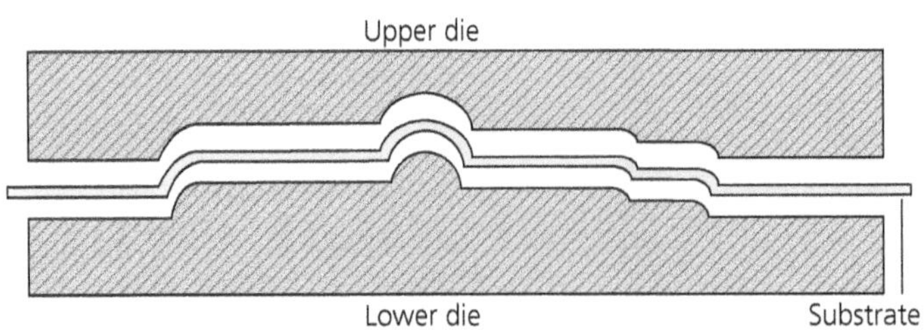

Figure 1.5.3 Embossing dies with substrate material.

Debossing

Debossing is the opposite of embossing, and produces an imprinted depression which sits below the surface of the paper or card. The debossing process is carried out in the same way as embossing, with two dies used to press and create the impression. Debossing is also used on greetings cards, invitation cards and packaging products which show the text as a depressed section of the paper or card.

Foil blocking

Foil blocking involves the application of heat and pressure to a metallic paper (foil) to create areas of depth and texture to add aesthetic impact to the product. Foils are available in different types such as metallic, gloss, matt, clear and holographic, and the foil blocking is used to enhance business cards and company logos on letterheads or seals.

Figure 1.5.4 Foil blocking on embossed lettering.

Foil blocking uses a special machine with a heated die, similar to the embossing process, except the design is stamped onto the material through the foil, which is transferred as it is pressed into the paper. A foil sheet is placed between the die and the paper on which the foil is to be applied. The die is pressed into the foil sheet and the heat allows the foil to stick to the paper. Foil blocking requires accuracy and consistency. If the heat is too high or the pressure is held for too long, the foil will bubble and peel. If the heat is not high enough or the pressure is not held for long enough, the foil will not adhere to the paper substrate.

Varnishing, UV varnishing and spot varnishing

Varnish is a clear, non-pigmented ink used on pre-coated papers and boards to enhance the colour, as well as offer some protection against dirt, fingerprints and water. Gloss, matt, satin and tinted inks are available and all are applied as a flood coat on the product. Varnish is only suitable for pre-coated papers because these will not soak up the ink and swell and/or bleed colour.

UV varnishing provides a very smooth finish but the ink must be completely dry before the finish is applied. Specific machinery is required that will instantly cure the UV varnish to provide the high-quality finish, which is usually high gloss or matt, and abrasion and chemical resistant. The varnish is applied via a set of rollers and the sheet is then passed under a UV light to cure.

ACTIVITY

Gather a range of different paper and board products that show a variety of finishing processes. For each product, state the finish used and the benefits this brings to the product.

Spot varnishing is varnish that is applied in specific areas or spots rather than to the whole surface area. Rather than flood coat the material, a plate or set of plates presses on to the surface and applies the varnish to that spot. Spot varnishing can be used with both varnish and UV varnish.

Paper and board printing processes

Printing processes allow the designer and manufacturer to attract a consumer to a product or product feature, as well as communicate information to the consumer. Colours, images and text can provide aesthetic appeal. Information such as barcodes, safety warnings and product ingredients can be printed to communicate specific data to the consumer and/or retailer.

Printing processes often refer to the term 'substrate', which is the material on to which the print ink is applied, such as paper or boards.

Screen printing

Screen printing is a simple and effective printing process used for small print runs of items such as posters, display boards and textile t-shirts. Screen printing can be carried out in a workshop and can be achieved with minimal setup costs due to the use of basic screens and printing inks.

The substrate to be printed sits on the screen print machine base and the upper section secures the screen. The image to be printed is created on a screen (or stencil), which is a mesh held by a frame. The screen has open areas for the ink to pass through. A different screen is required for each colour. This can increase the time taken to produce the print as well as the overall production cost per item. The screen is placed over the substrate on the machine bed.

The pigmented printing ink used is placed on the screen. A squeegee (a flexible polymer blade, held in a rigid handle) is then used to force the ink to flow through the mesh onto the substrate, for example paper or board. Once each colour is complete, the ink dries or cures to complete the print. Screen printing prints one colour at a time.

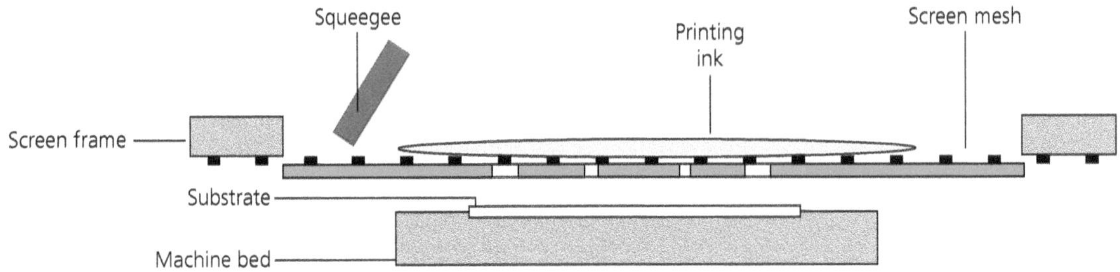

Figure 1.5.5 Screen printing process.

Flexographic printing

Printing processes such as flexography, offset lithography and digital printing use a four-colour process. The four-colour process uses four ink colours: cyan, magenta, yellow and key (black) (CMYK). These four colours are printed on top of each other in various quantities on to the substrate surface to create the print colour required.

To ensure an accurate and non-blurry print, the colours must line up precisely with each other. If one colour is slightly out of position, the printed image will appear blurred. In colour printing, a registration system is used to ensure the final image is consistent and of high quality. The registration mark is a set of precision marks on the final print substrate, outside the print area, which is used as a quality control (QC) check to ensure all four colours have printed in the correct place. These registration marks are often seen at the side of printed newspapers as a bar of CMYK colours.

Flexographic printing is the least expensive of the printing processes due to the simple operation and use of fast-drying water-based inks. It is used for products such as newspapers, comics, catalogues, folding packaging cartons, labels, carrier bags and continuous pattern products such as wallpaper and gift wrap. As flexographic printing uses rolls of substrate rather than sheets, this allows large continuous print runs to be completed with ease.

Flexographic printing uses flexible printing plates wrapped around rotating cylinders on a web press. Water-based or UV curable ink is fed from the first ink roller to the anilox roller. The anilox roller makes the ink a uniform thickness and transfers the ink to the plate cylinder. The substrate moves between the plate cylinder and the impression cylinder. The impression

Figure 1.5.6 Flexographic printing.

cylinder applies pressure to the plate cylinder to transfer the image on to the substrate. The web, which by now has been printed, is fed into the overhead dryer so that the ink is dry before it goes to the next print unit.

Table 1.5.1 Advantages and disadvantages of flexographic printing.

Advantages of flexographic printing	Disadvantages of flexographic printing
• High print speed. • Ideally suited for long runs. • Prints on a wide variety of substrate materials, both porous and non-porous • Low cost of equipment and consumables. • Low maintenance.	• The cost of the printing plates is relatively high, but they last for millions of print runs. • Takes a large amount of substrate to set up the job; excess material may be wasted. • Time consuming to change for any alterations to the print content.

Offset lithographic printing

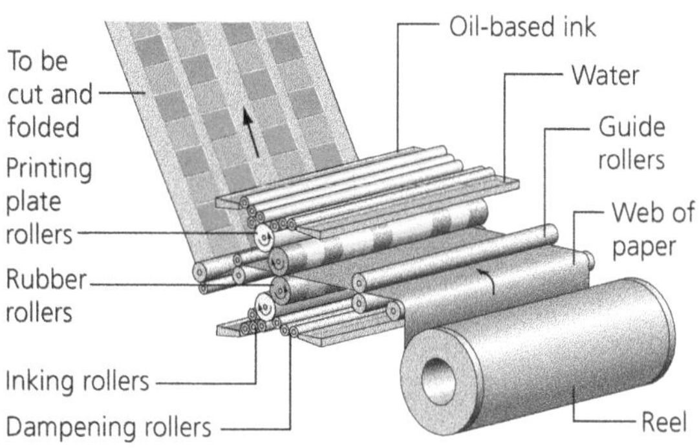

Figure 1.5.7 Offset lithographic printing.

This is an extremely versatile printing process capable of producing one colour (single roller), five colours (CMYK plus an additional metallic colour on a five-roller machine) and a ten-feature machine (CMYK, metallic, varnishing, spot varnishing and duplex (both sides printed) on a ten-roller machine). Offset lithography is used for printing medium and long print runs of products such as books, business forms and documentation, magazines, posters and packaging.

Lithographic printing is based on the principle that oil and water do not mix. Lithographic plates are chemically treated to make the image area oleophilic (absorbs oils easily) and therefore receptive to oil-based printing inks. The non-image area is treated to be hydrophilic (absorbs water easily).

During printing, fountain (dampening) solution, which consists primarily of water, is applied in a thin layer to the printing plate cylinder and is attracted to the hydrophilic non-image areas of the printing plate. Ink is then applied to the plate and is attracted to the oleophilic image areas. Since the ink and water essentially do not mix, the fountain solution prevents ink from entering the non-image areas of the plate.

The printing process is described below.

1 Printing substrate is either sheet-fed or web roll-fed into the printing machinery.
2 Printing plates are produced by a computer to plate (CTP) process via a laser which etches the image onto the plate. The plates are then attached to the machine and dampened by a damping roller.
3 The ink is applied to the printing plate. The plate cylinder rotates onto a blanket roller which becomes coated with ink. As the substrate is fed through, the image on the blanket is transferred to the substrate to produce the printed product.

Table 1.5.2 Advantages and disadvantages of lithographic printing.

Advantages of lithographic printing	Disadvantages of lithographic printing
• Consistently high image quality. • Suited to higher volume print runs of 1,000 or more. • Quick and easy production of printing plates. • Long life of printing plates because they only come into contact with the printing blanket, which is softer and less abrasive than the substrate.	• Expensive setup and running cost for small quantities. Only suitable for porous substrate materials.

Digital printing

Digital printing is becoming a far more popular and widely utilised resource for printing products, due largely to its speed and efficiency. Digital printers look similar to large photocopiers and produce full colour, highly detailed print runs with the option of different designs on each page, both front and reverse sides. Digital printers apply the colour in a similar way to laser printers, where the ink sits on top of the paper and is allowed to dry or be laminated. Digital printers can be used for both low and high volume print runs, and are very popular for printing promotional materials such as business flyers and business cards. The use of digital printing is ideal for mass customisation, such as the printing of specific names on drinks cans. This is due to the fast drying inks similar to those used in flexography.

ACTIVITY

Collect a range of paper and board products that have been printed using different methods. Note the particular features and quality each finish offers to enhance or protect the product.

Polymer finishing

Polymers are considered to be self-finishing materials because they require no additional finishing process once manufactured. Surface textures are produced from the mould and the addition of pigments during manufacture provides colour, making the polymer self colouring as well as self finishing. As such, polymers are considered finished and ready to use when the manufacturing process is complete. For example, the tomato ketchup dispenser shown in figure 1.5.8 is ready to use once the green pigmented top (injection moulded) and the red pigmented base (blow moulded) have been removed from the machines and assembled. Surface finish from the mould provides any textures or product features.

Figure 1.5.8 Pigments added during manufacture to produce the red and green colour.

Adding colour in the moulding process

Pigments are tiny particles which are added to the polymer to create a particular colour. The pigment particles can be added to the polymer during the manufacturing process (for example, polymer granules for injection moulding can have pigment particles added into the hopper) or, more commonly, pigments are added during the manufacture of the polymer stock form. For example, the manufacturer would buy the pre-pigmented granules used in injection moulding from the original granule manufacturer.

Smart pigments such as thermochromic or phosphorescent pigments can be added in the same way, producing ready-finished smart polymer products. For example, 'glow in the dark' polymer stars for a child's bedroom would be injection-moulded from a polymer with a phosphorescent pigment additive.

Pigments can also be added during other processes such as the lay-up process for producing laminated GRP. Ready-coloured boat hulls can be produced by using a pigment in the gel coat. The polyester resin gel coat is bought in ready-to-use pre-pigmented colours and can be applied to the mould before the GRP matting is added during the lay-up process. This eliminates the need to add individual pigments to the resin during the lay-up process or carry out an additional finishing process post manufacture.

Acrylic spray paints

A level only

Although polymers are self finishing, additional finishes are sometimes applied to enhance aesthetics or improve product function. Acrylic paint is a fast-drying, water soluble paint that becomes water resistant when dry, and is used on many polymer applications from model kit finishing to the automotive industry.

When acrylic paint is used on polymers it can provide improved aesthetics and additional protection against the effects of UV light and weathering, as well as allowing for mass customisation, particularly in the automotive industry. Mass customisation allows products to be tailored to the individual consumer's preference, such as colour coding bumpers and wing mirrors on cars either to match the body work colour or be a stand-alone colour. Car manufacturers make parts such as bumpers, wing mirror covers, sills and door handles in a base colour. The consumer can choose a different colour option when buying the car, and these parts are then individually colour coded and sprayed with acrylic paints to produce the final component. Although polymers can be pigmented during manufacture, it would be impractical and not cost effective for the manufacturer to keep changing the granules on the injection moulder to produce small numbers of differently coloured components.

Overmouldings

A level only

Overmoulding is moulding a second polymer over specific parts of a product. TPE is often used, and overmouldings are primarily used to provide areas of grip or texture, or to highlight different colours for different component parts on products. For example, a toothbrush made from PP would have a TPE overmoulding to provide an area of grip on the handle.

Figure 1.5.9 PP toothbrush body with TPE overmoulding grip.

There are different ways of producing an overmoulding:

- Use two injection moulding moulds: One mould is made for the product, for example toothbrush body, and one mould for the grip areas. The toothbrush body is injection moulded and then the product is placed into a second mould and the grip is injection moulded on to the body.
- Twin shot injection moulding: This process uses an injection moulding machine with a mould designed to produce the product in one cycle. The mould has two separate component cavities and can be rotated through 180° so that they line up with the twin injection points. First, the moulded part, for example the toothbrush body, is created in the first mould cavity. The mould then opens slightly and rotates 180° to the secondary position. The mould closes again and the second injection applies the overmould, for example the grip on the toothbrush body. Twin shot injection moulding is a continuous cycle – as the overmoulding is taking place, another toothbrush body is being made in the first mould cavity.

ACTIVITY

Find examples of products at home or in the workshop that make use of overmoulding. Explain the benefits that the use of overmoulding brings to the product.

Figure 1.5.10 Low carbon steel showing layers of rust on an outdoor handrail.

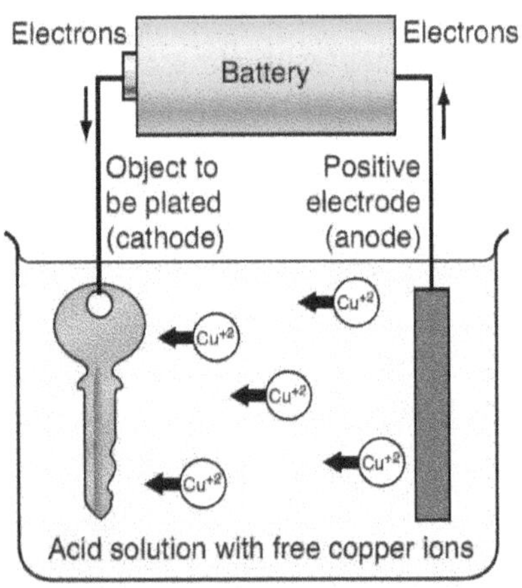

Figure 1.5.11 Electroplating process.

Metal finishing

Most metals have an oxide layer which provides some barrier against the effects of the environment. Copper originally has a bright reddish colour, but over time an oxide layer or patina forms to protect the metal from corrosion and environmental attack. Copper is often seen on older buildings for roofing; one example would be the green patinaed dome of St Paul's Cathedral in London, United Kingdom.

Steels (but not stainless steel) are the exception to this rule because they have an oxide layer which is porous, allowing moisture to penetrate the metal, leading to rust. The porous oxide layers continue to allow moisture to penetrate, leading to layers of crumbling rust that can be seen on many unfinished steel products such as gate posts or railings.

Finishes are applied to metals to enhance their appearance and help provide a barrier against corrosion.

Cellulose and acrylic paints

Paints provide colour and sometimes a textured finish to metals, while also providing a barrier against corrosion. Paints are primarily used on low-cost metals such as steel. The surface of the metal must be cleaned and degreased prior to paint application; this ensures that the primer coat has a sound surface to 'key' (or grip) to. A suitable primer coat that is often used is red oxide primer, followed by an undercoat in a similar colour to the final top coats. Paints, either cellulose or acrylic, can be applied by brush or spray.

There are an increasing number of specialist paints on the market that provide texture or special graphic effects such as pearlescence. There are some specialist paints that do not require a primer coat and are available in hammered or smooth textures.

Electroplating

Electroplating involves using a metal to coat a (usually cheaper) base metal, to provide both a protective layer and to give a greater aesthetic appeal; for example, metal teapots are silver plated to provide increased aesthetic appeal as well as a barrier to corrosion from the contained liquid. Also, kitchen taps are chromium plated to provide protection against the water and to give a shiny, hardwearing finish. Commonly used metals for electroplating include silver, gold, zinc, copper and tin.

The product and donor material are placed in a container with electrolyte solution. As the direct current is applied, the product attracts the donor metal and the product is electroplated.

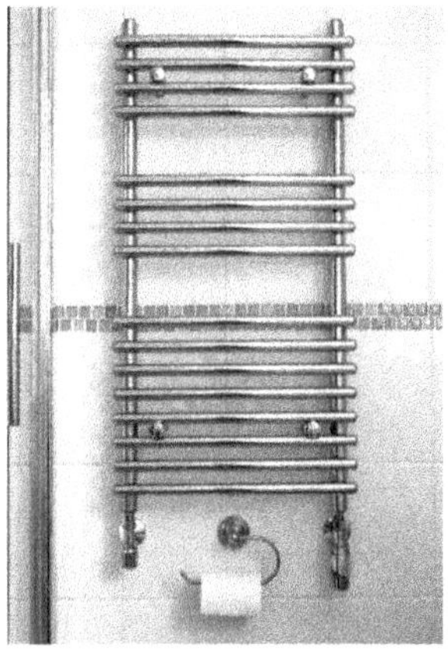

Figure 1.5.12 Chromium-plated bathroom wall heater.

Figure 1.5.13 Silver-plated teapot.

Dip coating

Polymer dip coating

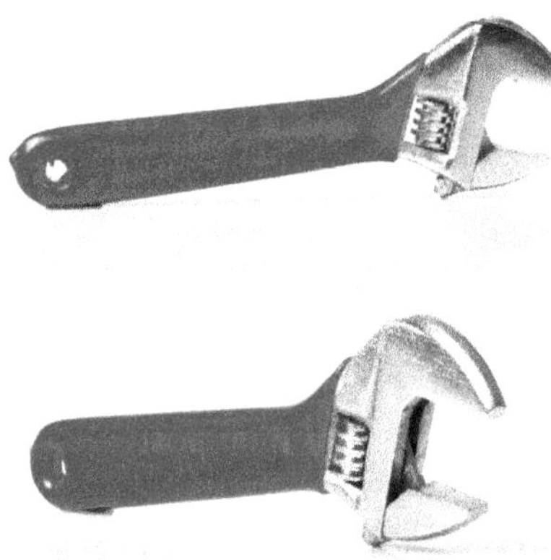

Figure 1.5.14 Polymer dip coating.

Polymer dip coating is used on a variety of products such as, wire coat hangers, kitchen dish drainers, coat hooks, dishwasher racks and outdoor play equipment frames.

The metal product is heated to approximately 230 °C. The hot product/component part is then dipped into a tank of fine polymer powder which has air blowing through it (fluidisation bath). The fluidisation of the plastic powder is an aid to providing an even coating on the product. The retained heat from the product allows the polymer powder to melt over the product, which is then simply air-cooled to allow the coating to set evenly.

Metal dip coating

Metals can be coated with other metals by dipping the product to be coated into a tank of molten plating metal (the donor metal). Metal dipping and plating is often used for inexpensive metals such as low carbon steels, primarily to provide a barrier against corrosion.

Prior to the dipping process, the metals must be cleaned and degreased.

Tin plating: Pass sheets of steel through a tank of molten tin at approximately 320 °C. This process is often used to provide a non-corrosive coating to food cans.

Zinc plating: This process is also called galvanising, and involves dipping steel into molten zinc at approximately 460 °C. Galvanising is used for many industrial and agricultural applications such as beams, gates and animal pens. It is also commonly used as the first protective layer in car bodies made from low carbon steel.

Powder coating

In **powder coating**, the product to be coated is initially statically charged (negative). Thermoset polymer resin (positively charged) is then sprayed through an airgun. The use of charge results in a strong attraction between the powder and the product. The product is then baked in an oven; the heat melts the powder over the product to give an even coating that is much more hardwearing than painting.

Figure 1.5.15 Powder coating

Many domestic white goods such as washing machine bodies, dishwasher bodies and refrigerator bodies are finished in this way. It is also widely used for applying a coloured finish to metal gates, fencing and railings.

Varnishing

A level only

Metal varnishing is a method that provides a clear finish to protect the metal and allow the colour of the base metal to show through. It is primarily used on more expensive metals such as aluminium, brass and copper, which all have good aesthetic properties.

The metal should be polished to a shine and any surface grease removed. The varnish is then applied by either a spray or with a fine brush to coat the metal with the protective layer.

Sealants

A level only

Metal sealants are tough polymer-based sealant coatings that protect polished surfaces from decay and tarnishing. Silicon-based clear sealants are often used in the automotive industry as an additional bodywork barrier against the effects of weathering, road salts and insect attack. The sealant is generally applied with a cloth or machine pad to produce a film that is then allowed to cure for up to 15 minutes and subsequently buffed with a cloth to a shine. Sealants can be used for automotive bodywork, freshly polished exhaust pipes and metal trims.

Preservatives

A level only

Metal preservatives can provide temporary in-process or final post-processing, medium- to long-term protection for metal surfaces. Preservatives are often used on moulds and dies for preventing fingerprints showing on surfaces, minor atmospheric corrosion or condensation build up. They can be applied by wiping on with a cloth, spray or immersion.

Silicon sealants can also be classed as preservatives and can be used on machine beds to provide a smooth, snag-free surface to allow materials to be pushed across the bed more easily and safely.

Anodising

A level only

Anodising has a finishing process commonly used for aluminium products which enhances the natural oxide layer, resulting in the aluminium having increased hardness and toughness. Anodising can also provide colour to the product and is often used in the outer casings of torches, external hard drive covers and casings for mobile phones or digital music players.

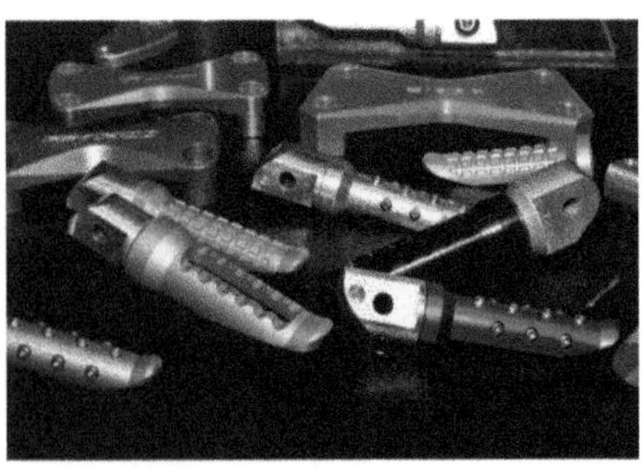

Figure 1.5.16 A range of anodised motorcycle components

Anodising has a similar setup to the electroplating process. The electric current passes through the sulphuric acid electrolyte solution from the part to be treated to a negative cathode. As the current flows from the part to be treated (positive anode) to the negative cathode, the aluminium oxide layer builds up on the treated part, producing the anodised finish. Anodised products can then be finished with a clear lacquer or varnish to seal the finish and provide further protection against surface scratches.

Cathodic protection

A level only

All metals have a natural voltage, and when two different metals are joined together, there is an electrical current. If water is present at the join, this results in the formation of an electrochemical cell and one of the metals will corrode. The reaction is basically current flow through the water from the anode (which is the more actively corroding metal) to the cathode (the less actively corroding metal).

For example, two aluminium sheets held together with brass screws will form an electrochemical cell in the presence of rainwater. An electric current will flow between the anode (aluminium) and the brass (cathode) due to the voltage difference between the metals and the aluminium will start to corrode around the join.

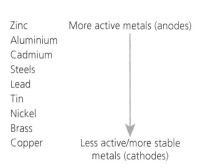

Zinc — More active metals (anodes)
Aluminium
Cadmium
Steels
Lead
Tin
Nickel
Brass
Copper — Less active/more stable metals (cathodes)

Figure 1.5.17 Electrochemical activity of metal in salt water.

The list in figure 1.5.15 shows metals arranged in order of electrochemical activity in a salt water solution. Metals at the top of the list provide cathodic or sacrificial protection to the metals below them. For example, aluminium protects brass due to the faster corrosion rate of aluminium compared to brass.

Cathodic protection is a method used to control the rate of corrosion by making the metal the cathode of an electrochemical cell. The anode of the electrochemical cell is a sacrificial metal which is more easily corroded. The anode (sacrificial metal) corrodes while the base metal is protected.

The main use of cathodic protection is to protect steel structures buried in soil or immersed in water, such as pipelines, ship hulls, jetties and offshore platforms.

There are two methods of cathodic protection:

- Impressed current: The component to be protected is connected to an electrical power supply. The impressed direct current flows from an inert electrode through any liquid to the component to be protected. For example, a buried pipeline receives direct current from an electrode buried in the ground. The pipe becomes the anode and the electrode is the cathode. The cathode corrodes over time and such corrosion is monitored so that the pipeline remains intact.
- Sacrificial anodes: This is a more basic method because it uses a sacrificial metal to protect the metal product of value. A more electrochemically active metal is wrapped around or joined to the less active metal to provide resistance to corrosion. On large structures, the sacrificial metal is monitored for signs of corrosion and replaced when the base metal is no longer protected.

Wood finishing

The purpose of finishing wood or wood-based products (such as manufactured boards) is to:

- prevent the wood from absorbing moisture
- protect against decay
- protect against insect attack
- enhance the appearance of the final product.

Natural wood can primarily degrade in the following ways:

- Alternating wetness and dryness means that moisture is absorbed into the wood and, although the surface may dry out, moisture remains below the surface. This wet rot leads to breaking down of the wood fibers and a fungus, which thrives in damp conditions where there is little air circulation, to spread through the wood. This converts the wood into a

soft, wet, crumbly state. A different, fast-spreading fungus causes dry rot, which can quickly turn wood into a dry, soft, powdery state.
- Insects, for example, furniture beetle (woodworm) which attacks softwood and hardwood, or deathwatch beetle, which usually only attacks hardwood, lay their eggs in cracks in the wood. The hatched larvae eat into the wood creating tunnels. When the beetle forms, it eats its way out of the wood via a flight hole.

Prior to applying a finish, good surface preparation is essential. Wood surfaces should be made smooth either by planing and/or sanding with glass paper using progressively finer grades. All sanding should be done in the direction of the grain to avoid scratching the wood surface. Finally, a cloth should be used to wipe away any remaining dust before the finish is applied.

Table 1.5.3 Methods of wood finishing.

Wood finish	Method of application	How it enhances appearance or prevents decay
Polyurethane varnish Acrylic varnish	Available in three types of surface finish: gloss, satin and matt. Available in clear or colours. Apply a thin coat with a brush in the direction of the grain. Several thin coats can be applied, rubbing down between coats with fine grade glasspaper.	Provides a hard, tough, heatproof, waterproof finish. Clear finish allows the grain of the wood to be seen through the finish.
Water-based paints	Available in different surface finishes such as gloss, satin, matt and metallic effects. Apply with a brush, roller or spray. Before paint is applied, wood knots should be treated with knotting to prevent the resin oozing out and spoiling the finish. The surface should then be primed and an undercoat applied prior to the paint finish.	Provides both protection and colour. Paint can be applied for both indoor and outdoor use, and is predominately used for softwoods. Paint makes surfaces more aesthetically pleasing, tougher, harder and resistant to moisture.
Stains	Available in many different colours and types such as mahogany, walnut or antique pine. Apply with a brush, roller or spray. Surfaces should be grease-free prior to application.	Used to help colour and enhance the grain. Stains are water or spirit based. Stains can be used to colour an inexpensive wood to make it look like a more expensive timber, or merely to add colour while allowing the grain to show. Stains do not have any specific protective properties.
Colour wash	Available in many different colours. Apply with a wet sponge.	Used to help colour the wood while allowing the grain to show through. Colour washes can be used to colour a wood to make it more aesthetically pleasing, or to add a vintage style while allowing the grain to show.
Wax	Available in clear and coloured waxes for indoor products. Apply with a brush or a stockinet cloth then, once dry, buff with a clean, lint-free cloth.	Can be buffed to provide a high gloss finish. Wax increases the surface hardness and toughness. Clear finish allows the grain of the wood to be seen through the finish.
Pressure treating	Wood is placed in a pressure vessel containing a solution consisting of copper sulphate and other preserving salts or preservatives. Vacuum and pressure are controlled to force the preservative deep into the fibres of the wood and then the wood is steam dried.	Helps protect wood for up to 50 years from rot, insect and fungal attack and weathering. Suitable for products such as decking, overhead cable poles, harbour or bridge construction. Some decking and outdoor cladding applications.

Wood finish	Method of application	How it enhances appearance or prevents decay
Yacht varnish	Available in high gloss and satin. Apply with a brush or sprayed directly on to the wood.	Suitable for woods that are outside all year round such as doors, window frames or boat parts. Increases toughness, hardness and weather resistance.
Danish oil	Available in clear and colour tints. Apply with a lint-free cloth, rub the oil into the surface of the wood in the direction of the grain. Leave the wood to absorb the oil for 5 minutes then rub away the excess oil with a clean cloth. Lightly sand or rub with wire wool to open up the grain between coats.	Maintains the natural appearance of the timber, soaks into the wood and enhances the grain. Although oil is not a very hardwearing finish, regular application and subsequent coats build up a matt, water-resistant finish.
Teak oil	Available in clear tint. Apply with a lint-free cloth, rub the oil into the surface of the wood in the direction of the grain. Leave the wood to absorb the oil for 5 minutes then rub away the excess oil with a clean cloth. Teak oil is used for naturally oily woods such as teak.	Primarily used for outdoor wood products to improve weather resistance and resistance to fungal and insect attack. Aesthetically pleasing as grain is enhanced.

ACTIVITY

Use notes and sketches to describe how to prepare and apply a polyurethane varnish finish to an oak table.

ACTIVITY

Find a range of metal products with different applied finishes. Create a table to show the product, base material and specific finish used. For each product, give two reasons why the specific finish is used.

KEY TERMS

Substrate: the material on to which the print ink is applied, such as paper or boards.

Embossing: the process of creating raised images or text on paper or card by placing the paper or card in between two dies under pressure.

Debossing: produces an imprinted depression that sits below the surface of the paper or card.

Screen printing: a printing process in which ink is forced onto the surface of a material through a mesh screen to create a picture or pattern.

Electroplating: using a metal to coat a (usually cheaper) base metal.

Powder coating: thermoset polymer powder is positively charged and attracted to the negatively charged product. It is then baked to melt the powder and given an even coating.

Anodising: commonly used for aluminium to enhance the natural oxide layer, resulting in increased hardness and toughness.

KEY POINTS

General aims of an applied finish:

- prevent the material from absorbing moisture
- protect against decay or corrosion
- protect against insect attack
- enhance the appearance of the final product.

Check your knowledge and understanding

1 Explain the meaning of the following polymer finishing terms:
 (a) self finished
 (b) self coloured.

2 (a) Describe how polymers degrade.
 (b) What methods could the manufacturer of a polymer outdoor garden slide use to ensure that the polymer does not degrade?

3 Suggest a suitable finish for the following products and suggest a reason why the finish is suitable:
 (a) oak coffee table
 (b) spruce telegraph pole
 (c) aluminium torch body
 (d) farm field gate made from low carbon steel.

Further reading

The timber trade federation's website contains some interesting information about sustainability and responsible sourcing: www.ttf.co.uk

The British Plastics Federation website has a 'plastipedia' of polymers: www.bpf.co.uk

Although quite advanced, the metal finishing information on the Materials Today website explains the various finishes: www.materialstoday.com/metal-finishing

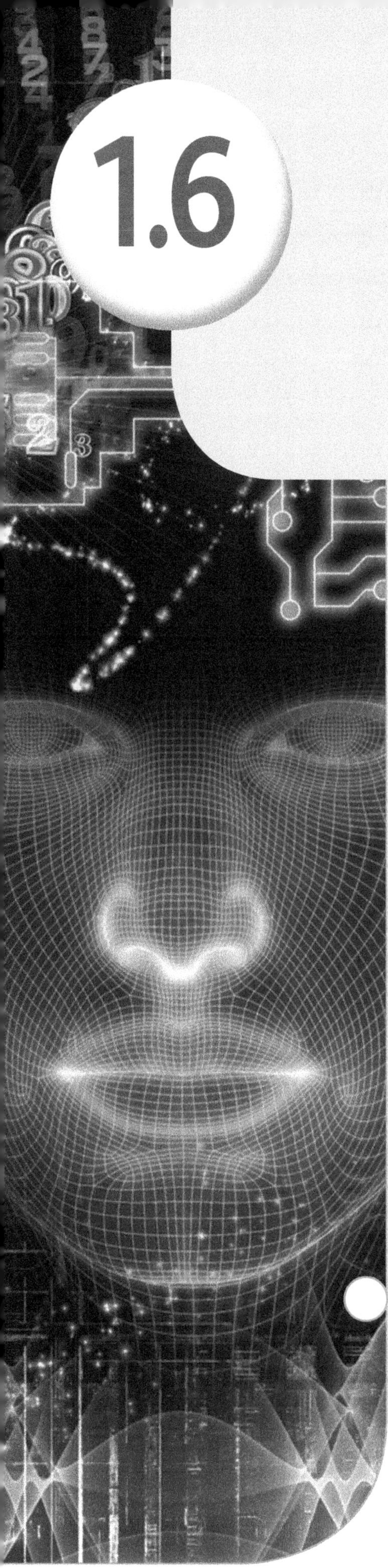

1.6 Modern and industrial scales of practice

LEARNING OUTCOMES

By the end of this section you should have developed a knowledge and understanding of:

- the different scales of production: one-off, batch and mass/line production
- the use of computer systems in production, distribution and storage
- the efficient use of materials: the relationship between material cost, form and manufacturing processes, and the scale of production
- sub-assembly.

If you are studying at A-level you should also have developed a knowledge and understanding of:

- unit production systems (UPS)
- vertical in-house production
- modular/cell production
- flexible manufacturing systems (FMS).

Designers and manufacturers have to be conscious of the **scale of production** required for any of their intended products, since this will have an important bearing on decisions that are made about its design and manufacture.

The growing impact of computer controlled systems in production, distribution and storage has, however, facilitated much greater flexibility for companies in responding to their customers' requirements.

Manufacturing efficiency is often improved by incorporating sub-assemblies, which are pre-made elements of a product, in the final assembly process.

Scales of production

One-off, bespoke production

Sometimes a product is required for a unique situation where it would be inappropriate to use anything other than an individually designed and manufactured, bespoke version. This is sometimes call 'job' production.

Examples include:

- a luxury custom-made yacht for a wealthy businessperson
- a chair for a child with a serious physical disability
- a personalised wedding cake
- a violin for a virtuoso player with specific requirements.

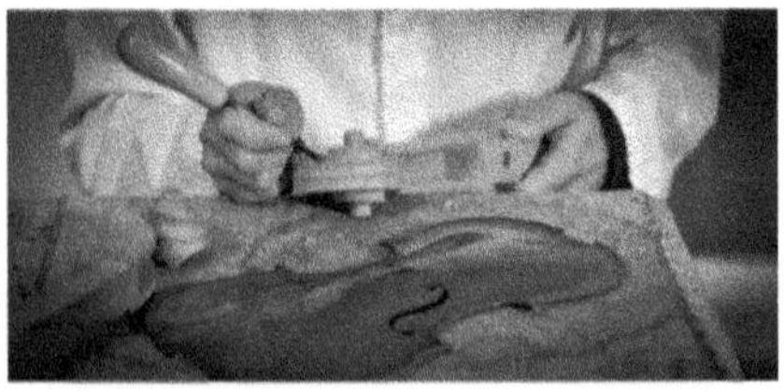

Figure 1.6.1 Hand crafting bespoke stringed instruments.

All of these products will be more expensive and are likely to be more difficult to produce than products made on a larger scale. This is because:

- Individual client consultation and design work is required.
- More skilled workers are likely to be required for manufacture.
- Each manufacturing process has to be set up individually for each of the unique components.
- Greater time is required for manufacture.
- There will be no economies of scale for materials as there would be with large-scale production.

Batch production

In 1700, the word 'batch' meant 'bake', as in the practice of making a number of bread loaves simultaneously. We now use this term to describe the manufacture of any products where the manufacturing processes are carried out simultaneously on a number of products in order to make the process more efficient and cost effective.

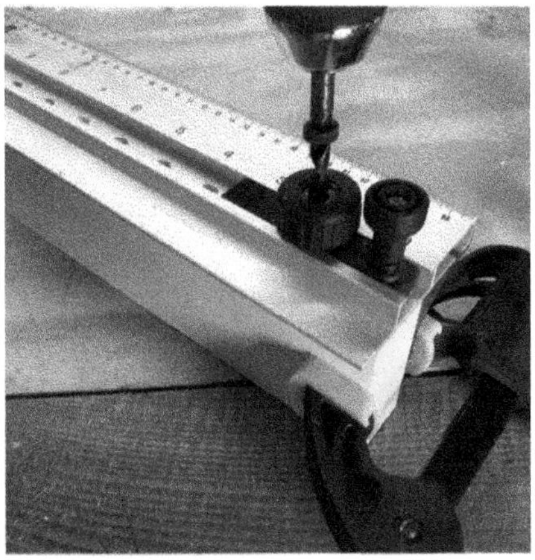

Figure 1.6.2 Drilling jig to facilitate batch manufacturing of furniture.

Examples include:

- pottery
- wooden furniture
- high-quality loudspeakers
- jet engines.

Jigs, fixtures and other aids will be used alongside manual and CNC machines in the manufacture of many of these items, since this will facilitate more accurate and faster production. To appreciate the improved efficiency of **batch production**, consider how long it would take to make, for example, one loudspeaker cabinet compared to making 24. Massive time efficiency results from machining and assembly operations being carried out simultaneously.

> **ACTIVITY**
>
> Design a simple stool or table to be made of dowelled pieces of beech, incorporating four legs and a square frame. Produce a list of manufacturing stages and calculate the time efficiencies that could be achieved by carrying out batch production processes in parallel.

Mass/line production

'**Division of labour**' is the term used to describe dividing up the tasks needed in order to mass produce a product in large quantities, using a sequence of stages as it proceeds through the manufacturing process. Henry Ford is often credited as the first person to create a large-scale **line production** system in the Piquette Avenue car plant in Detroit. Here, the famous 'Model T' could be produced in 1.5 hours rather than the 12.5 hours it had taken prior to the introduction of **mass production** using a production line.

Cars are particularly good examples of products whose manufacture can be automated to optimise the efficiency of manufacture, and there are many other products that can be manufactured on a large scale. The introduction of the mass production of PP chairs, such as the Robin Day design shown in Figure 1.6.10, made a significant impact on the furniture market, particularly in the context of the relatively inefficient batch production

Figure 1.6.3 The Model T production line in Detroit, 1913.

methods associated with many traditional, timber based chairs that preceded it.

Other examples of products that are mass produced include:

- clothing
- bicycles
- printed circuit boards (PCBs)
- mobile phones
- cameras
- TVs
- ready meals
- self-assembly furniture.

Unit production systems

A level only

Unit production systems (UPS) are used extensively in the manufacture of textile products and are based on the principle of using an overhead transporter system, incorporating appropriate hanging carriers, to convey the components that are required for the manufacture of a garment between workstations. This movement is sometimes manually operated, but is often computer controlled, and requires very careful organisation and synchronisation to ensure that it works efficiently. It does, however, reduce the number of handling operations and facilitates the optimal use of labour and equipment, while improving the quality of products.

Quick response manufacturing

The amount of time taken to respond to orders for a part or a product is known as the lead-time, and this needs to be as short as possible in order to compete successfully. One of the methods for reducing lead-time is known as **quick response manufacturing (QRM)**, which is a strategy for drastically reducing the time between a product being a concept and being completed.

The whole QRM strategy is focused on going through the design process and developing products to meet a customer's needs quickly. As the design process is a creative one that needs plenty of time, however, this is a significant change in approach. The needs of the customer are given very high priority throughout the design and production process, and quality is still important.

Vertical in-house production

A level only

Vertical in-house production (also known as **vertical integration**) is an arrangement in which the supply chain of a company is owned by that company. For example, companies that manufacture relatively complex products such as wind-turbines need to decide to what extent they will use external suppliers for parts or sub-systems that are incorporated in their products (often referred to as outsourcing). Those that opt for vertical in-house production set up their manufacturing in such a way that they minimise the need for external suppliers and organise their factories to include the facilities required to manufacture what is required.

Advantages include:

- reduction in the risk of unexpected price increases for components
- less susceptibility to suppliers going out of business

ACTIVITY

Compare the manufacture of a one-off 'thrown' clay mug with mass-produced mugs that sell in millions.

- protection of their brand and improved security of intellectual property rights (IPR)
- easier implementation of quality assurance (QA) strategies.

Disadvantages include:

- reduction in specialisation, potentially leading to dilution of expertise
- increase in administration
- reduction in flexibility.

Efficient use of materials

The efficient use of materials and the most appropriate selection of manufacturing methods is essential if manufacturers are to ensure that their production is cost effective and ultimately profitable. There is also an increasing awareness of sustainability issues.

Material cost, form, manufacturing processes and scale of production

When setting out to design and manufacture a product, deciding which material to use, with its inherent cost implications, is sometimes relatively straightforward since the specifications for the part may preclude other options. For example:

- A twist drill that is capable of penetrating low carbon steel in a commercial manufacturing environment would be better if made from high speed steel (HSS) rather than high carbon steel, since the ability to retain a sharp cutting edge at high temperatures is of paramount importance.
- A hinged lid for a freezer food container would preferably be made of PP rather than polyethylene, because it is the only thermoplastic that satisfactorily combines the ability to withstand low temperatures, has excellent fatigue resistance and displays very good mould flow characteristics.

However, cheaper versions of both of these products may be found, incorporating the lower cost, inferior material options. This is linked to product quality, since both of these cheaper products would be likely to suffer from premature failure.

The amount of material used for particular applications is another key decision that must be made, since this can have a knock-on effect in terms of the product's durability, weight, bending strength and other key factors. Consider, for example, making a decision on the appropriate wall thickness of a low carbon steel tube for a sports stadium crush barrier compared to a tube being used for aesthetic purposes in a shop window display.

In order to understand the relationship between material cost, form, manufacturing processes and scales of production, it is helpful to consider an example such as the main material used to manufacture the body of a range of different cars:

Table 1.6.1 The relationship between material cost, form, manufacturing processes and scales of production.

	Type of car	Price (£k)	Main material/form	Material cost (£/kg)	Manufacturing processes	Average annual production
A	Ford Fiesta (Economy)	13	Low carbon steel sheet	0.30	Presswork and automated spot welding	375,000
B	Jaguar F-type (Prestige)	50	Aluminium sheet	1.36	Presswork, fasteners and adhesives	9,848
C	Porsche 918 Spyder (Supercar)	625	Carbon fibre reinforced polymer pre-preg	6.38	Lay-up and pressure/ vacuum	500

There is a clear relationship in this table between the number of cars produced and the cost of the materials used, but there is the additional consideration of how the choice of material impacts on the manufacturing method:

Figure 1.6.4 Porsche 918 Spyder supercar.

A: Manufacturing economy cars can be highly automated with the use of presses that can simultaneously cut and form the panels required to make the body and, because low carbon steel parts are easily fused together, robotic arms can be programmed to spot weld them to produce the car body. Minimal human intervention is required in this process, and the dies used in the presses and the robot arm programming can be easily and quickly changed to facilitate variations in the model being manufactured. This facilitates the incorporation of **Just in Time (JiT)** manufacture to maximise profitability.

Figure 1.6.5 Robots welding low carbon steel economy car bodies.

B: Prestige vehicles increasingly make use of the more expensive metal, aluminium, which is lighter than low carbon steel and is also more corrosion resistant. The improved power-to-weight ratio makes the car's performance and economy much better. A disadvantage of using aluminium is that it is more difficult to weld, particularly using spot welding, so an ever-increasing range of mechanical fastenings such as rivets are being used to assemble these cars, often in conjunction with adhesives such as epoxy resin. This means that the level of automation cannot be as great as with the economy vehicle, since mechanical fastenings usually need to be fitted and locked in place by using more labour-intensive methods.

C: Supercars – the very expensive limited edition sports cars made by luxury vehicle manufacturers – are usually made using carbon fibre reinforced plastic (CFRP), since it offers an unparalleled strength-to-weight ratio (for this reason it has also been universally adopted in the manufacture of racing cars such as Formula 1). The disadvantage of using CFRP is that it requires high precision moulds, similar to the dies required for presswork, but the layers of pre-preg (carbon fibre, pre-impregnated with epoxy resin) usually have to be cut and also positioned in the mould by hand, and then air pressure systems have to be manually connected before curing in a special oven called an autoclave. Some manufacturers are, however, starting to automate this process, so CFRP may become more commonly used in everyday cars.

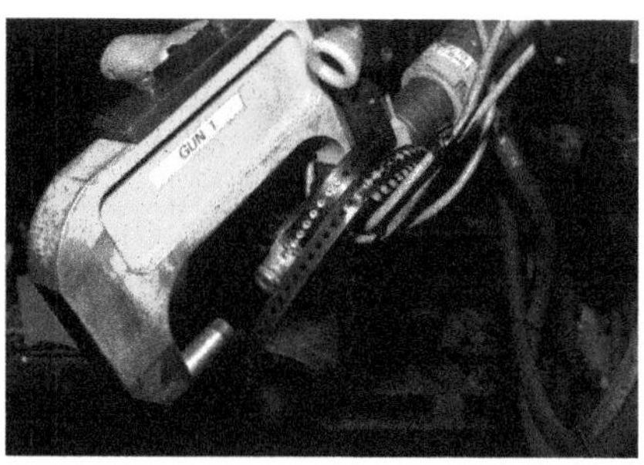

Figure 1.6.6 Labour intensive mechanical fastening assembly method required for aluminium car body.

Additional factors need to be considered such as the quantity of material being bought by the manufacturer, because clearly the larger manufacturers will be in a much stronger position to negotiate bulk purchasing discounts, unlike smaller, artisan manufacturers making one-off or small batch products that do not warrant large orders of materials.

Figure 1.6.7 CFRP car part being inspected after removal from the autoclave.

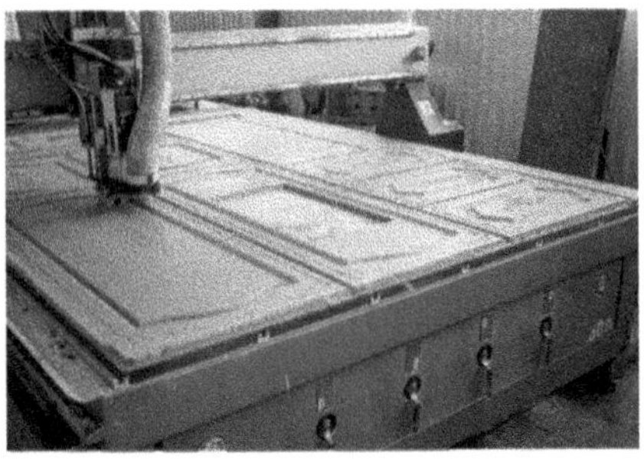

Figure 1.6.8 Nested components being machined on a CNC router.

Figure 1.6.9 Cellular beams making optimum use of material.

ACTIVITY

Research an alternative product to the car to use as an example in the scale of manufacture section of this chapter. Find out how the materials and manufacturing methods differ for each of the scales of production.

Design and the economic use of materials

Using materials economically can range from making sensible use of a standard size sheet of manufactured board for a one-off shop fitting project to the sophisticated incorporation of computer technology in the designing and manufacturing of reduced weight glass packaging.

Professional woodworkers often use software to plan how they will cut out multiple parts from a piece of MDF or plywood. This will ensure that the parts are nested economically and a large number of unusable off-cuts are not left over. Even without software, it makes sense to plan projects around standard sizes of board, timber and other materials, since it avoids wastage and additional machining.

A designer intending to make a number of batch produced storage boxes with sides specified to be 410 mm × 410 mm, would find it worthwhile to establish if this dimension could be modified to 400 mm × 400 mm, since this would avoid the potential wastage of approximately one-third of each of the standard 2,440 mm × 1,220 mm boards.

Many mass-produced children's toys are now made by rotational or blow moulding, which result in a hollow structure that keeps the amount of material to a minimum, while ensuring that the overall product is strong due to the coherent, one-piece moulding that results from these processes. Injection moulded hollow toys need to be made of at least two parts, which would require additional welding, integral, mechanical or adhesive joining methods, along with more expensive moulds.

Careful selection of appropriate metal sections can be used to minimise the amount of material used. The flexural (bending) strength of a beam, for example, depends on ensuring that the material in the beam is concentrated in the areas that will be subject to the most stress. This explains the widespread use of I beam girders in construction and is particularly well illustrated by cellular beams which have large holes along the central section, often referred to as the web. This type of girder makes optimum use of this principle by cutting and welding a standard girder to produce a particularly economical and light but strong beam for a roof.

This principle is often applied to the design of aluminium products such as ladders by using tubular or I beam extrusions, since it is essential that they are light while maintaining excellent flexural strength.

Furniture designers often make use of lightweight frames and suspended seats that capitalise on the elasticity of materials to provide comfort, rather than traditional upholstery methods, which require the use of a number of materials that have to be combined using labour-intensive methods. This movement towards economy of materials, which also facilitated easier mass production, was pioneered by the Bauhaus, whose designers made widespread use of tubular low carbon steel.

ACTIVITY

Produce an A4 page showing, with the use of notes and diagrams, how you planned the manufacture of components for your project in the most economical way possible from the raw materials you were supplied with.

ACTIVITY

Choose a product that you are able to study safely and produce a series of annotated, analytical sketches to explain how strategies for saving materials have been used in its design and manufacture.

Glass bottles are becoming thinner thanks to advances being made in the blow moulding process that is employed in their manufacture. They are now over 30 per cent lighter than they were in the 1980s, which makes a significant difference to the cost of materials for the product, as well as the associated energy costs and environmental effects. More sophisticated computer control and monitoring of blow moulding machines, as well as the introduction of finite element analysis (FEA) in the design process, has helped in achieving these reductions.

Manufacturing processes which increase accuracy and reduce waste

Manufacturers are keen for components in their products to be as accurate and uniform as possible in order for them to function effectively and integrate seamlessly with other elements of the design, therefore facilitating the interchange of parts.

For example:

- The engine of a car requires parts that fit and work perfectly together as a pre-requisite of effective function and serviceability.
- The positioning of the holes for KD fittings in flat-pack furniture has to be precise in order for the furniture to be assembled without difficulty.
- PET and glass soft drinks bottles must be accurately made in terms of material thickness, as this will affect their capacity, and the precision of the screw opening must be made to very strict tolerances, since this will impact on the seal of the bottle top.

In order to ensure the accuracy of manufacturing, particularly in large-scale production, automated machinery is used, since it would be far too expensive and time consuming to rely solely on human input. Automated screw-making machines work by means of cams and levers rather than relying on computer control, since they are set up to carry out the same process millions of times, so the flexibility offered by CNC machinery is inappropriate. Cams usually take the form of specially shaped discs that revolve and press against other parts of the mechanism to make them move as required. Mass production of products using automated machines like this dates back to 1805 when Marc Isambard Brunel introduced machines for manufacturing pulley blocks for the navy.

CNC machines tend to be used for automated production when a range of different parts needs to be programmed and manufactured in a relatively short time, so their flexibility is invaluable. Some companies are now moving towards making greater use of CNC rather than dedicated machines.

Figure 1.6.10 Mass produced Robin Day PP chairs.

Comparing bulk and one-off production

One-off production is the **bespoke** design and manufacture of single products in accordance with individual specifications, often requiring a high proportion of manual labour and skill. This is reflected in the high cost of such products.

Bulk production, whether at line, mass or batch level, makes it viable to use automated manufacturing techniques, and also means that materials can be bought more cheaply in larger quantities.

Tens of millions of the Robin Day PP chair have been sold around the world because it is an excellent solution to the problem of providing a practical, easily cleaned, stackable and economical seat for use in offices, schools and other locations. The injection moulded PP shell and welded tubular low carbon steel frame are highly compatible with high volume production in a range of sizes and make it possible for it to be sold cheaply.

Figure 1.6.11 Ergochair Adapt® chair with multiple individually configurable features.

This contrasts with bespoke office chairs, such as the Ergochair Adapt® range, which can cost thousands of pounds when assembled to suit individual anthropometric specifications.

The adoption of computer systems for design and manufacture has provided the capacity for manufacturers to produce products in bulk that can have bespoke elements incorporated in their design. Mass customisation like this makes it possible for customers to have access to personalised goods without paying the premium that is normally associated with one-off products.

Examples include sports shoes, cars and computers. Orders for such products are placed using online retail websites which link to the manufacturer's computer-based production system.

Some personalised products simply modify the name on a packaging label or randomise elements of the manufacturing process, such as the addition of colour to glass bottles during moulding. These techniques result in products with variations, but cannot be considered to be truly bespoke.

The use of computer systems

Computer technology has revolutionised the way that products can be made, ranging from the large-scale use of robotic welding and machining in a car manufacturing plant to a hobbyist using a 3D printer at home to create parts for various projects. As we will see in this section, however, the influence of computer technology goes well beyond this and impacts on the design, manufacture, marketing, storage, distribution, sales and disposal of products.

Computer systems for planning and control

Figure 1.6.12 RFID tag.

Sophisticated planning and control software uses mathematical calculations to plan the schedules necessary to organise a complex manufacturing facility such as a car plant. Staff involved in operating and monitoring the software and resultant schedules can obtain an instant overview of their operations by accessing constantly updated spreadsheets, charts and graphs. Barcodes and radio frequency identification (RFID) tags are used to facilitate the transfer of information relating to components and stock to and from computer systems. Barcodes are more reliably read than text when scanned from various angles and RFID tags contain a small integrated circuit (IC) that can communicate information in both directions, when used with a scanner, as well as providing security against theft in retail environments.

Modular/cell production

A level only

Figure 1.6.13 AGV being used to transfer car parts.

Figure 1.6.14 Unwanted, stockpiled cars.

Computer controlled manufacturing cells are set up to combine a number of CNC machines in a group. This group of machines, known as a CNC module or cell, can be programmed to carry out a sequence of operations on a number of CNC machines, such as lathes, milling machines, drills and grinders.

The job of loading the material at the start of the process, transferring the work piece between machines and removing the finished part is carried out by a robotic arm. Robotic arms can carry out a range of tasks depending on the type of gripping mechanism fitted and can have up to six axes of movement.

If longer distances are involved when moving materials or parts, automatic guided vehicles (AGVs) are used, which are programmed along with the manufacturing cells that they are servicing.

For example, in order to manufacture a drive shaft for a machine or vehicle, the following stages give an indication of how the CNC cell would operate:

1. Low carbon steel stock is delivered to the cell by AGV and loaded by robot into the chuck of the first machine.
2. The part is turned (CNC lathe).
3. A robot transfers the part to the next machine.
4. Grooves and flats are machined (CNC milling machine).
5. A robot transfers the part to the next machine.
6. A perfectly smooth finish is achieved by grinding (CNC grinder).
7. The finished part is removed by robot and is ready for transfer by AGV to another cell or assembly facility.

JiT manufacture

JiT production is a method of organising the manufacture of products in such a way that they are made to order rather than being produced in the hope of them being sold at some point in the future. Embracing and developing JiT manufacturing was a key element in the success of Toyota, the car manufacturing company, in the 1970s. Prior to this, it was the practice of most manufacturers to stockpile their products until they were bought, with corresponding financial burdens due to factors such as materials costs, storage requirements, damage and redundancy of stock. The stockpiling approach to manufacturing is sometimes referred to as 'Just in Case'.

Taiichi Ohno, an executive at Toyota, was an influential figure in the development of JiT manufacture, who developed a system that enabled the company to react flexibly to customers' requirements rather than trying to predict them. Vehicles could, therefore, be manufactured according to the orders received, and the quality of them was vastly improved in the process. This was particularly successful due to the adoption of steps to reduce waste, continuously monitoring and improving their methods of manufacture and using techniques to reduce errors on the production line so that products are 'right first time'. 'Lean production' is an alternative term that is often used in this context, since it conveys the idea of reducing waste and maintaining stock levels at the minimum possible.

This type of production is particularly appropriate for products that can be bought in multiple configurations, such as vehicles and computers, but can also be seen in action in fast food restaurants, where customers expect their meals to be freshly prepared rather than containing ingredients that have been pre-cooked earlier in the day.

The stockpiling of products is a potentially problematic strategy, particularly for goods such as modern digital devices, since new models with updated and improved features will quickly make older devices obsolete and result in stock that is difficult to sell. The unsold stock has to be stored in warehouses, which involve considerable costs, as well as rendering goods vulnerable to damage and theft.

As many of the modern industrial practices were first developed in Japanese manufacturing, they are after referred to by their Japanese names. Some of the strategies used to implement JiT in manufacturing include:

- increased consultation, discussion and co-operation between management and workers
- regular feedback meetings to ensure continuous improvement (Kaizen)
- careful selection of suppliers, often in relatively close proximity to the factory where assembly takes place
- reduction of waste and stock (Muda)
- optimisation of the configuration and layout of machinery
- stock labelling systems, which give instant visual and/or electronic indication of current levels and reordering requirements (Kanbans).

Figure 1.6.15 Kanban cards being used to control parts stock levels for JiT production.

There are also potential disadvantages to the JiT system, the main one being the dependence on reliable suppliers and transport infrastructure. Toyota lost millions of dollars' worth of business in 1997 due to a fire at a subsidiary company that supplied them with braking system valves.

In car manufacture, JiT meshes with a computer based process to ensure that customers take delivery of exactly the vehicle they require in the time scale that they agree when placing their order with a dealer. This is achieved by using the following simplified process:

1. A customer uses the manufacturer's interactive car configuration website to select all appropriate options for the vehicle they require, such as engine type/size, wheels, colour, trim, audio and accessories.
2. The information is digitally transferred to the car manufacturer's planning and control software that generates the order, organises the necessary parts and places the manufacturing process in an available time slot.
3. The lead-time that is generated by the software will enable the dealer to give the customer an anticipated delivery date, which may be longer if the car is being assembled abroad or if there are any particular issues with the supply of parts.
4. Manufacture of the vehicle commences, with an RFID tag being used as an electronic Kanban to ensure that the correct parts are available and appropriately assembled as the vehicle proceeds along the assembly line.

Figure 1.6.16 Car configurator being used by a customer at a manufacturer's dealership.

5 Once completed, the car can be delivered to the customer's local dealer on a transporter that has been loaded in accordance with the computer software's optimum schedule for economical delivery.

Manufacturing vehicles in this way can result in potentially millions of different option combinations, so it is sometimes referred to as mass customisation.

Quick response manufacturing

QRM makes extensive use of CAD and rapid prototyping with 3D printers in the product development stage, and flexible manufacturing cells and multi-skilled workers in product manufacture. It also incorporates and builds on JiT, Six Sigma and total quality management (TQM) strategies. More detailed information on Six Sigma and TQM can be found in Chapter 2.9 Design for manufacture and project management.

This type of production is particularly appropriate for companies that produce relatively low volumes of products that incorporate elements of customisation.

A typical example of a product that would be produced using QRM principles is a machine for packaging food. QRM is appropriate because:

- These machines have to be designed and built to carry out a particular sequence of operations for the goods they are packaging; therefore customised requirements have to be accommodated.
- The use of QRM facilitates rapid generation of design proposals and prototypes as well as a smooth, fast transition into the manufacturing stage.
- Customers can take delivery of their machines more quickly than would otherwise have been possible.

Flexible manufacturing systems

A level only

Flexible manufacturing systems (FMS) make extensive use of **modular/ cell production** (see page 107), and other systems that facilitate easy transfer of production from one part or product that is being made, to another. This makes it possible to respond quickly to changes in demand, so factories can switch production to accommodate this.

A key development in this respect is the adoption of single minute exchange dies (SMED) that facilitate changing the dies in a press or moulding machine very quickly in response to demand for a different part. The term relates to the target of being able to change dies and moulds in less than ten minutes (single digit minutes) and is often facilitated by using a range of quick release locking devices and multi-purpose parts.

An example of SMED can be seen in factories manufacturing car radiators where presses are used to blank, pierce and form the many components that are required. Before this development, the presses would typically be idle for many hours while the dies were changed ready for the new parts to be manufactured. SMED methodology encompasses very quick transfer of the dies and moulds, resulting in reduced idle periods, known as downtime, for the machine. It also offers financial advantages since keeping expensive machines like this in operation for the maximum time possible is a key element in running a manufacturing facility efficiently. Moulds used in forming processes for plastics, such as blow moulding and injection moulding, are another example of a situation where quick changeovers are important.

Computer systems for production, distribution and storage

When manufacture is taking place, the stock of parts – known as the inventory – can be monitored by using barcodes or RFID, and most JiT systems rely on them to facilitate an efficient flow of parts for the production line.

Once products have been made, their storage, distribution and sales can be monitored by computer systems that also make use of these technologies. Examples of where computers have a major impact include using software to plan optimum loading patterns for freight containers and electronic point of sale (EPOS) systems that are used to record sales of products and feed this information through to warehouses for re-stocking purposes.

Figure 1.6.17 Worker scanning an inventory in a warehouse.

A term often used to describe the use of computer systems to control all the stages of bringing a product to market from design, through manufacture to distribution, is computer integrated manufacture (CIM). Achieving full integration of all these elements is a highly complex undertaking, involving a number of stages including CAD, prototyping, material costing and ordering, CAM, QC checks, warehouse organisation and distribution.

Warehouses increasingly employ computerised AGVs, forklifts and narrow aisle stacking machines. This type of arrangement, with no manual handling involved, makes it possible to use space very efficiently and facilitates round-the-clock operation.

Standardised and bought-in components

Standardised components

Standardised components are those that must be able to fit other elements of a design. Consider what would happen if you bought a punched notepad only to find that the holes were in the wrong position for the clips in your folder. This has been addressed by stationery manufacturers adopting a universally accepted standard spacing of 80 mm.

One of the most significant breakthroughs in standardisation was the British standard Whitworth screw thread that was introduced in 1841. Before this development, threads made by different manufacturers had varied angles, depths and pitches, which meant that replacement of parts and maintenance were very problematical, as threads made by one manufacturer did not match those made by another one. A metric international standards organisation (ISO) system has now been adopted, which makes compatibility even more effective as it has been adopted in most countries. The USA and Canada still use a thread sizing system based on fractions of an inch, but ISO metric threads are now becoming much more widely used.

Another example of standardised components that are vital to the success of a particular group of products is the system of connecting light bulbs to their fittings. Bayonet fittings, for example, were first used in lighting in the

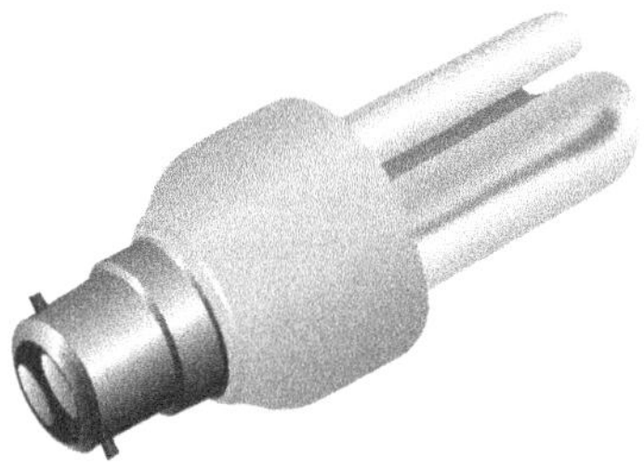
Figure 1.6.18 Low energy light bulb with bayonet fitting.

1870s, but there have been many subsequent developments, all of which are dictated by the British Standards Institute (BSI). In the case of bayonet fittings, the specification including aspects such as dimensions, tolerances, material properties, etc. is set out in BS EN 61184. Manufacturers of lighting products requiring the use of bayonet fittings need to comply with all aspects of this document.

Compatibility achieved by the adoption of universal standards makes it possible for product designers and manufacturers to specify third party **bought-in components** for their products. The following are examples of standardised components that are incorporated into products that are in wide use:

- tyres
- plumbing fittings
- door locks
- audio connectors
- batteries
- fuses
- fitted kitchen units
- printer cartridges
- biro refills
- memory cards.

Bought-in components

Bought-in components are used extensively where it would not be practicable or economical to manufacture those components, often due to their specialist nature and/or the high numbers involved. The most common include fastening devices such as screws, nuts, bolts and rivets, along with electronic components such as ICs, transistors and resistors.

Sometimes companies use bought-in components that take the form of **sub-assemblies**, such as electric motors, hydraulic valves and gearboxes. Suppliers of components and sub-assemblies make CAD files of their products available in standard formats so that manufacturers can more easily integrate them into their designs and therefore improve sales of their components.

By using bought-in components, companies can make major financial savings because:

- the components can be bought in bulk, which reduces the unit cost
- they do not have to set up their own manufacturing facility, reducing the need for specialist equipment and expertise which can be very expensive
- there will be a greater level of consistency in the components
- time will be saved.

Figure 1.6.19 Tyres ready for fitting on cars on a production line.

It is essential, however, that the quality of bought-in components is maintained, therefore many manufacturers use the accreditation of ISO 9001, which is an internationally recognised quality management system that ensures the required consistency of goods.

Figure 1.6.20 These camera battery chargers are being sold because they cannot be used with newer batteries.

Changing standards

When standards used by manufacturers change, it can cause major problems to customers who use their products, as it can sometimes make the products or the ancillaries used with them obsolete. It is, however, sometimes necessary for companies to make major changes to their products to ensure that they stay competitive, and it is in this situation that brand loyalty is paramount. Some changes are adopted universally as a result of technological advancement. Examples of changing standard elements of products that have caused consumers compatibility problems include:

- Apple lightning connector replacing 30-pin type and the removal of the 3.5 mm headphone socket from smart phones
- introduction of HDMI connectors instead of SCART to facilitate HD TV
- Windows 10 installation causing software compatibility issues
- camera manufacturers changing the type of battery used in new models, so older batteries and chargers can no longer be used
- removal of optical disc capability from some new models of laptop, making it difficult to use DVDs, etc.
- new printers that will not accept earlier types of cartridge, so they may have to be discarded.

Sub-assembly

Sub-assemblies are elements of a product that are created as a unit prior to being combined into the final product. An example of this is a bicycle, which would incorporate sub-assemblies such as:

- gear cassette
- derailleur gear changing mechanism
- braking system
- lighting system
- mudguards
- saddle
- frame
- chain.

Most bicycle manufacturers concentrate on producing their own individual frame designs, while incorporating bought-in sub-assemblies from a range of specialist suppliers in the particular configuration required for their finished products. It would be uneconomical for companies to make most of these sub-assemblies themselves, since a great deal of specialist knowledge and high levels of investment would be required.

There are also situations in which sub-assemblies simplify and speed up the manufacturing process in industries where it is normal practice for all parts to be made in the same factory. An example of this is a furniture manufacturer, who would make drawers as a sub-assembly prior to fitting them into a cabinet.

Figure 1.6.21 Some of the sub systems used in the final assembly of a bicycle.

KEY TERMS

Scale of production: the consideration of the number of products to be made, using particular manufacturing methods to suit a particular market.

Batch production: the manufacture of groups of products to increase efficiency and economy.

Division of labour: the efficient organisation of a workforce so that individuals specialise in particular manufacturing tasks.

Line production: the manufacture of large numbers of products in factories set up so that processes can be efficiently carried out by workers and/or teams organised in a specific sequence.

Mass production: the manufacture of large numbers of products in factories that are usually highly mechanised.

Unit production systems (UPS): the use of overhead transporters for component transfer between workers to improve factory efficiency.

Quick response manufacturing (QRM): the use of mainly computer-based technology to facilitate efficient, competitive production of low-volume, customised products.

Vertical integration/vertical in-house production: the organisation of manufacture to reduce dependence on externally sourced parts and sub-assemblies.

Just in Time (JiT) production: the manufacture of products as needed, in response to existing orders.

Bespoke (one-off) production: the design and manufacture of products to individual specifications.

Flexible manufacturing systems (FMS): production using work cells of CNC machines and robots that can be used to make a wide range of different products, typically one-off bespoke items or in small batches.

Modular/cell production: the use of groups of CNC machines, robots and AGVs to facilitate efficient, flexible manufacturing.

Standardised components: parts such as screws and light bulbs that are made to a common standard to ensure interchangeability.

Bought-in components: product parts that are sourced from external suppliers rather than being manufactured in-house.

Sub-assembly: a self-contained element of a product that is made separately and incorporated in the final assembly stages.

KEY POINTS

- The number of a particular product that needs to be produced, and the nature of the materials and manufacturing processes required, dictate the means of organisation of production to ensure commercial success, efficiency and appropriate quality standards.
- Division of labour, whereby workers specialise in executing particular production tasks, is a particularly significant element in facilitating the efficient manufacture of products in large numbers.
- Designers must consider the appropriate type, size and form of supply of materials they are incorporating in products and parts, in conjunction with the manufacturing methods to be used and the shape and form required. This will help to ensure that their products are effective in withstanding forces and satisfying other specification requirements.
- Wastage of materials can be avoided by carefully considering stock sizes at the design stage to avoid generating significant amounts of unusable surplus.
- Computerised systems are an increasingly indispensable factor in improving the quality, efficiency, flexibility and organisation of the design, manufacture, marketing and distribution of consumer products.
- CNC machines are often used, along with robots, in highly efficient and flexible manufacturing cells.
- Many products are made more efficiently by splitting them into sub-assemblies, which are sections of the product that can be pre-manufactured as a self-contained unit and then incorporated into the product at a convenient stage of final assembly.

Check your knowledge and understanding

1 Name all of the scales of production, choose an example for each one and explain why it is appropriate in each case.

2 Explain the term 'division of labour' and how it relates to the organisation of line production systems.

3 Why is low carbon steel used to manufacture the bodies of economy cars? What are the disadvantages of two alternative materials that result in them not being chosen for economy cars?

4 Explain three contrasting methods of reducing the amount of material used in the manufacture of products.

5 Explain the difference between CNC and dedicated machines, and give examples of the products or parts they would manufacture and why they would be appropriate in those situations.

6 Name and describe a specific design of chair that would be manufactured by using mass production methods, and explain how its design makes it suitable for this.

7 Explain how RFID tags are used in the manufacture and distribution of parts and products.

8 Explain the meaning of 'CNC manufacturing cell' and explain how it might be used to manufacture a typical part.

9 Contrast JiT and stockpiling production methods.

10 What is QRM and what type of products is it appropriate for?

11 Name two methods that could be used to enable manufacturers to be as flexible as possible in responding to varying demand for products.

12 How does EPOS help to maintain satisfactory product stock levels?

13 Name three standardised parts or products, and explain why their standardisation is important.

14 Why might a manufacturer make use of bought-in electric motors rather than making their own?

15 Name two sub-assemblies used in the manufacture of a bicycle.

Further reading

The Genius of Design by Penny Sparke (Quadrille Publishing)

Materials Selection in Mechanical Design by Michael F. Ashby (Butterworth-Heinemann)

The Machine that Changed the World by James P. Womack, Daniel T. Jones and Daniel Roos (Simon and Schuster UK Ltd)

The International Organization for Standardization website: www.iso.org/iso/home.html

This website explains the applications and potential of 'radio frequency identification': www.rfidjournal.com

This website gives information about single-minute exchange of dies: www.leanproduction.com/smed.html

This website explains the significance of the Portsmouth Block mills in the development of mass production: www.gracesguide.co.uk/Portsmouth_Block_Mills

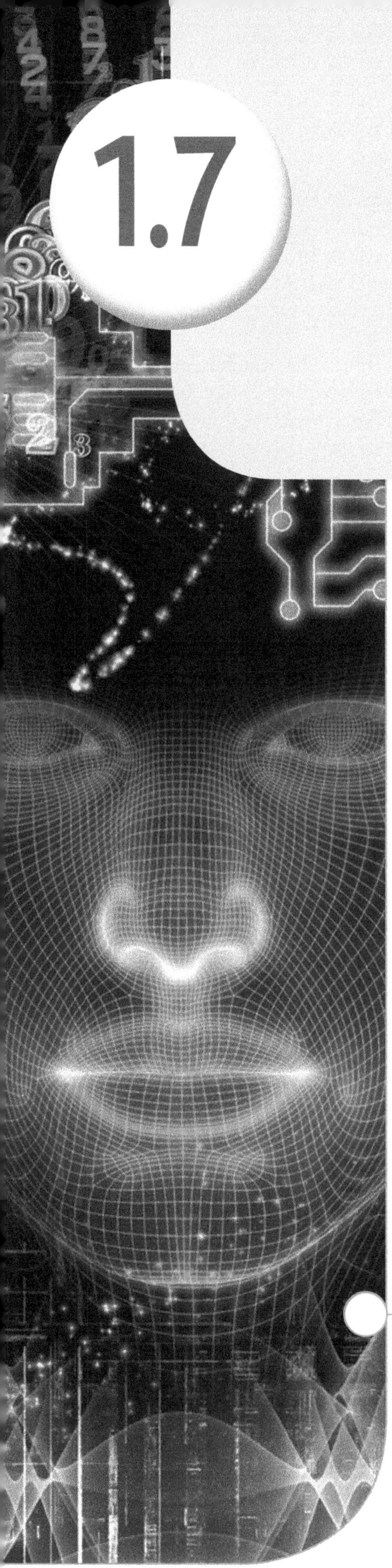

1.7 Digital design and manufacture

LEARNING OUTCOME

By the end of this section you should have developed a knowledge and understanding of:

- the use of computer aided design (CAD) to develop and present ideas for products, and the advantages and disadvantages of using it compared to manually generated alternatives
- how computer aided manufacturing (CAM) – including laser cutting, routing, milling, turning and plotter cutting – is used in the manufacturing of products.

If you are studying at A-level you should also have developed a knowledge and understanding of:

- virtual modelling/testing procedures used in industry
- rapid prototyping processes and additive technologies, and the benefits they offer to designers and manufacturers
- the use of EPOS for marketing purposes and the collection of market research
- the role of production, planning and control (PCC) systems in the planning and control of all aspects of manufacturing.

In the last 20 years, **computer aided design (CAD)** and **computer aided manufacturing (CAM)** has become one of the key methods of developing and manufacturing products both in schools and colleges, but particularly in industry. As software and hardware has developed, it has become both easier to draw designs and less complex to download designs to manufacturing equipment. Coupled with these advancements, there has been the growth of simulation software, and with it, the ability to plan production. Finally, advancements in software have also enabled the use of computers to control the flow of materials, components and finished goods through the factory.

Computer aided design

CAD includes the use of computers to produce either 2D working drawings or 3D full colour virtual models. In recent years, CAD software has become easier to use and more powerful, with a wider range of features to help in the design and development of products, and the ability to download drawings to CAM equipment.

Figure 1.7.1 CAD software has a wide range of features to help in the design and development of products.

Advantages and disadvantages of CAD

The main advantages of using CAD include:

- CAD speeds up the product design and development process. The ability to edit and develop existing drawings that are stored on a computer makes it faster than having to redraw items by hand.
- Using CAD makes it easier for teams of designers to work collaboratively. Such teams may be located in different offices and even different countries. The use of CAD and web conferencing makes collaborative working possible, and reduces the need for people to travel.
- Completed CAD drawings can be downloaded to CAM equipment such as laser cutters, routers, lathes and milling machines. Alternatively, they can be sent to 3D printers to make models or prototypes.

The main disadvantages of using CAD include:

- The initial setup cost can be quite expensive, especially when staff have to be trained in the use of software.
- Software has to be updated on a regular basis, which can be expensive.
- CAD software is not very easy to use for quick sketches.

The use of CAD to develop and present ideas for products

3D CAD software allows designers to produce high quality 'artist impressions' of their designs. This can include full colour rendering, with light, tone and texture to give a real impression of what a product would look like when made. These drawings can be rotated and viewed from any angle. When a designer can produce such drawings, they are able to show them to clients or marketing teams for feedback. This allows the design to be modified and developed more quickly than if models or prototypes were made.

Other ways in which CAD is used to develop products include:

- Original art work can be scanned and copied into CAD software, so that the drawing can be digitally developed. Editing tools such as 'copy', 'mirror', 'rotate', 'scale' and 'array' allow designers to change drawings quickly.
- Libraries of standard component drawings can be used to complete designs. This is particularly useful for design engineers who might need to do circuit designs in electronics, pneumatics or hydraulics.
- Designers can use 'layers' to draw on. This allows complex designs such as electronic circuits to be built up layer by layer, or for architects to overlay services such as plumbing, electrics and air conditioning over the layout of rooms.

CAD and CAM software often has the facility to run a simulation of what will be machined when a drawing is downloaded to a piece of equipment such as a laser cutter, router or 3D printer. Such simulation will highlight any potential problems and indicate if the item being made will turn out as expected. In the case of a laser cutter, simulation will often show if an outline of an object has gaps that would result in the product not being able to be removed from the stock material. In the case of a product being laser cut and engraved, simulation can show how long the item will take to be machined, as well as ensuring that the cutting and engraving is done in the correct order (usually engraving first).

Simulation can be used to show the tool path (the direction in which tooling will move across a piece of stock material). This will highlight any problems such as potential clashes with clamps, vices or the tool moving beyond the X, Y or Z limits.

Simulation is used in 3D printer software to show what the model will look like and how long it will take to machine. It is also used to indicate where the support material is required, so that the model will not collapse under its own weight when being printed.

Finally, simulation can be used to ensure that multiple copies of the same item are machined in the most efficient way. When cutting items from sheet material, this is known as 'nesting'; the tooling would be set to start in one corner of the sheet and there would be the minimum amount of space left between each part to be cut, so that the material is not wasted.

ACTIVITY

Using the internet, carry out research on the use of CAD and find real-life examples of how it is used in industry.

CAM processes

CAM usually involves downloading completed CAD drawings to machines that will cut and shape materials. Most CAM equipment uses software to convert CAD drawings into a machining programme, and only requires the operator to select the appropriate speed and power settings for the material they are machining. (Usually a slower speed and higher power is selected for denser material, and with increased thickness.) This ease of use has made it possible to use CAM processes for one-off production.

Figure 1.7.2 CAD drawing can be converted to a CNC program, which controls laser cutters.

Laser cutting

Laser cutting uses a high-energy laser beam to cut or vaporise materials and compressed gas or air to blow the waste material away, leaving a clean edge. Two-dimensional CAD drawings are downloaded to laser cutters, which convert the drawing into a **computer numerically controlled (CNC)** program. This program controls the movement of the laser in the X and Y axes, and the power and speed can be adjusted to cut or engrave different materials. Laser cutters in school or college are usually powerful enough to cut materials such as woods and plastics, but in industry, more powerful lasers are commonly used to cut sheet metals.

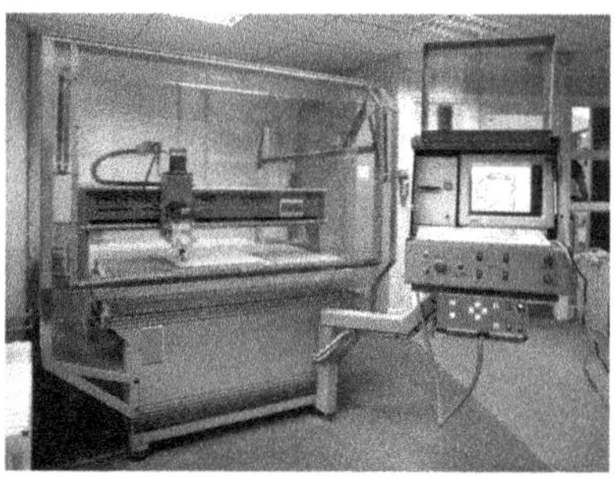

Figure 1.7.3 CNC routers are used to cut materials that are too thick for laser cutting.

Routing

Both 2D and 3D CAD drawings can be downloaded to CNC routers. They are usually used to cut sheet materials that are too thick for laser cutting, such as 9 mm MDF or thicker materials such as blocks of timber and Styrofoam, or high density modelling foam to make moulds or 3D models.

Milling

Milling machines work in a similar way to routers, and can be fitted with tools to drill holes, cut slots or shape the surface and edges of a range of materials (see Figure 1.7.4). Milling machines can move the 'bed' of the machine in the X and Y axes, but the cutter can also be moved in the Z axis to vary the depth of cutting. Some milling machines can also angle the cutting head and the machine bed. These are called 5-axis machines and they are ideal for complex engineering components. Modern CNC milling machines often have the ability to automatically change tools to suit different machining jobs and different materials.

Figure 1.7.4 Milling machines can drill holes, cut slots or shape a range of materials.

Figure 1.7.5 CNC lathes can be used to reduce metal bars or plastic rods, and for thread cutting and boring.

CNC milling machines are usually totally enclosed, which improves safety in the work area. They will be fitted with automatic coolant which floods the tooling and workpiece with a lubricant, resulting in improved tool life and surface finish. The machines are fully enclosed, so workers are protected from potential health and safety problems from flying debris and exposure to the coolant.

CNC milling machines can obviously operate much more quickly and accurately than human operators, and they can continue to machine for long hours with the exception of stoppage for maintenance.

Turning

Turning involves the use of lathes, usually to machine metals in bar form or plastic rods. This could be to reduce the diameter and length of a bar, or for more complex operations such as thread cutting and boring (see Figure 1.7.5). Like milling machines, modern CNC lathes are able to change tools automatically. Industrial CNC lathes might form part of a flexible manufacturing system (FMS) together with milling machines, laser cutters or punches, all served by robot arms loading material and transferring parts between machines.

Figure 1.7.6 Plotter cutters convert 2D CAD drawings into a CNC program to cut out from sheet materials.

Plotter cutting

Plotter cutters convert 2D CAD drawings into a CNC program that will use X and Y coordinates to cut out what has been drawn, usually on to thin sheet materials such as self-adhesive vinyl for graphics applications (see Figure 1.7.6). Most plotter cutters pull the sheet material in and out of the machine (Y axis), while the cutter moves from side to side in the X axis.

> **ACTIVITY**
>
> Using examples of CAM, make a list of the advantages of using such equipment when compared to machining by hand.

Virtual modelling

A level only

Virtual modelling can be used to test products before they are manufactured. By creating 3D CAD drawings, the product or a component can be put into a virtual model of where it may be used. For example, an engine component can be 'connected' to other parts in a virtual engine to check it 'fits' and that it could be removed for maintenance.

Virtual modelling can be used to simulate a manufacturing process, for example to check that a product will be machined correctly. In industry, it might also be used to plan a whole production line before investment is made in purchasing new equipment. An example of this is in the planning of a car assembly line. The equipment needed is highly specialised and costs millions of pounds to purchase and install. Using simulation software, designers and engineers can design the layout of production cells, including the positioning of robots, people, tools and workstations, and how the car and components will arrive at the correct place along the production line.

Figure 1.7.7 Virtual modelling is used to test products before they are manufactured.

> **ACTIVITY**
>
> Find three specific examples of where virtual modelling and testing is used in industry. Where possible, try to find examples of software that might be used to model and test products.

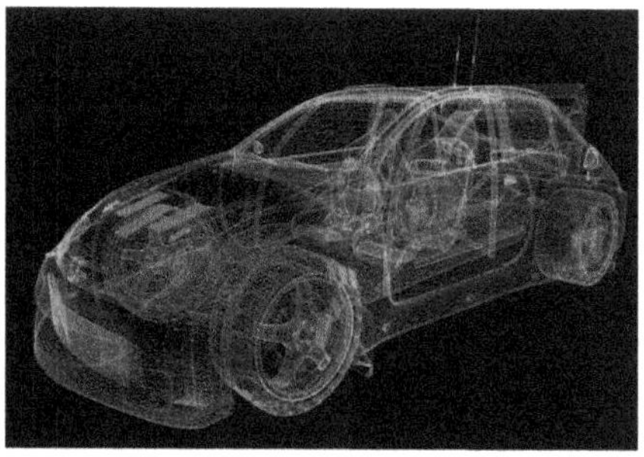

Figure 1.7.8 CFD results on a CAD drawing of a car. The red parts on the drawing indicate the areas that offer more wind resistance.

Computational fluid dynamics

Computational fluid dynamics (CFD) is a tool available in some 3D CAD packages, and is used to simulate the flow of fluids or gases in or around a product. For example, a vehicle design can be tested in a simulated wind tunnel to evaluate the drag (wind resistance) created by air flow over the external surfaces. Designers will then use the test results to develop and improve the aerodynamics of the vehicle in order to reduce wind resistance.

Another example might be a design for a boat, which could be tested in a simulated wave tank to evaluate how efficiently water flows around the surfaces of the boat.

By using CFD, designers can save time and money which would be needed to produce physical models and prototypes, and the expense of using specialist wind tunnels or wave tanks.

Finite element analysis

Finite element analysis (FEA) uses computer modelling to carry out component stress analysis. A designer or engineer can use FEA software to simulate things such as vibration or shock loads on products or components that they have drawn. This can highlight any potential weak points in a product which might need further development, before the design is actually manufactured. Using such software can save a lot of time and expense that would be required to make prototypes for physical testing.

Figure 1.7.9 A beam is being tested to check how the component will perform under static loading.

In Figure 1.7.9, a beam is being tested to check how the component will perform under static loading. Here you can see that the centre is the area that will be under the most stress and therefore it needs further welded reinforcement.

Rapid prototyping processes

A level only

Rapid prototyping is the process of downloading a 3D CAD file to a machine that will produce a 3D model of the drawing. The models can be made in a range of materials, and the modelling process can also vary because several types of rapid prototyping machines are available.

Regardless of the type of machine used, rapid prototyping is a fast way of making realistic 3D models of product designs. Prior to rapid prototyping, modelling would have to be done by hand, which is a highly skilled and time consuming process which, in industry, leads to delays in developing products and increased costs.

An example would be in the development of ceramic tableware. Traditionally, this would have been modelled by a skilled potter who would throw or cast one-off items, fire them, glaze them, fire them for a second time, decorate and fire them for a third time. This process would take days or even longer for complex items. With rapid prototyping, this time can be reduced to under 24 hours. Today, modern pottery companies use rapid prototyping machines such as stereolithographic modelling. This process uses a vat of photopolymer resin, which is solidified layer by layer, using an UV laser.

Stereolithography is one of the most accurate 3D printing methods capable of producing plastic models of products with a high level of accuracy and surface quality. Returning to the example of making models for the pottery industry, a designer can take a stereolithography model, spray paint it to simulate glazing and apply transfer decals, to give the impression of decoration which would be added in a production version. Models produced in this way can then be evaluated by potential customers, and by engineers who might suggest improvements or modifications.

Fused deposition modelling/3D printing

One of the most common types of 3D printers uses fused deposition modelling (FDM). These machines build or 'print' a 3D model from 3D CAD drawings by using processing software which 'slices' the drawing and guides an extruder in paths to build the product from the bottom up, layer by layer.

FDM rapid prototyping machines soften a thermoplastic filament and extrude it to build the model. They sometimes use a support material that acts like a scaffold while the modelling material hardens. The support material, which is extruded in a less dense pattern, is broken away when the model is finished.

Rapid prototyping allows products to be modelled quickly, accurately and cheaply. Such models can be used to show clients and potential consumers what the product will look like, to obtain feedback. If made full size, rapid prototype models can be used to make moulds for casting and making production dies.

In addition to modelling, 3D printers are starting to be used for one-off production of bespoke items. An example is in the production of cast jewellery, where a 3D printed pattern is used to then make moulds for casting. Another example is in the development of custom made cutlery grips for someone with arthritic hands, or to adapt products such as mobile phones or cameras for someone who has limited dexterity due to injuries such as burns or loss of fingers.

Recently, manufacturers have developed 3D printers capable of printing in metals. These are ideal for one-off bespoke production or limited production runs. 3D printing metals can be used to produce highly complex engineering components or structures that would be impossible to make in any other way. By 3D printing in metal, there is no need to fabricate (join) parts which creates weaknesses, or machine parts from solid which wastes material. Typically, metals such as titanium, stainless steel and aluminium might be used with such 3D printers.

Metal printing 3D printers generally use either electron beam melting (EBM) or selective laser melting (SLM). An electron beam or laser beam fuses metal powder layer by layer to make a product or component.

Metal printing 3D printers are also used in producing medical products. Examples include custom-made skeleton joints, prosthetics such as bone replacements or other medical devices.

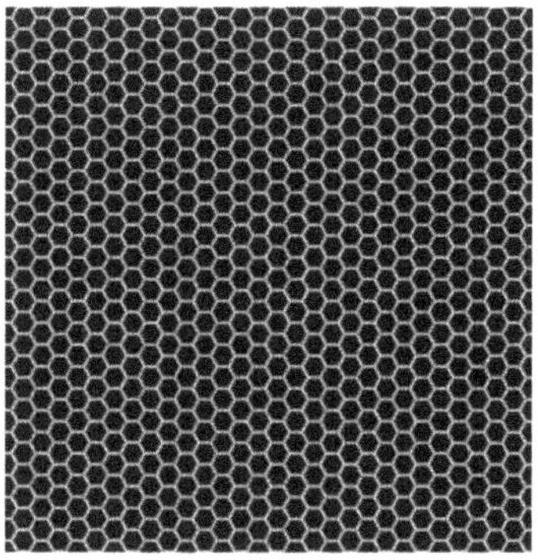

Figure 1.7.10 Honeycomb structure made from metal 3D printing.

Figure 1.7.11 A knee joint made by 3D printing bio-compatible cobalt-chrome alloy.

Electronic data interchange

A level only

Electronic point of sale

Electronic point of sale (EPOS) is a system of capturing data when consumers purchase goods. Many items are labelled with barcodes, and when they are scanned at the till, the details of the sale are recorded. This information is normally used to control stock levels; 'limits' can be set so that replacement stock will be reordered from distributors when a certain number have been sold. In turn, as distributors remove stock from warehousing, replacement products will be reordered electronically from the manufacturers.

In addition to controlling stock, the data is used by sales and marketing teams to monitor how many items are being sold. They can then use this information to target their marketing of products to help improve sales, or even determine when a new or improved product has to be launched if existing products seem to be in the decline stage of their life cycle. Sometimes, when consumers purchase goods at the till, they may be asked to provide some details such as an email address. This will be used to send the consumer a receipt, but it can also be used to advertise additional products to the consumer.

Production, planning and control networking

A level only

Computers are used in industry to plan and control production, availability of materials and components, and to co-ordinate suppliers and customers.

In modern manufacturing and a global market, it is vital for companies to operate an efficient supply chain network (SCN). The flow of materials and components required to manufacture products is essential for companies and most operate on **Just in Time (JiT)** production.

A **master production schedule (MPS)** is a plan that determines how many products will be made in a given time period. This planning is done using software that networks with other parts of the business, including personnel. For example, MPS software will determine the required amount of labour to carry out production and check the availability of labour. MPS software will also network with suppliers to order materials and control stock inventory.

Many manufacturers today only produce products to order – this means that the item is already sold before it is made. Customer orders are placed electronically via EPOS, and this is transmitted automatically to the manufacturer. The customer's order is then placed in the MPS software. This software will determine the appropriate timing for production of the customer's order, and trigger the ordering of materials and components from the supplier.

In the automotive industry, customers can specify particular requirements for their car, for example colour, two- or four-door, engine size, manual or automatic, diesel, petrol or hybrid. These details are transmitted to the manufacturer and their suppliers, so that the required materials and components are ordered in time for JiT production.

JiT production requires suppliers of materials and components to deliver stock only when it is needed for production (usually just a few hours before it is actually used). This has several advantages, such as less warehousing needed to store stock, and companies are not left with unused materials and components.

Many of the modern manufacturing systems that we use today were originally developed by the Japanese car manufacturer Toyota (known as the Toyota Production System). For this reason, much of the terminology is derived from Japanese words and phrases.

In manufacturing, the flow of materials and components through a factory is often controlled using a '**Kanban**' system. Small consumable items such as fasteners (nuts, bolts and screws) are kept in reusable plastic bins, and a barcode is used on these bins. As each bin is drawn from stores, the barcode is scanned and an order is placed to reorder them from suppliers.

Figure 1.7.12 Car assembly line. Note the gantry/track system used to move the car along the line.

Kanban is Japanese for 'signal card' and originally, as materials or components were used on a production line, a card would be put into a holder by the relevant workstation. These were collected at several points in the shift and the items reordered. These have now been replaced with barcodes and scanners or **radio frequency identification (RFID)** tags, and orders are sent automatically using computer networks. The Kanban system is also used to ensure materials and parts arrive at the correct point in a production or assembly line at the correct time. In the car industry, this is typically achieved using overhead gantries and conveyors that deliver parts to work cells at the precise time they are needed. As a car under assembly moves down the line, it will be tracked using barcodes or a telemetry device such as RFID. This feeds back to the MPS computer so that the progress of the assembly can be monitored.

MPS software can also be used to control manufacturing equipment. For example, in the production of cars, robots and other equipment can be automatically reconfigured to produce different models of cars or even completely different vehicles on the same production line. Jaguar Land Rover is an example where the production line can switch between making Jaguar cars and Land Rover Discovery models in order to respond to customer demand. This is an example of QRM; the use of flexible manufacturing equipment is an essential part of such manufacture.

On a smaller scale, FMS typically consisting of computer controlled lathes, milling machines, laser cutters and punches can make a wide range of different engineering components. The part programs that contain the machining code will be downloaded from the MPS network. The flow of materials coming into the flexible manufacturing cell can be automated by using robots and **automatic guided vehicles (AGVs)** – also scheduled by Kanban and MPS software. Because some processes take longer than others, there will be 'buffer zones' where items will be temporarily held in order to allow time for the slower processes to be completed. Finished items can be transported to the warehouse using AGVs, and the warehouse itself can often be computer controlled and networked to the MPS system.

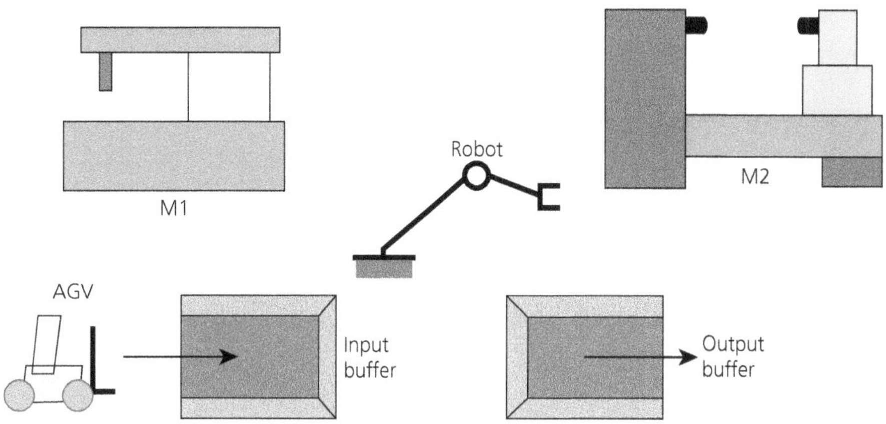

Figure 1.7.13 Diagram illustrating a simple FMS cell. In this example, there is a CNC milling machine (M1), a CNC lathe (M2) and a robot moving material from the input buffer, between the machines and to the output buffer.

ACTIVITY

Make notes on why rapid prototyping might be used to model the following products:

- a handle to be used on cutlery by someone with arthritis
- a prototype mobile phone.

KEY TERMS

Computer aided design (CAD): using computers to draw in 2D or 3D and develop products.

Computer aided manufacture (CAM): using computer controlled equipment to machine materials.

Computer numerically controlled (CNC): using a program to convert CAD drawings to drive CAM equipment.

Computational fluid dynamics (CFD): using software to virtually tests products in liquid or gases.

Finite element analysis (FEA): using software to test components and products to identify potential stress or weak points.

Electronic point of sale (EPOS): shop tills or similar that use barcodes or computer entry to transmit customer orders to suppliers and manufacturers.

Just in Time (JiT) production: a production method that relies on materials and components arriving at the point of manufacture, just in time for production.

Master production schedule (MPS): a plan used to determine how many products will be made in a given time period.

Kanban: a system used to control the movement of materials and components from suppliers and through a factory.

Radio frequency identification (RFID) tags: small electronic devices that can be scanned and used to track items as they move through a factory and into dispatch.

Automatic guided vehicle (AGV): robots, similar to forklift trucks, used to carry materials and finished goods around a factory.

KEY POINTS

- CAD software is used to quickly produce accurate 2D and 3D drawings.
- CAM equipment is used to accurately manufacture products directly from CAD drawings. The use of CAM reduces waste due to manufacturing errors, and consistent, high-quality products can be reliably made.
- Virtual modelling allows products to be tested before they go into manufacture.
- Computer software can be used to test products virtually to simulate loads and stresses.
- Rapid prototyping is used to produce 3D models of products, directly from CAD drawings.
- Rapid prototyping speeds up the process of developing products, and can be used to evaluate items before they go into full production.
- QRM involves the use of computer controlled, flexible manufacturing equipment that can respond to changes in customer demand.
- The use of computers in manufacturing to control the movement of materials, people and equipment allows for the most efficient use of resources and helps companies stay competitive.
- JiT production allows manufacturers to reduce their overheads. They do not have to carry stock and are not left with unsold products.
- Flexible manufacturing cells are not dedicated to making one particular product. They are 'flexible' because they can be used to manufacture a wide range of products.

Check your knowledge and understanding

1 What are the benefits of using CAD in the product development process?

2 What are the advantages of using rapid prototyping in the product development process?

3 Using examples, describe how CAM can ensure the quality of products.

4 How do manufacturers use computers in production planning, the movement of materials and control of manufacturing equipment?

5 Explain how computers are used in the testing of products in the development process.

Further reading

This website gives details of the Toyota Production System and explains the origins and philosophy behind modern manufacturing systems such as Just in Time: www.toyota-global.com/company/vision_philosophy/toyota_production_system/

This website gives a good overview of how CFD is used in the development and testing of products. Key terms are explained with useful real-life examples: www.solidworks.co.uk/sw/products/simulation/computational-fluid-dynamics.htm

This website explains what CAM is. It lists the advantages of CAM and there are useful links to related areas such as CAD: www.plm.automation.siemens.com/en_gb/plm/cam.shtml

This website is packed with examples of projects that have been made using 3D printers. It is useful for gaining inspiration for your own projects and there is the facility to download files to 3D print yourself: www.thingiverse.com

This website has several useful case studies that illustrate how 3D printing has been used in engineering projects and in medical applications. It shows examples of state of the art 3D metal printing and explains the process in easy to understand terms: www.renishaw.com/en/metal-3d-printing-32084

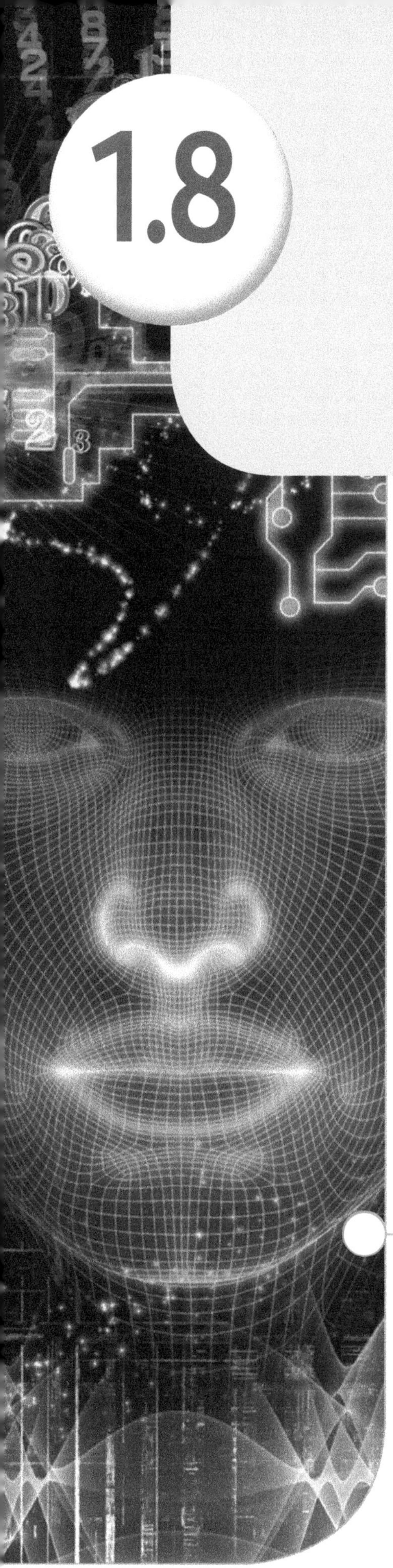

1.8 The requirements for product design and development

LEARNING OUTCOMES

By the end of this section you should have developed a knowledge and understanding of:

- the design, development and manufacture of products to meet specification criteria and fitness for purpose
- how the critical assessment of products can lead to the development of new designs
- how to critically assess products and develop new design proposals
- how to work with a variety of materials, including 2D and 3D forms, to produce creative and original products that satisfy the demands of the target market, and consider accurate and efficient manufacture
- the development of products that are inclusive in their design so that they can be used by a wide range of users including the disabled, children and the elderly.

If you are studying at A-level you should also have developed a knowledge and understanding of:

- accuracy of production
- how to consider aesthetics, ergonomics and anthropometrics when designing products.

Successful product design and development requires an understanding of the product and its intended purpose. By studying existing solutions to the problem or similar problems, designers can formulate accurate product design specifications (PDS) to guide them within the development of future solutions.

Product development and improvement

The concept of good design can be confused with aesthetically pleasing, but not necessarily functionally good design. We therefore need criteria to specify what good design is – criteria that can be used by all to remove the subjectivity of aesthetic appearance. The criteria we use in product development take the form of a PDS, and all products are assessed against their own specification. We can therefore compare products against their own specifications to decide whether they are 'good designs'.

For example, the Citroen 2CV car by André Lefebvre and Flaminio Bertoni was designed 'to carry four people and 50 kg of potatoes or a keg at a

Figure 1.8.1 The Citroen 2CV prototype.

maximum of 60 k.p.h with a consumption of 3 litres of petrol per 100 kilometres'. It successfully achieved all of the specification points set for it, and therefore is categorised as a 'good design'.

If you wish to assess the quality of a design beyond its specification, you can use a range of criteria such as the ten principles of good design created by Dieter Rams – a famous German industrial designer. When developing a PDS, we must assess existing solutions to the problem. This may take the form of a product analysis.

Specification criteria and fitness for purpose

The development of a design specification requires a clear set of testable criteria based on appropriate research that will allow the final product to be assessed.

Stuart Pugh, a highly influential design engineer and founder of the term 'total design', refers to the PDS as the key document that is used throughout all subsequent design activities. He devised a wide range of sub-sections that a PDS must address – known as Pugh's plates. A reduced list of the most common elements is shown below, and can be used to guide project research and development.

- Environment: consider the impact of the product on the environment during production, use and disposal/reuse.
- Testing: consider how each part of the product will be tested and the measurements that will mean success.
- Safety: consider all relevant safety standards and legislation.
- Product life span: consider how long the product should last before replacement is necessary (consider technology push, market pull and built-in obsolescence).
- Materials: consider the key material properties required for each component within the product.
- **Ergonomics**: consider how the product will be interacted with, and how the design will assist the ease of use for the client/user.
- **Aesthetics**: consider any appearance characteristics of the shape and form to meet the client's requirements.
- Performance: consider how well the design must function in comparison to existing solutions.
- Product cost: consider all of the cost restrictions upon the design regarding materials, manufacturing and labour costs, etc.
- Time scale: consider the project deadline.
- Size and weight: consider all size and weight restrictions on the product, including relevant **anthropometric** data.
- Maintenance: consider how the product will be maintained. Consider whether there are any consumable parts and how these will be replaced/changed.

A designer must ensure that a product meets the product specification to ensure the quality of the design, and that the end product is fit for purpose. All aspects of a PDS must be objectively tested; this means that quantitative data is key to allow consistency in testing.

Critical assessment for new design development

Design is constantly evolving. David Pye states:

'Everything we design and make is an improvisation, a lash-up, something inept and provisional.'

In this quote, Pye is referring to the fact that designers must assess and improve upon what has come before to avoid making the same mistakes. Designers must be willing to accept that they will never fully solve a problem, but can only devise a temporary solution that, with current technology and knowledge, addresses the specification they were given.

By critically assessing current products, we are able to use the experiences of consumers to guide the development of further products.

Critical analysis

Designers and manufacturers evaluate current products to help them develop ideas for new or improved products, by carrying out various analyses:

- Product analysis: an effective product analysis can only be completed on a product that you have direct access to, so you can interact with it. You can only assess how well a product succeeds in performing its perceived function if you have used it for the task it was designed to fulfil.
- Function (perceived): to analyse the function of a product, you must perform a **task analysis**, using the product and assessing its success in performing each stage of the task.
- Task analysis: the idea of a task analysis is to break a specific task down into individual stages. We often forget to include what we would regard as 'common sense' or obvious stages in the task. By breaking the task down further, we can look more closely at how a product is suited to a particular task and if there are any elements that may be improved.
- Ergonomics: to analyse the ergonomics of a product, you must consider how the user interacts with it (see page 132).
- Aesthetics: when analysing the aesthetics of a product, you should be factual and avoid using subjective generalisations such as 'it looks good'. The use of geometric or natural forms and the design inspiration for the product should be referred to, considering the aim of the designer and possible designers or movements that may have influenced the product. See Chapter 2.2 for further information.
- Materials: the choice of specific materials within a product is dependent on key mechanical and physical properties. You should consider the importance of each of these properties and whether the design has compromised on material selection for other reasons. The material selected may also be influenced by the method of manufacture and the scale of production.
- Manufacture/process: the manufacture of each component within the product should be assessed. Consider how the components have been assembled and in what order.
- Scale of production: the number of products produced has a huge impact on the method of manufacture and materials used. You should consider the suitability of the processes used and why they are appropriate.

- Environmental impact: you should evaluate the impact of the product on the environment – whether this is in the extraction of raw materials, the use of the product or what happens to the product at the end of its life. See Chapter 1.11 for further details.

Critical analysis: user-centred design

The concept of **user-centred design (UCD)** is that all products we design should be focused around the end user and the potential end user should be involved at all stages of the design process to guide the product's development. The reality is that the degree of end user involvement varies greatly within design.

Product concepts and models must be tested in real world situations with potential users. This is essential because the designer's opinion of how the product should be used may be different to the user's opinion, and customers should not be forced unnecessarily to change how they perform a task to accommodate a new product.

ACTIVITY

1 Perform a product analysis of the chair you are sat on, using the criteria listed on page 126.
2 Consider how you change the ringtone on a mobile phone. Write down a list of all the individual button presses required from the main screen to do this. Compare this with the procedure on another model of phone. How could the task be simplified?

Critical analysis: task analysis

When performing a product analysis, it is often useful to perform a task analysis. A task analysis involves monitoring potential users of a product performing specific tasks with that product. This can be done with existing solutions to the problem to identify issues or with working prototypes to iron out issues in development.

Working with a variety of materials

Concept modelling

During the design process, it is essential to model design ideas/concepts. This can be done in a wide variety of ways using a nearly endless number of materials. The selection of an appropriate modelling method is crucial, and this selection is dependent on the purpose of the model. All modelling is done to help make decisions or prove/disprove predictions.

Figure 1.8.2 Styrofoam prototype model.

Initial concept generation

Initial 2D or 3D sketch modelling can be used to gain feedback from clients prior to investment in physical prototypes. These sketches serve a clear purpose early in the design process; however, the feedback from a client is mainly restricted to comments on the form/aesthetics.

Block modelling

After sketch modelling, designers may progress into block modelling concepts to prove theories on mechanisms or show ergonomic features of design in a physical form. The use of lightweight compliant materials such as Styrofoam allows simple shaping of 3D forms using hand tools, but when moving parts are required, more resistant materials may need to be included to test reactions to forces.

Figure 1.8.3 Clay visual appearance model.

Visual appearance models

Visual appearance models are used to demonstrate the aesthetic form and appearance of the design concept. The model is not usually made from the same materials as the final product because it will be a one-off product so investment in complex moulding equipment would not be appropriate.

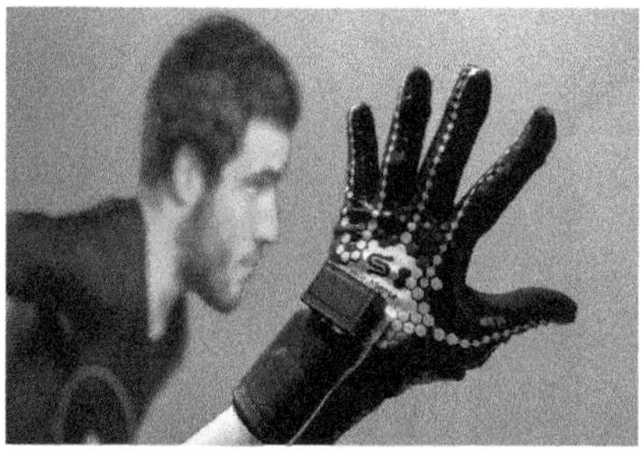

Figure 1.8.4 Working prototype model.

Figure 1.8.5 Rapid prototyping model.

Working prototypes

Working prototypes are used to test mechanisms and technical principles. They do not often resemble the final product and may be constructed from a range of materials and components; mechanisms can be adjusted with temporary fixings to finalise dimensions and movements.

Rapid prototyping

Due to the increased availability of rapid prototyping technology, these 3D block models are often created from 3D CAD models, which can be 'printed' in a wide range of materials including thermosetting polymers and resins. These CAD models can be tested prior to the printing of the model and, once saved, can be replicated several times following feedback from focus groups. This gives a range of iterations for the same product in a fraction of the time required to produce the same number of hand-made iterations.

Summary

The choice of block modelling method largely depends on the time and facilities available to the designer because investment in CAD software and rapid prototyping technology can be cost prohibitive. Although the cost is a major concern, there are clear benefits to both the manufacturer and client – the design can be tested using a range of computer-based facilities such as FEA and CFD (covered in Chapter 1.7), which also allow the manufacturer to simulate tooling situations and possible component production issues.

Accuracy of production

A level only

During design development, concepts must be suitable for production. The use of simple geometric forms within design increases the ease of manufacture, reducing complexity in machining and aiding the ability to produce accurate repeatable products by hand or machine.

Although the use of standard geometric forms is not critical, the ability to produce accurate, fully dimensioned drawings is necessary to ensure components can be combined effectively.

The level of dimensional accuracy used within design development must be in line with the production facilities available. When producing components within a school or college workshop, it is possible to draw CAD drawings to an accuracy of eight decimal places; however, the machinery used to produce the component will probably not be able to replicate this accuracy.

Considering aesthetics, ergonomics and anthropometrics

A level only

Aesthetics

The term 'aesthetics' refers to the features of a product that make it visually appealing. The appearance is affected by shape/form, size and proportion, colour and texture.

Donald Norman states in his book ***Emotional Design*** that products and systems which make you feel good are easier to deal with, and produce more harmonious results.

The balance between aesthetics and function is key to the commercial success of a product. How we assess aesthetics is complicated because it is a very subjective area. During design development, we must consider a range of aesthetic aspects.

Shape and form

Figure 1.8.6 Morgan Aeromax.

Shape and form refers to the 2D and 3D physical appearance of a product. The shape of a product refers to the 2D profile, whereas the form refers to the 3D object. We can see how the use of geometric shapes and forms have enabled modern design to develop producible and functional products when we study the work of design movements such as art deco and modernism.

The use of tangential alignments (lines and curves meeting at a single point) is essential to produce minimal and aesthetically pleasing forms without distracting blemishes on surfaces. This is seen in the transition between curves shown in Figure 1.8.6, which is essential for aerodynamics within the design.

Symmetry and asymmetry

The use of symmetry within design can aid inclusivity for handheld products aimed at both left- and right-handed users. Symmetry is also used to add visual balance to a product; however, the use of asymmetry can add interest to design, and can be used to focus attention on specific aspects of a design.

Proportion

Proportion refers to the relationship of size between different elements of an object, for example the width compared to the height. This can be seen as the overall proportion; however, the position of key features, controls or elements within a design can also be seen as proportional. The aesthetics of a product are largely affected by the proportions used in its design. One of the most important considerations within this area refers to the use of the 'golden ratio' – a term referring to a ratio of 1:1.61803398875 units, which when used to construct a rectangle, produces the most visually pleasing proportions to the human eye. This proportion has been used in architecture, art and design for hundreds of years. Below are some recognisable examples where this proportion has been used.

Figure 1.8.7 Acropolis facade.

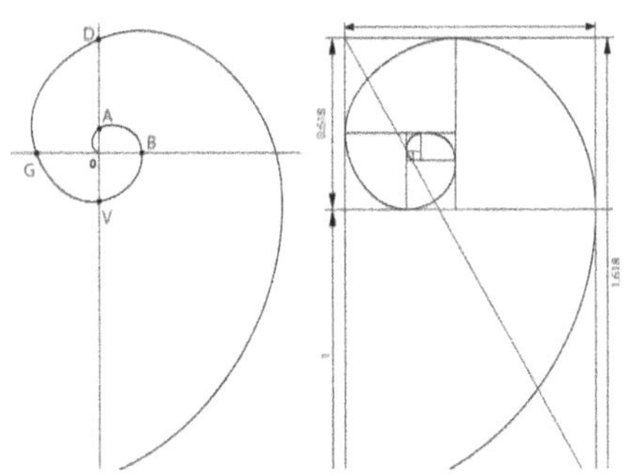

Figure 1.8.8 The 'golden ratio'.

Anthropomorphism

The use of anthropomorphism within product design refers to the design of products that reflect proportions associated with the human body and face. This can be seen in the design of the Anna G corkscrew by Alessandro Mendini, and in a subtler approach with the Coca-Cola bottle of 1915, which differed vastly from bottle design of the time.

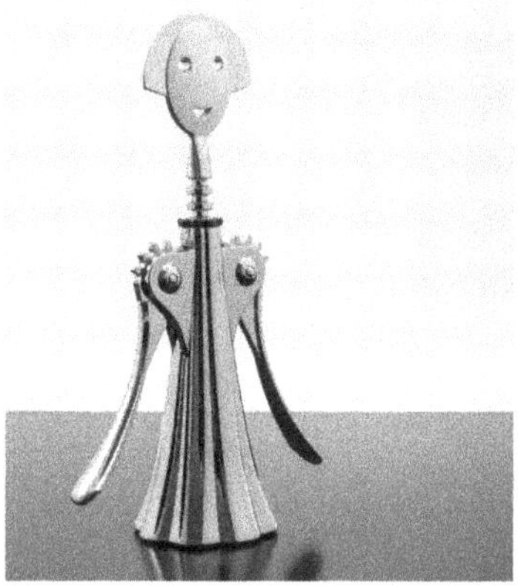

Figure 1.8.9 Alessandro Mendini Anna G corkscrew.

Figure 1.8.10 VW Beetle front view.

Anthropomorphism within design development is used to provide an emotional connection with the product; the association of the character shown in the Anna G corkscrew with a playful approach to design adds fun to a mundane task. The addition of facial expressions or perceived facial expressions to products can be calming or add 'personality' to products as seen in figure 1.8.10.

The addition of these characteristics can often be unintentional, and designers must appreciate the ability of people to associate forms with human characteristics.

ACTIVITY

Find a range of example products where you can see human facial characteristics. How do the facial expressions change your opinion of the product?

Colour and texture

The use of colour within product design is important in the aesthetic appeal of the product. When designers combine colour, they use a variety of selection processes; two of the main techniques are: combining colours from opposing sides of the colour wheel (these colours are known as complementary) and combining colours that are next to each other on the wheel (this is known as an analogous colour combination).

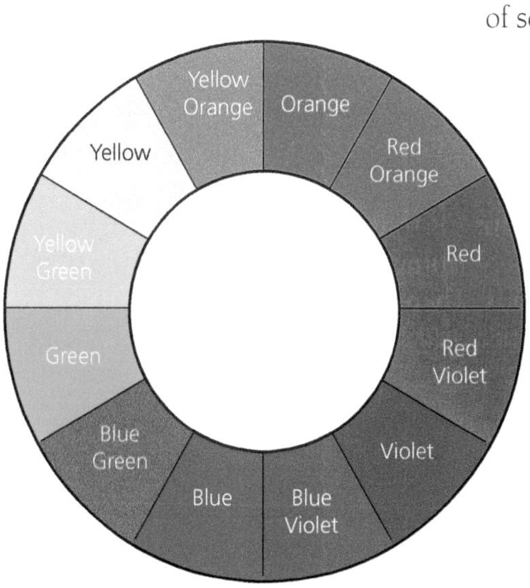

Figure 1.8.11 The colour wheel.

Symbols and ideograms

The use of instructive symbols is growing as it becomes more important to appeal to a more diverse global market; designers must ensure that they avoid language barriers within the use of products. The use of standardisation within symbol design also means that children will have more chance of understanding the product.

Along with the use of standardised symbols, we use colour associations to aid with the intuitive nature of products. For example, it is understood that red will indicate stop, danger, heat or warning, whereas the use of green indicates go or environmentally friendly, and blue indicates cold.

Combining colours can also give an even greater range of associations such as yellow and black, which is seen in nature as a warning, and therefore is used in signage to indicate danger (see figure 1.8.13).

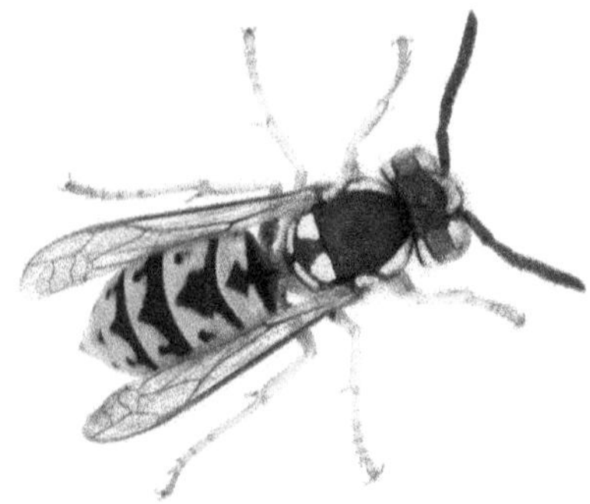

Figure 1.8.12 A wasp.

Figure 1.8.13 Corrosive warning sign.

Ergonomics

The term 'ergonomics', also referred to as human factors, is defined as the scientific study of people and their working conditions. When considering the ergonomics of a product, situation or environment, we must consider how we interact with it using all of our senses. When we design a product, we have to consider how interaction with it will affect the user – will the method of operation cause discomfort or undue stress?

For example, if we look at the design of a pair of scissors, we will see that there are all sorts available, each designed for specific tasks. The largest pair are designed for performing long cuts through fabric and, due to this, they have longer blades (Figure 1.8.14). In Figure 1.8.15, the handles are angled to allow the user to rest the scissors on a flat surface and therefore prevent undue movement of the wrist during cutting.

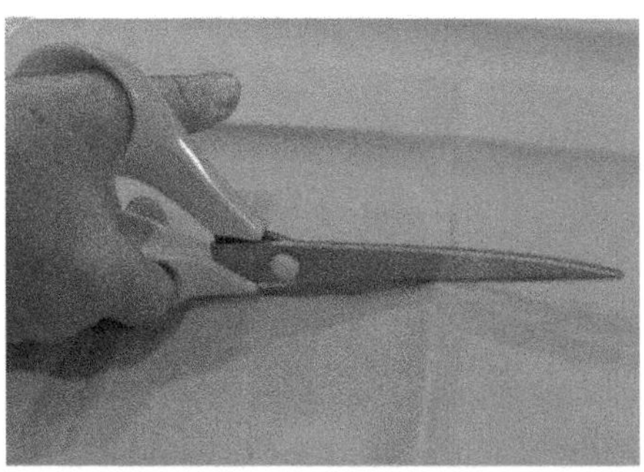

Figure 1.8.14 Using large fabric scissors.

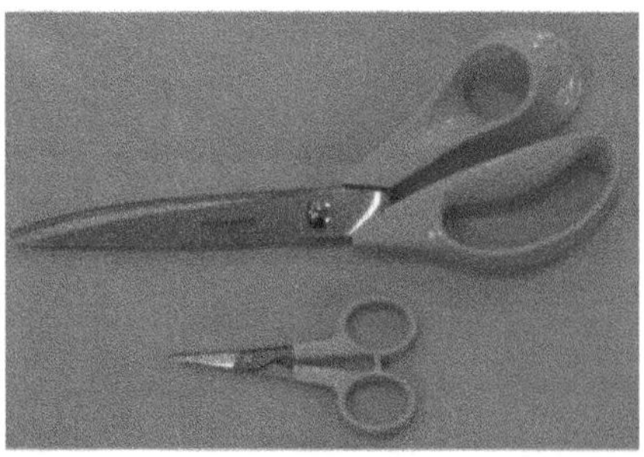

Figure 1.8.15 Ergonomics in scissors.

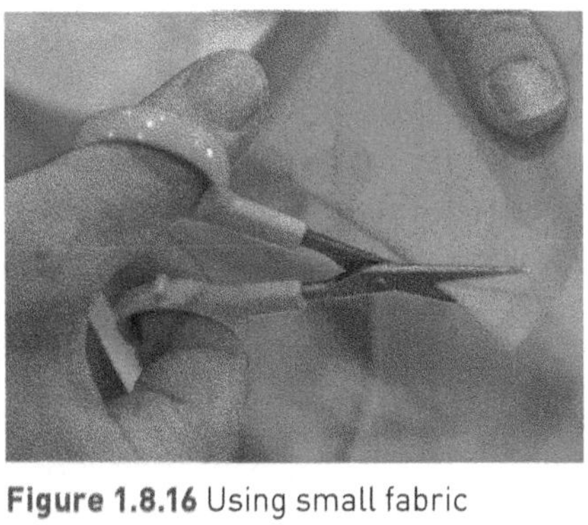

Figure 1.8.16 Using small fabric scissors.

The smaller pair of scissors in Figure 1.8.16 is designed for removing strands of a woven material and performing intricate cuts, where the user will hold the material in their hands. Therefore, the blades are small to reduce weight, and no support is necessary. Without support, the handles are symmetrical around the central axis.

There are many things we must take into account when considering the ergonomics of a design, including the:

- range of sizes of possible users (anthropometric data)
- range of differing abilities displayed by possible users (concerning all senses)

- movements and forces used in relation to the product
- shape and form of the areas that a user will interact with
- use of appropriate materials for the product
- use of appropriate colour schemes
- use of appropriate symbols on the product.

The ergonomic considerations required for products vary considerably. For example, if we consider the ergonomics of a dashboard within a car, the appearance is mainly driven by the need to maintain the driver's focus on the road during travel. While maintaining this focus, a driver needs to be made aware of important changes, such as the need to refill the fueltank. The method used to attract attention to the fuel level is such that both audible and visual stimuli are given.

The design of an office chair is highly dependent on anthropometric data and adjustable heights to accommodate a range of users. The product has a high focus on functionality, with the aesthetic appearance being driven largely by its ability to keep the user comfortable for long periods of time.

Design considerations for control interfaces

The following are things you should consider when designing control interfaces, such as the control panel for a cooker or the dashboard of a car interior.

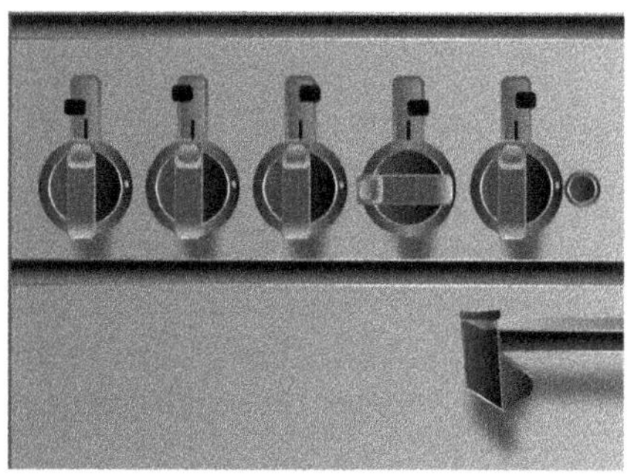

Figure 1.8.17 Cooker controls.

Figure 1.8.18 Metalwork lathe controls.

Guidelines for control design:

- Users should not have to move unnecessarily to reach any controls.
- All controls should be easy to grip to reduce the strain put on joints from bending fingers and hands. This may mean making grips larger.
- Control surfaces should prompt increased grip. This may be done through the inclusion of elastomer overmouldings or in mould texture application.
- Controls should be easy to access and operate; this means that although creating flush surfaces improves aesthetics, users may not be able to locate the control in the dark.
- The pressure required for operation of controls must also be considered and optimised for the task and user.
- Varying shapes and sizes of controls mean that identification can be made by touch.
- Large, clear labels of a contrasting colour to the background are easier to see.
- Tactile markings need to be easy to locate, so as to guide partially sighted and blind people.
- Important information should be shown in different ways; wherever possible, a variety of sounds, symbols, lights and textures should be used.

Anthropometrics

Anthropometrics is the use of scientific measurements of the human body in the design and construction of a product. Everyone differs in size and ability, and so designs need to take this into consideration. Anthropometrics involves collecting statistics or measurements relevant to the human body such as height, weight, shoe size, arm length, grip strength and head circumference.

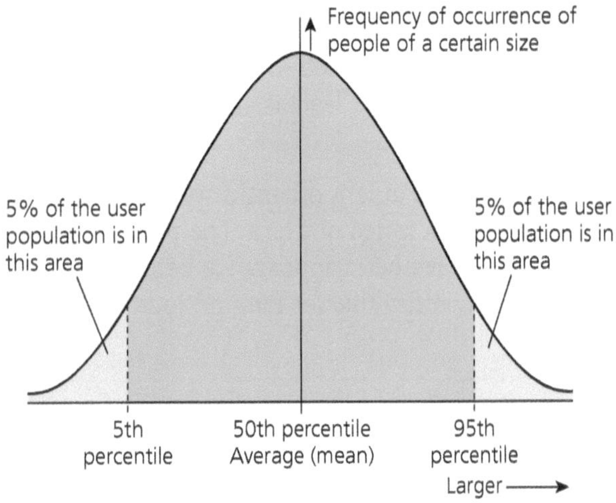

Figure 1.8.19 Percentiles.

The data gathered for each measurement is often presented in a graphical format. Although, as designers, we aim to be inclusive, the extremes of human measurements can often make it impossible to design products suitable for all.

The graph shows the distribution of human heights from a sample of results; the 50th percentile shows the average results for this measurement. If we designed a door frame for a 50th percentile height, everyone above this would have to crouch down when entering the room. This is not inclusive, as 50 per cent of the population would have issues with the design. If we designed the frame for 95th percentile height humans, we would only cause problems for 5 per cent of the population. This is the optimal solution – the largest 5 per cent of the population are accommodated with specialist products because it is not cost effective to mass produce products for this market.

Inclusive design

Inclusive design is the design of products that are accessible to, and usable by, as many people as reasonably possible without the need for special adaptation or specialised design. It incorporates the concept of UCD and means designing a product suitable for the widest number of people, taking into account variables such as user size, age, gender, background and ability.

As a designer, it is extremely difficult to design for users with differing abilities to yourself without the opportunity to empathise with the target user. Within the idea of **empathic design**, the designer is encouraged to take part in task analyses while emulating possible restrictions faced by the user. While developing the Ford Focus, Ford utilised a 'third age suit' to allow designers to experience the prototype from the view of an older user who may have mobility issues.

Figure 1.8.20 Using a third age suit.

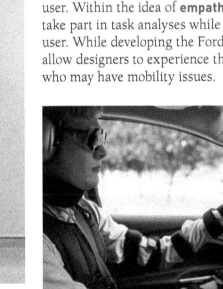

Figure 1.8.21 Driving while wearing a third age suit.

The suit has been designed to restrict movement in joints, reduce the user's ability to grip objects and also to simulate visual impairment, giving designers the opportunity to empathise with specific age-related disabilities.

Hand tool design requires a clear understanding of inclusive design; bad design can cause undue stress on joints and cause the user to work more slowly, or even to stop the task altogether.

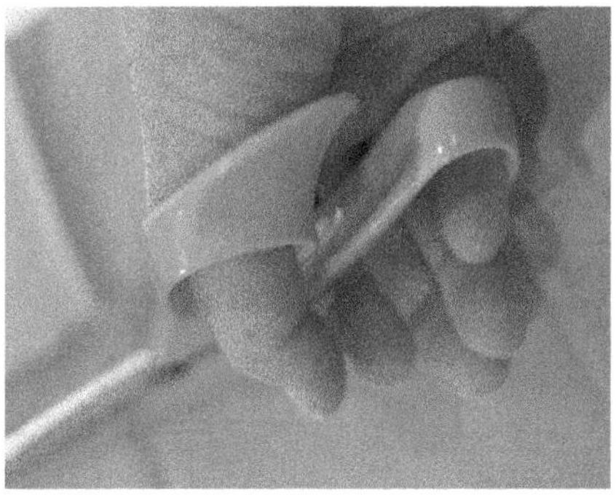

Figure 1.8.22 Holding scissors in the right hand.

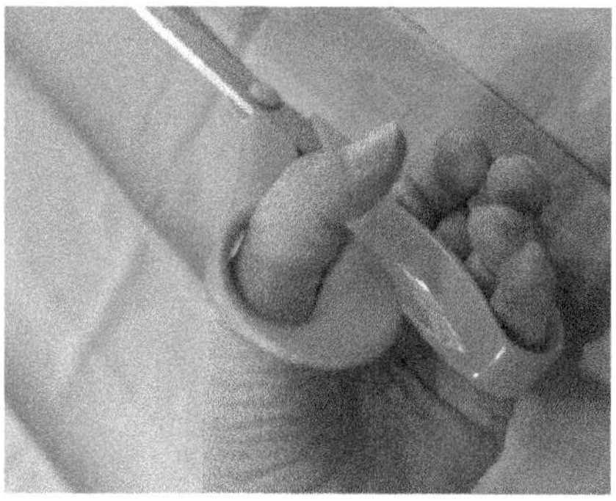

Figure 1.8.23 Holding scissors in the left hand.

For example, in Figure 1.8.22 you can see the scissor handle has been shaped for right-handed use, guiding the user in the way they 'should' hold the handle. In Figure 1.8.23, you will see the use of a smooth finish rigid polymer for the handle; this means that the grip could be improved, possibly by adding texture or an elastomer overmoulding on contacting surfaces. The thermal insulation properties of the polymer mean that the scissors will not react to changes in environmental temperature, which may cause discomfort when using the product in cold conditions.

Within the design of hand tools, there are two major grip types – the precision grip shown in Figure 1.8.24 and the power grip shown in Figure 1.8.25.

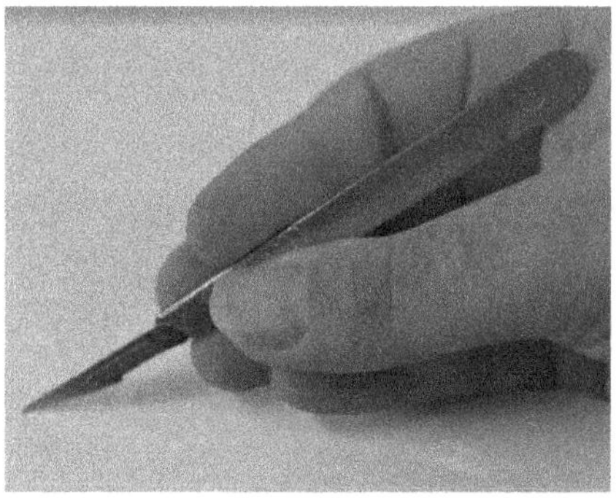

Figure 1.8.24 The precision grip.

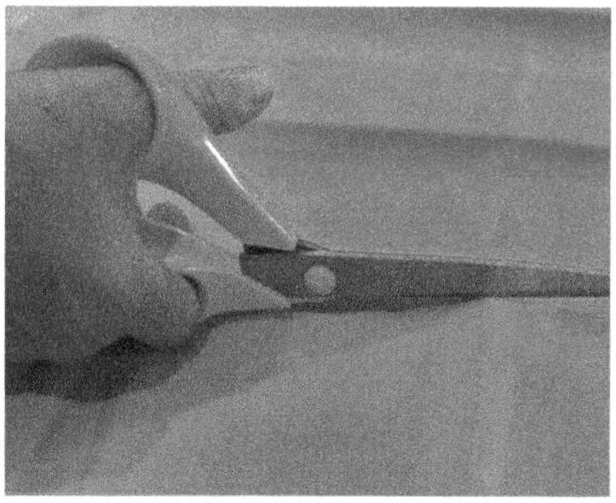

Figure 1.8.25 The power grip.

The precision grip is used for small intricate movements where the force applied is small, so weaker finger muscles are utilised. The tool is held between the thumb and index finger. You use this grip with pencils, paintbrushes and scalpels. The ideal diameter for a precision grip is 8–16 mm, although the shape may vary depending on the use and direction of force.

The power grip is used when large force is required; the hand wraps around the handle and force comes from the forearm. This is used with screwdrivers, hammers and also utility knives. The ideal diameter for a

ACTIVITY

1 The task of fastening a necklace is fiddly but achievable. It becomes more problematic when the fastener is behind your neck. How could the fastener be modified to make this easier?
2 Try to type an email on a computer keyboard while wearing gloves. How does this affect your ability to complete the task?
3 Place one hand behind your back and try to cut a straight line with a pair of scissors.
4 Find five examples of products where you can see anthropomorphism.

power grip is 30–50 mm, although the shape may vary depending on the use and direction of force.

The use of ridges or indentations in hand tool design can improve grip, which is especially useful for screwdrivers. This improved grip can also be achieved with the inclusion of a TPE overmoulding.

As the length of hand tools is generally fixed, designers use average measurements to ensure that they are more inclusive. The use of indented finger grips is avoided here because they can be too prescriptive and uncomfortable for users with small or large hands. To accommodate the widest range of users, gentle curves are used, giving multiple gripping positions for the user.

KEY TERMS

Ergonomics: the scientific study of people and their working conditions.

Aesthetics: the features of a product that make it visually appealing.

Anthropometrics: the use of scientific measurements of the human body in the design and construction of products.

Task analysis: the study of individual tasks performed by consumers to allow for successful product development.

User-centred design (UCD): the use of the end user throughout the design process, to ensure the developed product is focused on their actual needs and not the perceived needs determined by the designer.

Inclusive design: the concept of designing products that will allow the maximum number of possible users to successfully use the product without discomfort.

Empathic design: the process of simulating conditions experienced by potential users of products, to assist with the development of successful products.

KEY POINTS

- All aspects of a PDS must be objectively tested; this means that quantitative data is key to allowing consistency in testing.
- It is vital that the potential end user is involved at all stages of the design process to guide the product's development.
- Task analysis allows a designer to identify issues with existing products, or to iron out issues in prototype development.
- All modelling is done to help make decisions or to prove/disprove predictions.
- The appearance of a product is affected by shape/form, size and proportion, colour and texture.
- As a designer, it is extremely difficult to design for users with differing abilities to yourself without the opportunity to empathise with the target user. Empathic design encourages the designer to take part in task analyses while emulating possible restrictions faced by the user.

Check your knowledge and understanding

1 When a design concept is 3D modelled purely for aesthetics, what is this known as?

2 If hand tools/controls are designed without consideration of ergonomics, how can this affect the user?

3 What does the term 'analogous colour combination' mean?

4 Why might a designer use anthropomorphism in their designs?

Further reading

For fantastic examples and explanations of why some things work well and others do not: *The Design of Everyday Things* by Donald A. Norman

A book full of anthropometric drawings and essential measurements of the human body: *The Measure of Man and Woman: Human factors in design* by Alvin R. Tilley, Henry Dreyfuss Associates

A book looking at how attractive things really do work better: *Emotional Design* by Donald A. Norman

A video showcase for prototype design ideas: *Producttank* YouTube channel

A website giving further information about and examples of anthropomorphism in design: www.nextnature.net/2011/12/11-golden-rules-of-anthropomorphism-and-design-introduction/

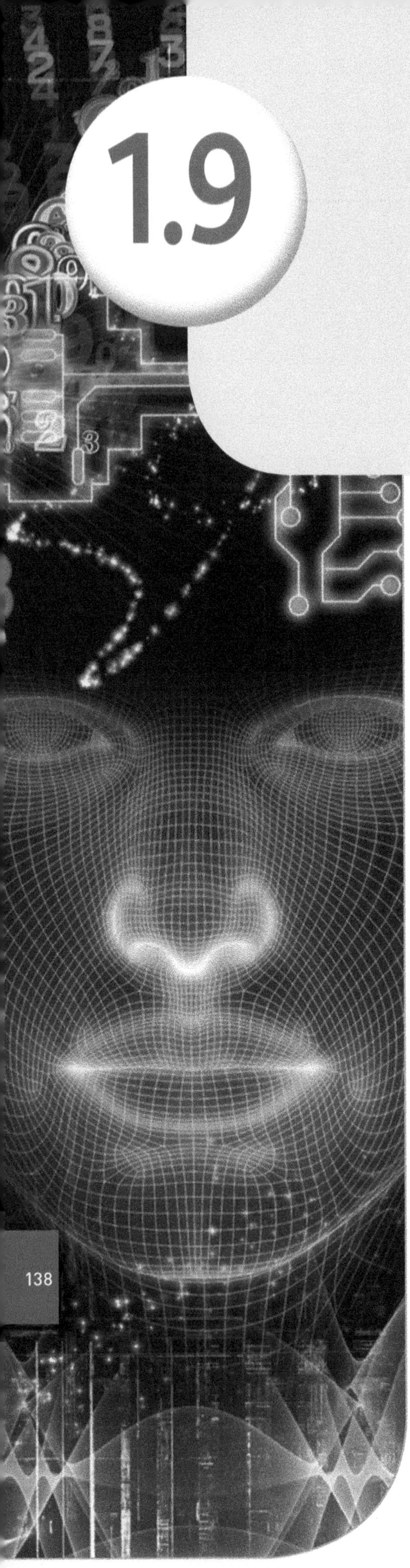

1.9 Health and safety

LEARNING OUTCOMES

By the end of this section you should have developed a knowledge and understanding of:

- safe working practices, including identifying hazards and the need for risk assessments
- the Health and Safety at Work Act (HSWA) 1974 and how it influences the safe manufacture of products
- COSHH and safety precautions that should be taken with relevant materials
- safe working practices in school or college workshops and in industry
- safety precautions taken with specific manufacturing processes
- how designers and manufacturers ensure products are safe for consumers to use
- legislation used to protect consumers, and its impact on product design and manufacture
- the BSI and how products might be tested to meet safety standards
- measures to ensure the safety of toys
- advice given to consumers, including manufacturer's instructions, safety warnings and aftercare advice.

All employees have a right to work in places that are safe. Employers have a duty of care to ensure that procedures are implemented to keep employees safe, and to prevent them from being harmed or becoming ill through work. Health and safety is everyone's responsibility, and employers need to consider a number of different safety procedures in the workplace. The Health and Safety Executive (HSE) is a national independent watchdog for work-related health, safety and illness. The HSE helps companies and employers to meet health and safety obligations. They produce guidance documents and also investigate accidents in the workplace.

To protect employees in the manufacturing industry, there is specific legislation that the employer should comply with. The main legislation will be discussed in this chapter and includes HSWA, **COSHH** and the Personal Protective Clothing Regulations 2002.

In general, employers should ensure the following:

- The workplace is safe and free from **risks** to health.
- Welfare facilities such as first aid are provided.
- Dangerous items and substances are used and stored safely, and removed from use if necessary.
- Machinery is safe to use, maintained and in good working order.
- Training, information and adequate supervision is provided as necessary.
- Appropriate safety signage is provided.

Employees also have to consider health and safety, and should ensure they:

- take reasonable care for their own health and safety, as well as the health and safety of those in the vicinity
- use the personal protective equipment (PPE) provided, and use machinery in accordance with the given training instructions.

Safe working practices

Safe working practices are a set of guidelines that an employer may produce as part of an overall health and safety management system. The guidelines should help to ensure the safety of the employee, as well as others who may be affected by their actions.

Safe working practices usually cover the following aspects:

Training

Employees should be trained in the safe use of work machinery and equipment and trained to carry out safely any process that is part of their job. Training may be done in-house or may require formal certification, for example to prove competence in a particular machine process.

Machine maintenance and guarding

All machinery should be part of a regular maintenance programme in which parts are inspected for wear, blades are replaced as required and items such as bearing or gears are lubricated. This is vital, not only ensure the safe and effective operation of the machine, but also to prove the machine has been adequately looked after in the event of an accident investigation. Machinery should also have the correct guards to protect the employee. Guards can include blade covers for band saws and circular saws, quadrant guards on disc sanders as well as infrared beams or micro-switches, which will stop the machine immediately if the beam is broken or the switch is enabled. Micro-switches are often used on CNC machinery to ensure that the machine will not operate if the door or guard is open, and that the machine will stop immediately if the door or guard is opened mid cycle.

Extraction systems

If a workshop process produces any dust or fumes, then extraction must be provided. The extraction system ensures that any **hazards** such as timber dust or welding fumes are extracted at the source, thus preventing the employee from inhaling any potentially harmful airborne particles. Extraction systems must also be regularly maintained, and this is usually carried out by an external company on a 12- or 14-month schedule.

Provision of personal protective equipment

Employers should ensure that personal protective equipment (PPE) is available to any employee who may be subject to a health or safety risk at work. PPE can include items such as safety helmets, eye protection, dust masks, respiratory protective equipment (RPE), high-visibility clothing, safety footwear and gloves.

Accident reporting

Employers must keep a record of any accident, and must report this via RIDDOR (Reporting of Injuries, Diseases and Dangerous Occurrences Regulations, 2013) to the HSE. RIDDOR covers any accident, near miss or dangerous event which has resulted in injury, death or occupational disease, such as a respiratory issue that has been induced by working in very dusty conditions.

Health and Safety at Work Act (1974)

The **Health and Safety at Work Act (HSWA)** is the main piece of legislation for British health and safety law. To comply, all employers must ensure (so far as is reasonably practicable) that all employees and visitors to the workplace are protected in terms of health, safety and welfare.

The key phrase for employers is 'as far as is reasonably practicable', which means that although they cannot ignore health and safety responsibilities, the employer does not have to take measures to reduce or avoid the risk if the cost is not justified by the reduction in risk that the measure would produce.

By complying with the HSWA, assessing risks and adopting appropriate safety measures, employers are providing a safe environment for product manufacture to take place. If appropriate and safe manufacturing procedures are followed, the manufacturing process will be deemed as safe as is reasonably practicable.

Control of Substances Hazardous to Health regulations

To comply with COSHH (Control of Substances Hazardous to Health) regulations (2002) employers need to prevent, reduce or control their workers' exposure to substances that may be hazardous or cause ill health. They have a duty to protect both the user of the hazardous substance as well as any worker in the vicinity who may be exposed to the substance.

Hazardous substances include:

- paints, varnishes, cleaning agents and adhesives that may be used during work
- fumes that may be created as a result of a process such as soldering or brazing
- airborne particles such as dust created by cutting or sanding timber.

If hazardous substances are not controlled, this may lead to workers becoming hurt or ill. This could be from skin irritation such as dermatitis, eye irritation or damage from chemical splashes, the development of lung disease or asthma, or in extreme cases even death. As a result, this could lead to loss of productivity, an increased cost to society through medical expenses as well as the potential for legal costs for the employer as a result of prosecution.

There are eight principles to consider under COSHH regulations:

1. Plan and carry out processes and activities to minimise the release and spread of emissions and substances hazardous to health.
2. Consider all of the relevant routes of exposure, such as inhalation, skin absorption and ingestion, when developing control measures.

Acute toxicity, Very toxic (fatal), Toxic, etc

Gases under pressure

Harmful skin irritation, serious eye irritation

Flammable gases

Explosive, self reactive, organic peroxide

Harmful to the environment

Oxidising gases, oxidising liquids, oxidising solids

Respiratory sensitiser, mutagen, carcinogen, reproductive toxicity, systemic target organ toxicity, aspiration hazard

Corrosive (causes severe skin burns and eye damage), serious eye damage

Figure 1.9.1 CLP hazard pictograms.

3. Control exposure by measures proportionate to the health risk.
4. Choose the most effective and reliable control options, which minimise the escape and spread of substances hazardous to health.
5. Provide, in combination with other control measures, suitable PPE.
6. Check and regularly review the control measures for their continuing effectiveness.
7. Inform and train all employees on the hazards and risks from the substances with which they work and the use of control measures to minimise the risks.
8. Ensure that the introduction of control measures does not increase the overall risk to health and safety.

Hazardous substances will usually show a symbol that meets the classification labelling and packaging (CLP) and/or the Globally Harmonized System of Classification and Labelling of Chemicals (GHS) regulations. The CLP/GHS hazard pictograms (see figure 1.9.1) allow easy identification of the hazard as corrosive, flammable, toxic, etc.

Health and safety in product manufacture

All manufacturers must be able to demonstrate safe working practices and abide with any legislation, so that persons involved in the manufacturing process can be kept safe. Manufacturers not only need to consider worker safety but also, in the event of an accident, be able to prove to an accident investigation team that safe working practices have been adopted.

Safe working practices in the school or college workshop and in industry

Safe working practices are primarily the same for schools, colleges and industry; they all serve to keep the person doing the work, as well as those in the vicinity, safe and free from harm. These safe working practices may include a set of guidelines for the person(s) in the workshop to follow.

In industry where there is a movement of goods around a factory floor, for example in a car manufacturing plant, the employer may have designated safe zones or walkways that must be kept free from machinery. Forklift trucks or vehicles will usually be fitted with a flashing light and/or reverse signal alarm to alert those in the vicinity of the vehicle movements.

Job rotation may also be used so that workers are not at risk of repetitive strain injury (RSI) or injury as a result of a lapse in concentration from doing the same task with no variation.

In school and college workshops, the guidelines may be produced as workshop safety rules such as:

- Do not operate any machinery or equipment unless you have been trained in its use.
- PPE must be worn when using machinery.
- All guards and safety measures should be utilised when using machinery.
- Work areas should be kept neat, clean and free of hazards. Spillages should be reported to a teacher.
- All students must be aware of the emergency procedures, such as fire evacuation routes and muster points, emergency phone numbers, locations of fire extinguishers, emergency stops, etc.

- Walkways should be kept clear and free from trip hazards.
- Students must report to a teacher whenever they become ill or injured or in the event of an accident.

Safety precautions

Any action carried out in advance that protects against a possible danger or injury is called a safety precaution. **Safety precautions** are usually determined by the specific process being undertaken, for example when cutting timber on a band saw, the safety precautions would include wearing eye protection, the use of guards and the use of a dust extraction system.

Safety precautions an employer may take could include the use of signage such as:

Smoking and naked flames forbidden

Figure 1.9.2 Prohibiting an action likely to cause danger.

Industrial vehicles

Figure 1.9.3 Warning of a danger or hazard.

Eye protection must be worn

Figure 1.9.4 Requesting a specific action to be taken, e.g. wear eye protection.

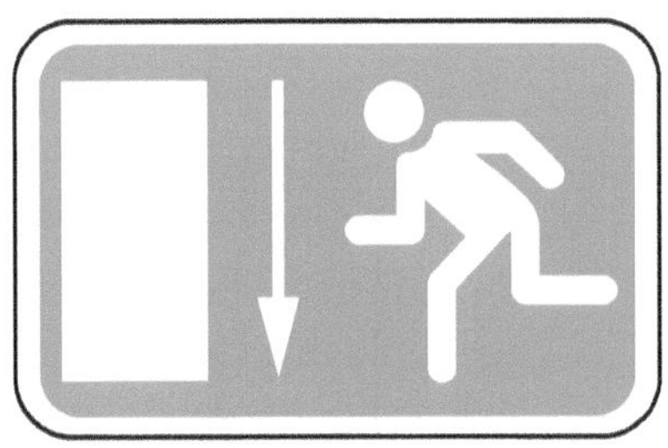

Figure 1.9.5 Emergency escape.

Figure 1.9.6 Location of first aid.

Risk assessment

As part of health and safety management, any risk in the workplace must be controlled. **Risk assessment** is something that an employer is required to carry out by law, and should be recorded in any establishment where there are five or more employees. The risk assessment must consider what might cause harm to people, and if reasonable steps are being taken to prevent that harm.

It is important to note that not all risks can be eliminated, and the law does not require them to be. The employer must however take all reasonably practicable measures to identify potential hazards and minimise the risk.

A hazard is anything that could cause someone harm. The risk is the likelihood or chance that someone may be harmed as a result of being exposed to the hazard.

For example, if there is a water spill on the floor, there is a slipping hazard. If the spill is not cleaned up, the risk of slipping is high for everyone walking on the floor. If a sign is put up or the area is cordoned off, the risk of people slipping is low.

Risk assessments are extensively used to identify potential hazards within manufacturing processes where not only the worker but also others in the vicinity may be potentially harmed by the process.

HSE recommend five key steps for carrying out a risk assessment:

1 Identify the hazard: hazards can be identified by walking around the workplace and noting any potential hazards. Referring to manufacturers' instruction booklets and data sheets can also help to identify hazards – both immediate, for example cut risk from a circular saw, or long term, for example long-term exposure to timber dust.
2 Identify who might be harmed and the nature of the harm: this includes the worker, people visiting the workplace, as well as anyone in the vicinity such as members of the public during external building works.
3 Evaluate the risk and decide on the control measure: when the hazard has been identified, the risk assessor must decide how likely it is that harm may occur, and how to remove or lessen the level of risk. For example, if the hazard is from falling debris on a building site, the control measure may be to cordon off the area to those not directly involved in the process, and to issue PPE to workers in the risk area.
4 Record findings and implement actions: a risk assessment document should be written to show that checks to identify risks were made, the people at risk were identified and the precautions put in place to reduce the risk.
5 Review the risk assessment after a set period of time and amend if required: risk assessments should be regularly reviewed. The review should identify if the risk remains the same or if there have been changes that have been implemented.

An example of a basic risk assessment for a manufacturing process is shown in the table below. The employer would carry out this type of risk assessment and keep a written document for every process required for product manufacture.

Table 1.9.1 An example of a basic risk assessment for a manufacturing process.

Process	Cutting wood on a bandsaw
Hazard(s)	Cut risk Inhalation of wood dust particles Particles in the eye
Potential harm	The machine operator Persons in the immediate vicinity
Risk and control measure	Risk level is medium. Provide training prior to machine use. Provide access to machine operating manual. Provide machine guards and push sticks. Use dust extraction system during machine operation. Provide good ventilation in the workplace. Provide eye googles. Post safety signs on machine and in vicinity. Mark out a safe zone around the machine which only the operator can enter.
Review date	The risk assessment should be signed and dated by the person assessing the risk. The review date should also be noted, such as the date of the six- or twelve-month review.

ACTIVITY

1 Complete a risk assessment for a manufacturing process carried out in your school or college workshop.
2 Make a list of five substances used at home or school that carry a CLP pictogram label. Note the substance, the symbol and any special precautions to be taken during use.

Safety in products and services to the customer

Designers and manufacturers need to consider the safety of the potential user of their product, and should ensure that the product is safe to use.

To help protect the user, many products have specific legislation and standards that the designer must meet. These specify certain requirements that the product must meet, suitable safety levels, and the right for consumers to get their money back if goods are found to be faulty or do not perform as the manufacturer claims.

Legislation to protect consumers

Consumer Protection Act 1987

The UK Consumer Protection Act 1987 is designed to safeguard consumers from products that do not reach a reasonable level of safety, and give them rights when buying goods and services.

The Act aims to help protect consumers in the following ways:

- Product liability: if the product is defective and causes injury, the manufacturer will be held responsible.
- General safety requirement: all goods for domestic use must be reasonably safe. Any product deemed unsafe may be removed from sale for up to six months while safety checks are carried out. If the product is found to be faulty, it may be destroyed.
- Price indications: the Act prohibits sellers giving consumers a misleading price indication, such as in an advertisement, price ticket or in a price comparison. The Act is enforced by trading standards officers who have the power to seize goods and business records if they have reasonable grounds to suspect a misleading price indication has been given or displayed.

The manufacturer or supplier to consumers must ensure that products are safe. If the product is not safe and the consumer suffers injury or damage to their property because of using the unsafe product, the manufacturer could be sued or face legal action, which could result in fines or imprisonment.

To ensure products are safe, manufacturers should:

- warn consumers about potential risks
- provide information to help consumers understand the risks
- monitor the safety of products
- take action if a safety problem is found.

When a consumer buys a product, the risks and health and safety information are contained in an instruction booklet, either provided with the product or available online.

Nintendo – the games console manufacturer – includes health and safety information not only relating to potential risks from aspects such as faulty batteries (product risk) but also potential health risks such as RSI caused by excessive play.

When products are found to be faulty, the company has a duty to repair or recall the product. Product recall usually occurs when the same fault is

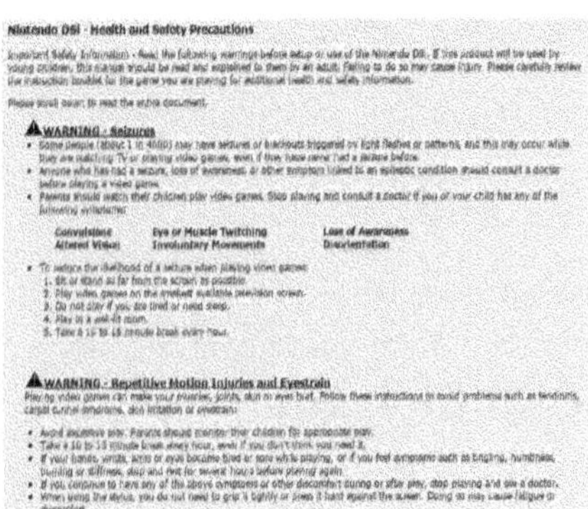
Nintendo DSi · Health and Safety Precautions

WARNING - Seizures

Convulsions	Eye or Muscle Twitching	Loss of Awareness
Altered Vision	Involuntary Movements	Disorientation

WARNING - Repetitive Motion Injuries and Eyestrain

Figure 1.9.7 Nintendo DSi – health and safety precautions user manual.

found in several of the same model or make of product. For example, in the period 2009–2010, Toyota Motors recalled over 8.5 million vehicles following driver reports of unintended acceleration due to a fault with the accelerator pedal sticking. Toyota kept records of any customer who had reported the fault through a service department or company hotline. The volume of reports made Toyota act by recalling the vehicles to investigate the potential fault and take appropriate repair action.

The Trade Descriptions Act 1968

Figure 1.9.8 Electronic bear toy seized by trading standards. Upon inspection, the eyes on the toy fell off. This not only formed a potential choking hazard but also revealed electronic wiring and components which were then accessible to the child.

This Act makes it an offence for a trader to make or apply false or misleading statements about goods or services. Under the Act, it is an offence to apply a false statement about a product or to supply goods that have a false statement.

Trading standards officers regularly check products to ascertain whether they meet any required legislation, and officers often attend street markets, stores and import facilities to check for counterfeit or unsafe products. Legislation covers a whole range of goods – from cosmetics to children's toys. Cosmetics and skin products must meet safety standards in relation to constituent ingredients, so that there is little likelihood of adverse reactions to the product.

The British Standards Institute

TM

Figure 1.9.9 BSI kitemark.

The British Standards Institute (BSI) was formed in 1901, and was the world's first national standards body. BSI works with industry and government, as well as business and consumer groups across the world, to produce a set of standards to assure consumers that products are well produced, safe and fit for purpose. A standard is a published document that shows an agreed way of doing something, and provides an assurance of consistently reliable and safe products.

The BSI kitemark is a quality mark, and is one of the most recognised symbols of quality and safety in the world. It is used on hundreds of products from cycle helmets to children's toys, manhole covers to electric plug casings. Companies can pay to have their products tested against the set standard. If the product and the manufacturing process meet the standard, the company will be awarded the BSI kitemark and can display the quality mark on their product. Companies registered with BSI are subject to ongoing testing of the product and manufacturing process, to ensure standards are maintained.

Figure 1.9.10 Child safety gate to prevent the child accessing the stairs.

Child safety gate

Child safety gates to prevent young children accessing stairs are a popular safety feature in many homes, but how can the consumer be sure the gate has been manufactured in a way that will keep the child safe?

Safety gates and 'safety barriers used in the home that limit access to the stairs of children under 24 months old' are manufactured to BS EN 1930, and any safety gate product not complying with this standard may not be safe for use.

The requirements for child safety gates cover the following areas:

- Structural integrity and alignment: the gate should fit securely in position and should not be easily dislodged. The gate must be difficult for a child under 24 months to open.

Figure 1.9.11 Standard UK 240 volt 3-pin plug.

- Footholds: there should be no structure that would give a child a foothold for climbing.
- Distance between adjacent members: vertical slats or rods should be no more than 5.5 cm apart.
- Holes: there should be no holes that could trap a child's fingers.
- Height from floor: the distance between the foot of the gate and the floor should be no greater than 5 cm.
- Snagging: there should be no sharp edges or protrusions that could catch on clothing or cut a child.
- Openings: the opening method should require either 'some force', two consecutive actions or two simultaneous actions of different types.

3-pin plug

The standard UK 240 volt 3-pin plug is a product used every day. It connects the electricity supply to the appliance, and needs to keep consumers safe from electrical shock and electrical fire. The 3-pin plug is packed full of safety features and must conform to BS 1363.

A plug is certified by the BSI and therefore safe if:

- It carries the words 'Approved by BSI', the kitemark and the kitemark licence number for the plug, for example, Licence No KM 123456.
- It is supplied with a removable card label, fitted over the three pins, which shows the wiring diagram.
- Inside the plug is a fuse, which is also certified (BS 1362). The fuse shows the BSI kitemark and BS 1362 printed on the fuse body.
- The earth pin is the longest pin. The live and neutral pins have a plastic cover at the pin base. The earth pin does not have a sleeve to ensure a proper connection of the earth circuit; this protects the user from getting an electric shock if a fault occurs.
- The wiring is coloured in accordance with the wiring diagram: earth (green and yellow), neutral (blue) and live (brown).

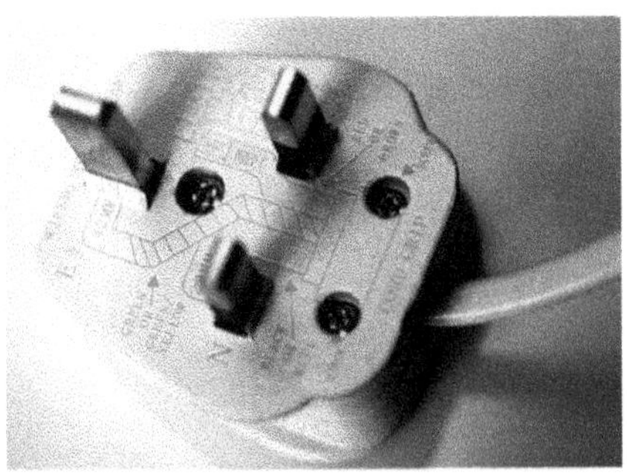

Figure 1.9.12 Card cover showing plug wiring diagram.

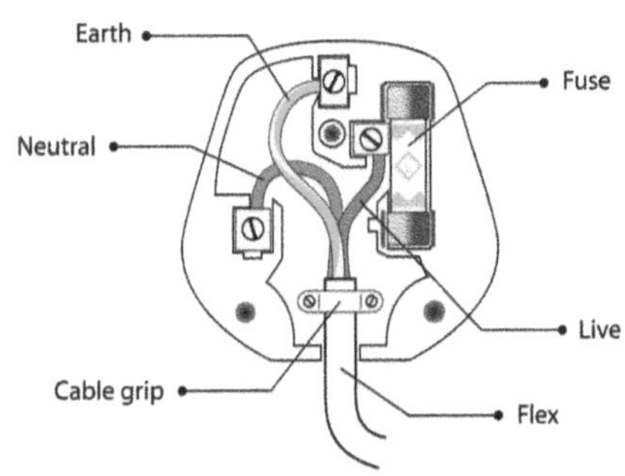

Figure 1.9.13 Wiring colours and positions within the plug casing.

The safety of toys

Manufacturers of children's toys must comply with the provisions of the Toys (Safety) Regulations 2011, which is part of the European Toy Safety Directive as well as British Standard BS EN 71.

Figure 1.9.14 Small parts cylinder used for assessing choking risk for toys intended for children under 36 months.

The standard deals with every aspect of toy safety from flammability and toxicity to their potential to trap clothing or injure a child physically. Toys intended for children under 36 months must not present a choking risk. To test for a choking risk, a 'small parts cylinder', which represents a child's throat and windpipe, is used and the toy or any component part is placed on the cylinder. Toys or parts of toys that can fit entirely inside the cylinder are identified as choking hazards and will be rejected for use.

Toy manufacturers must:

- make sure that the toy has been designed and manufactured to comply with the essential safety requirements during its normal period of use
- carry out a safety assessment of the toy
- demonstrate that an applicable conformity assessment procedure has been followed and affix the CE (Conformité Européene) marking
- make sure that the toy is accompanied by instructions for safe use and safety information where appropriate
- carry out sample testing of toys and bring non-conforming toys into compliance
- investigate and record any complaint made in relation to the toy, and keep a register of complaints, non-compliant toys and recalled toys
- draw up the technical documentation and keep it for ten years
- apply information identifying the toy and manufacturer on the toy or packaging.

The British Toy and Hobby Association (BTHA) developed the lion mark in 1988 to show consumers that the toy had been made with quality and safety in mind.

The lion mark indicates that the toy has been made by members of the BTHA, who adhere to a code of practice which includes rules covering ethical and safe manufacture of toys, a ban on any counterfeit goods, a commitment to improving sustainability and a desire to promote the value of all play.

Advice to consumers

Consumers who want a product that will be safe and fit for purpose, should look for recognised symbols such as the BSI kitemark. This proves that the manufacturer has taken steps to ensure the product has been checked, tested and inspected to meet the recognised industry standards. If a product carries a BSI kitemark, customers know that it has been independently tested on a regular basis, and that it will perform to that consistent quality every time.

Figure 1.9.15 The lion mark.

Manufacturers provide safety warnings and health and safety instruction manuals to help advise consumers of any risks, as well as inform them of safety precautions. To ensure safe product use and minimise the risk of injury or harm, consumers should read the provided information prior to using the product.

Manufacturers also have a responsibility to inform consumers how to care for the product during use, and how to dispose of the product at the end of its life. This information forms part of the extended producer responsibility, and covers aspects such as recommendations for non-damaging screen cleaning products for computer screens and methods for safe disposal of the

Figure 1.9.16 In 1991, the Toy Retailers Association (TRA) joined with the BTHA to launch the approved lion mark retailer scheme, indicating that members of the TRA only sell products confirming to EN 71.

product such as free product return or instructions to take the product to a local waste recycling facility.

ACTIVITY

Carry out an investigation into the specific details associated with the British standard for toy safety (BS EN 71).

KEY TERMS

COSHH: Control of Substances Hazardous to Health regulations 2002.

Risk: the likelihood or chance that someone may be harmed as a result of being exposed to a hazard.

Hazard: anything that could cause someone harm.

Health and Safety at Work Act (HSWA)1974: the main piece of legislation for British health and safety law.

Safety precaution: any action carried out in advance that protects against a possible danger or injury.

Risk assessment: consideration of what might cause harm to people, and if reasonable steps are being taken to prevent that harm.

KEY POINTS

- Safe working practices include carrying out risk assessments and taking action to ensure workers are as safe as is reasonably practicable. Areas covered include the provision of training, PPE, machine guards and extraction systems with consideration of COSHH.
- Legislation such as the Consumer Protection Act 1987 and the Trading Standards Act 1968 help to protect consumers from faulty goods and services.
- To ensure products are safe for consumer use, manufacturers should:
  - warn consumers about potential risks
  - provide information to help consumers understand the risks
  - monitor the safety of products
  - take action if a safety problem is found.
- The BSI is an organisation dedicated to producing an agreed set of standards for products, which helps provide assurance for consumers that the products they are buying are reliable and safe. CE marking shows that the manufacturer has checked that products meet and comply with European Union (EU) safety, health and environmental requirements. CE-marked products can be freely circulated within the European market.

Check your knowledge and understanding

1 Give an example of a substance that may show this CLP hazard pictogram.

Figure 1.9.17 CLP hazard pictogram.

2 List five specific safety precautions that a manufacturer may implement to keep workers safe.

3 Identify three specific risks and associated control measures for making a facing cut on an aluminium round bar using the centre lathe.

4 Give one example of specific health and safety legislation that protects workers in the workplace.

Further reading

The UK's independent regulator for work-related health, safety and illness: www.hse.gov.uk

The national standards body of the UK: www.bsigroup.com

The BTHA represent the interests of British toy manufacturers and aims to raise standards of practice in the industry: www.btha.co.uk

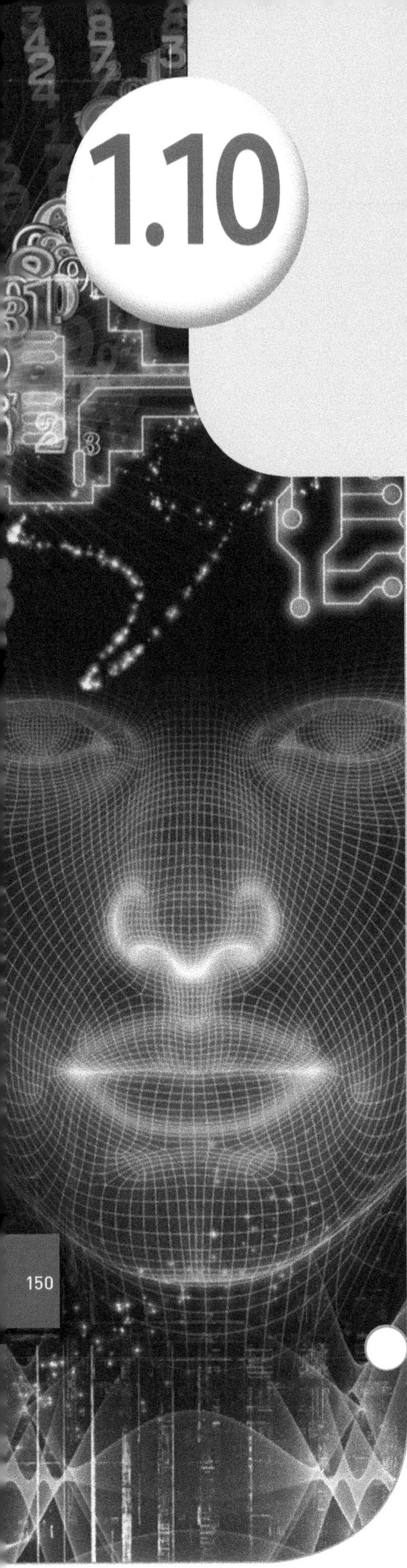

1.10 Protecting designs and intellectual property

A level only

This topic is for A level only. If you are studying at AS level you do not need to cover the content in this section.

LEARNING OUTCOMES

By the end of this section you should have developed a knowledge and understanding of:

- copyright and design rights, patents, registered designs, trademarks and logos, and their importance to the designer
- the concept of 'open design'.

Intellectual property (IP) is something that has been physically created and did not exist before, such as a book, a film, a camera lens mounting system, a handbag or a car manufacturer's name badge. Ideas alone for all of these things do not count as IP because they must be translated into a physical format in order for them to be recognised. Uniqueness is an essential element when seeking **intellectual property rights (IPR)**, since the key point of these rights is the protection of original ideas.

Although there is evidence that the idea of IP originated in Roman times, it is generally accepted that the need for the recognition of the rights of owners of IP began in the late-nineteenth century with the Trade Marks Registration Act, which began to address the growing problem of copying and stealing the original work or identity of others.

As it stands now, some rights are automatically endowed when a piece of work is created, and some rights have to be applied for. It all depends on the type of IP involved.

IP laws vary depending on the country; this section concentrates on the UK.

Copyright and design rights

Copyright is a legal right that provides protection for work such as books, photography, drama, music, films, television programmes, software, etc., and lasts, in most cases, for 70 years after the death of the creator of the work. It provides a framework that facilitates the management of copyright in such a way that others can be given permission to make use of it, such

Figure 1.10.1 Harry Potter author J.K. Rowling signing books.

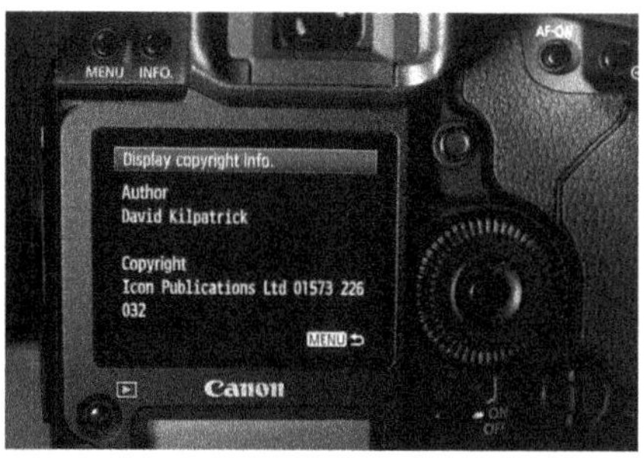

Figure 1.10.2 Professional camera being used to record copyright information in the metadata of digital images.

as when photographs are licensed for use in books and magazines or when a theatre company pays a fee to stage a musical. These are rights that do not need to be formally registered since they are automatically applicable as soon as the work in question has been created.

The Copyright, Designs and Patents Act 1988 provides the appropriate legislation in the UK, and gives 70 years of cover to literary, dramatic, musical or artistic works including films, and 50 years to sound recordings.

Unregistered design rights provide protection for designs so that they cannot be copied, and also covers unauthorised dealing in copied designs. Unregistered design rights apply for ten years and cover the appearance of a product, including aspects such as shape, material and ornamentation. They do not cover how the product works – this falls under '**patents**'.

Although registration is not required for either of these types of IP, in practice it can be very hard to prove ownership if a dispute arises. There are, therefore, organisations such as the UK Copyright Service that facilitate the provision of independent evidence of authorship. It can help to make use of the © symbol, the 'C in a circle', which is the accepted copyright symbol that can be applied to most types of work, and which also helps to reinforce that the owner of the work is serious about protecting it, and would be likely to follow up any infringements with legal action.

This symbol is usually accompanied by the name of the copyright holder and the year of first publication, and makes it easier to reinforce any subsequent claim that may be required to assert the status of the originator of the work, although it is not a requirement.

Additional steps may be taken to provide proof of original ownership by using strategies such as:

- adding metadata (encoded information) and watermarks to digital images
- keeping early drafts, sketches, recordings, etc. to show the development of the piece of work
- the incorporation of deliberate, unique modifications that can be cited in evidence, such as slight alterations or additions in maps.

Patents

Patents for an invention are granted by the government to the inventor, and are granted in order to give inventors up to 20 years' legal protection of their ideas from being stolen and used by other people without their permission. They are only applicable to novel inventions for the working parts of a design such as a:

- suspension system of a car
- circuit board in an electronic product
- sensing device in an automatic kettle
- sealing method used in push-fit pipe fittings.

If a design is to be patented, it is essential that it is a completely new invention or a significant improvement on previous versions. A patent attorney is often employed to establish the validity of an application for a patent prior to it being submitted, since there are often complex searches to be carried out and checks that need to be made on the validity of the application. Confidentiality is also essential, since premature exposure of

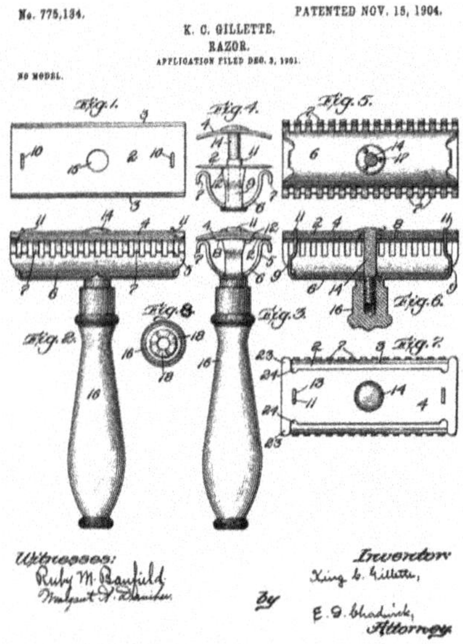

Figure 1.10.3 Patent drawings for the 1904 Gillette razor.

the idea through a demonstration or a newspaper article, for example, would render it invalid for patenting. The legal complexities that exist often result in conflict, such as that between Apple and Samsung with regard to phone and tablet patents. The associated claims, amounting to billions of dollars, clearly illustrate the vital commercial importance of establishing the ownership of IP.

Another high-profile patent case was the ongoing battle that James Dyson has had with rival companies in order to protect his patented ideas, such as the dual cyclone incorporated in his range of vacuum cleaners.

Patent submissions, which are made to the **Intellectual Property Office (IPO)**, are expensive and complicated. One of the most essential elements is the set of drawings illustrating the way that the invention functions. All of the key features in these drawings have to be numbered to cross reference with the explanation of the invention in the patent application. The encouragement of further innovation is often argued to be a useful by-product of the publication of the details of inventions through the patent system.

Registered designs

Registered designs cover the appearance and decoration of a product; therefore, in the case of an electronic device, they are applicable to the exterior casing rather than the circuit and other working parts (which would need to be patented, if appropriate). New designs can be registered for up to 25 years by submitting suitable illustrations of the design and paying the necessary fees to the IPO.

The international trade in fake designer label products is an ongoing problem, which still continues despite potentially massive fines for perpetrators and purchasers.

Figure 1.10.4 Fake designer handbags for sale on the streets of Rome, Italy.

Trademarks and logos

One of the most precious and eagerly sought aspects of commercial activity is achieving and maintaining an instantly recognisable combination of words, sounds, colours and logos that can be used for marketing and brand identity. A **logo** is a graphical symbol that is often an important element in producing an instantly identifiable and unique identity to make brands stand out from others. A **trademark** may incorporate graphics, but can often simply consist of words, such as the phrase 'dual cyclone'.

A trademark is registered by making an application and payment to the IPO, and has to be renewed every ten years. The TM® symbol can be used to warn others that it is a protected trademark.

Figure 1.10.5 The ® symbol being used on a label for a toy.

Registering a trademark entitles the registrant to the following:

- The ability to take legal action against counterfeiters and any other unauthorised users of their brand.
- The use of the TM® symbol alongside their brand.
- The right to sell and license their brand.

Many products have a combination of the main types of IP, so it is worth analysing a couple of products to see examples of this.

Table 1.10.1 The main types of IP.

	Mobile phone	Car
Copyright	Legal notice in information settings	Driver's manual
Patent	Central processor circuit	Hybrid transmission system
Registered design	Shape and form of bezel on edges	Shape and form of bodywork
Trademark	Company logo	Grille badge

Open design

Open design is an approach to designing and manufacturing which operates outside the protection of patents and other types of IP that are eagerly sought by most individuals and companies. Instead, people opt to develop physical products, machines and systems that can be freely accessed by everyone. Contributors, often unrewarded, who wish to donate some or all of their work for the greater good, carry out open design activities in a spirit of communal involvement that might be regarded as an evolution of the do-it-yourself (DIY) culture that gained momentum in the 1950s. Internet-based design and manufacture is usually at the heart of current developments in this field, and there are many free open source software (FOSS) projects facilitating this, along with open source hardware, which often take the form of 3D printed components and also printed circuit boards (PCBs). The open approach to contributing design skills for the interests of society is one that was enthusiastically promoted by Victor Papanek, the American designer and educator who disliked patents; he saw them as a restriction on the development of inventions and something that prevented urgently required designs being developed for the benefit of all.

Figure 1.10.6 RepRap Prusa i3 3D printers printing parts for more 3D printers.

The licencing of open designs is a complicated issue that designers have to consider to ensure their work can be used and developed in an appropriate way. A creative commons licence is often used for open designs, as a way of ensuring that the original designer retains the IP of the idea and the right to be acknowledged as the originator of the design, while conferring on others the freedom to build on their input.

The 'Italic shelf' system, for example, is one of many open designs produced by the Israeli designer Ronen Kadushin, who has embraced the capability offered by computer controlled machinery and standardised file formats such as DXF (Drawing eXchange Format), to make his creations freely available for others to manufacture and develop.

Figure 1.10.7 Open design aerial ropeway installed in Nepal by Practical Action.

The RepRap Prusa i3 3D printer can be made from parts that are printed on the machine itself from freely available 3D files. In this way, it is possible for users to work collaboratively to make more 3D printers and broaden the influence of the group.

Open designs free from IP rights are a key element in the work of Practical Action, an international non-governmental organisation (NGO) which promotes 'technology justice' to challenge poverty in developing countries. Their range of designs for tools, equipment and systems can be freely copied by individuals, communities and organisations in these countries, to enable them to lead healthier, safer and more productive lives. Examples of their designs include water pumping and purification systems, ploughs and aerial ropeways.

The current trend is to replace mobile phones about every 18 months on average. Fairphone is an open source software-based mobile phone that is being developed in such a way to challenge this trend. It will provide a device based on an operating system (OS) and modules that can be easily upgraded and repaired. This allows its serviceability and usefulness to become relatively immune from the need for complete replacement in order to keep abreast with technological developments in hardware and software.

Figure 1.10.8 Large-scale shipping of goods facilitated by standardised containers.

A contrasting but very influential historical example of designers shunning the protection offered by IP rights is that of the shipping container. In the 1950s, entrepreneur Malcom McLean and engineer Keith Tantlinger responded to the massive problems of delays, wastage, accidents and theft experienced with conventional cargo transport by developing the highly successful worldwide system of shipping containers, which is still used today. The patents for the design were leased to the International Organisation for Standardisation (ISO), free of royalties, so that they could allow others to use the design without hindrance and thus ensure the worldwide compatibility that is a prerequisite of the system.

ACTIVITY

1 Use the IPO website to search for the IPR applicable to a product design project you have worked on. For example, if you have made some furniture, see if you can find a registered design or patent for an existing design and write up what you discover.

2 Patent applications have to be accompanied by detailed drawings of the invention, along with labels of all the relevant parts and a description of how it works. Have a look at some of the patent drawings and descriptions that have been submitted on the IPO website, and produce some of your own for a simple workshop tool or piece of equipment, as if you had invented it.

3 Without looking them up, draw the logos of four product manufacturing companies. When you have finished, check how close you were to the actual logos, to see how effective the companies' trademarked brand identities are. This would be even better if you organised it as a group activity.

4 Draw a simple labelled diagram to illustrate a Practical Action project, and explain how its open design helps to facilitate improved lives for communities in developing countries. Visit the website www.practicalaction.org for more details.

5 Research a patent dispute that James Dyson has had with a rival manufacturer. Set out the argument with the aid of notes and simple diagrams, and explain how it was resolved.

KEY TERMS

Intellectual property (IP): a novel, physical creation that is entitled to protection for its originator in the form of copyright, design rights, patent, registered design or trademark.

Intellectual property rights (IPR): the legal protection of IP.

Copyright: unregistered rights that protect original works such as books, photographs and plays.

Unregistered design rights: rights that give automatic protection to prevent copying the appearance of a product.

Patent: legal protection for inventions relating to the way in which the products function.

Intellectual Property Office (IPO): the official UK government body responsible for IP.

Registered design: a product whose appearance or decoration has been legally protected to counter copying.

Logo: a graphical symbol that often serves as an important element of a trademarked brand identity.

Trademark: a unique combination of words, sounds, colours and logos used for marketing, and the legal protection of brand identity.

Open design: a design that has not been protected by IPR, so that it can be freely used and developed by others.

KEY POINTS

- Intellectual Property Rights (IPR) are an important aspect of product design, manufacture and marketing because they provide protection against the theft of ideas and other commercially valuable aspects.
- Copyright and unregistered design rights are automatically applicable to the work of people such as writers, photographers and designers, but steps often need to be taken to ensure that proof is available of original ownership of the work.
- Patents provide legal protection to the inventor for the way that their products function, but are only applicable to original ideas and must be registered with the IPO.
- Registered designs are legally protected original ideas for the appearance of a product that has been registered with the IPO.
- Trademarks have to be registered with the IPO and provide legal protection for the words, logos and other elements employed in the marketing and brand identity of products and companies. Examples include car grille badges and slogans used for advertising.
- Open designs are not protected by IPR because the originator of the ideas is keen for them to be accessed, used and developed by others, therefore an appropriate licence is normally set up to facilitate this.

Check your knowledge and understanding

1 What is IP?

2 Why do designers need to obtain IPR?

3 What are copyright and unregistered design rights, and what is the difference between them?

4 What aspect of a product design would be covered by a patent?

5 How is a registered design different from a patent?

6 What aspects of a company and the products they produce might need to be trademarked? How might a logo be utilised for this?

7 Consider a product such as a bicycle and make a simple chart showing examples of the types of IP that might apply to various aspects of it.

8 What is meant by the term 'open design'? Explain how open designs can benefit society.

Further reading

Legally Branded: Logos, Trade marks, Designs, Copyright, Intellectual Property, Internet law, Social Media, Marketing by Shireen Smith (Rethink Press Limited)

Copyright Law for Artists, Photographers and Designers (Essential guides) by Gillian Davies (A&C Black Publishers Ltd)

Against the Odds: An autobiography by James Dyson (Texere Publishing)

Design for the Real World: Human Ecology and Social Change by Victor Papanek (Thames and Hudson)

The Box: How the Shipping Container Made the World Smaller and the World Economy Bigger by Marc Levinson (Princeton University Press)

Website for the IPO: www.ipo.gov.uk

This website encourages the wider use of open design: www.opendesignnow.org

This website is for Practical Action – an organisation that uses technology to challenge poverty in developing countries: www.practicalaction.org

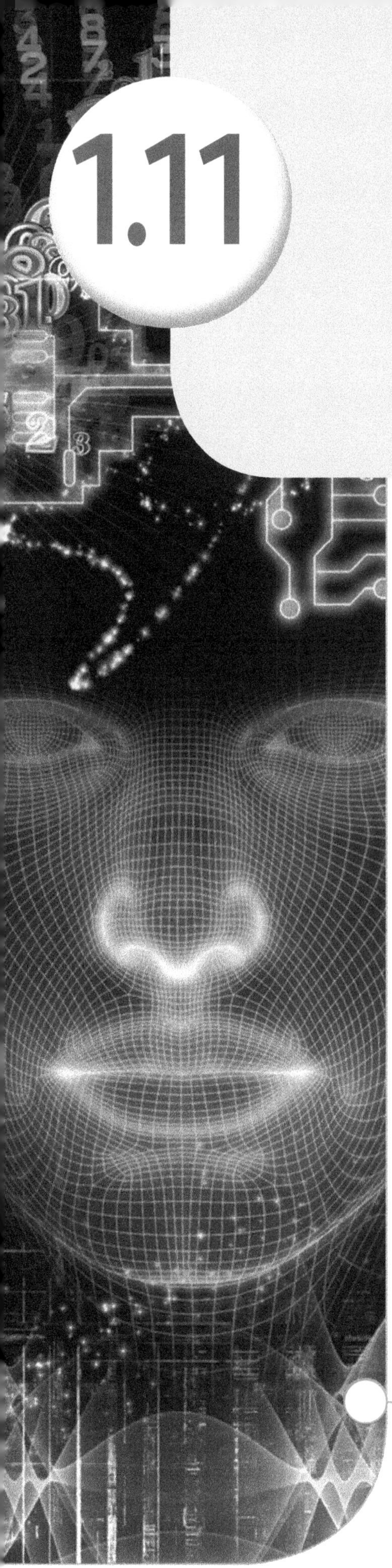

1.11 Design for manufacturing, maintenance, repair and disposal

LEARNING OUTCOMES

By the end of this section you should have developed a knowledge and understanding of:

- how the choice of materials affects the use, care and disposal of products
- the application of the **six Rs of sustainability** to product design and manufacture.

If you are studying at A-level you should also have developed a knowledge and understanding of:

- the need to modify designs to make them more efficient to manufacture and to reduce the number of manufacturing processes
- maintenance including temporary and integral fixings, the use of standardised parts, allowing for service and repair and the ability to upgrade with software downloads
- different ways in which a product can be designed to allow for more efficient manufacture
- how a product can be designed and manufactured with disassembly in mind.

Designers need to carefully consider the materials that they will incorporate into their products to ensure that they are appropriate from a financial, manufacturing, functional, aesthetic and **sustainability** point of view. Their designs need to be practical and fit for purpose, economical to manufacture accurately without generating excessive waste, as well as being easy to maintain and environmentally friendly. The material choice affects the care of the product, and it needs to be based on the life expectancy of the product. Consumer preference will be a factor too.

Choice of materials

The designer's choice of materials for a product is critical in ensuring that it is fit for purpose. An example of this is the selection of materials appropriate for drinking vessels depending on how they are to be manufactured, used and cared for, as well as fitting in with consumers' requirements such as cost:

Table 1.11.1 The choice of material for various drinking vessels.

Purpose of a drinking vessel	Material	Reasons for choice
Outside catering (disposable)	HIPS (high impact polystyrene)	Very inexpensive, recyclable, easily vacuum-formed at low temperatures.
Suitable for a young child	PP (polypropylene)	Relatively inexpensive, tough, resistant to heat, cold and fatigue, can be injection moulded.
General purpose	Soda-lime glass	Transparent, easily blow moulded, inexpensive, relatively thick and resistant to breakage, dishwasher safe.
Special occasion	Crystal glass (various oxides added)	Sparkling appearance (high refractive index), potential for ornate designs, likely to be hand washed, so lack of toughness is not an issue.

Figure 1.11.1 Polystyrene disposable cup.

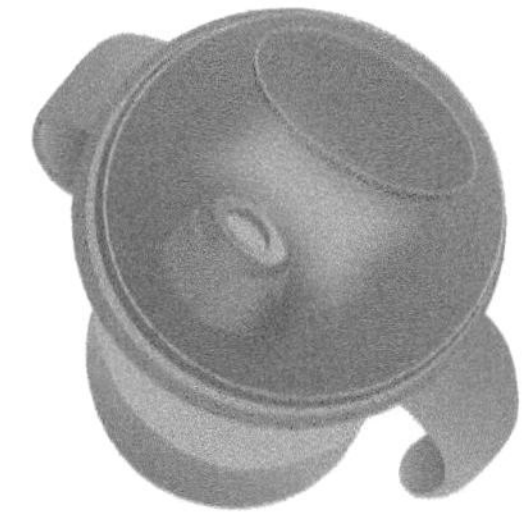

Figure 1.11.2 Child's PP drinking vessel.

Figure 1.11.3 General purpose glass.

Figure 1.11.4 Crystal wine glass.

The six Rs of sustainability

Sustainability is concerned with management of resources to ensure that a legacy of depletion and pollution will not be the result of human activities.

There are six principle ways in which this can be achieved in respect to product design.

Recycle		Recovering parts and materials to be used again when products reach the end of their useful life.
Reduce	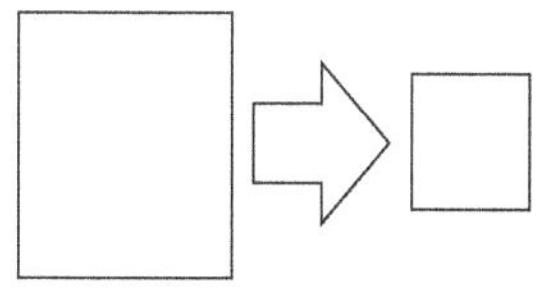	Minimising wastage, by using less material or eliminating excessive packaging, for example.
Refuse		Avoiding designs which, for example, use materials which are harmful to the environment or are difficult to recycle.
Repair	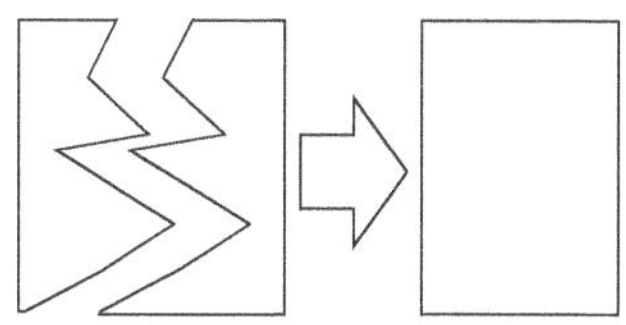	Designing products that can be repaired rather than those that have to be thrown out if they are damaged.
Rethink		Considering carefully whether there might be more sustainable alternatives.
Reuse	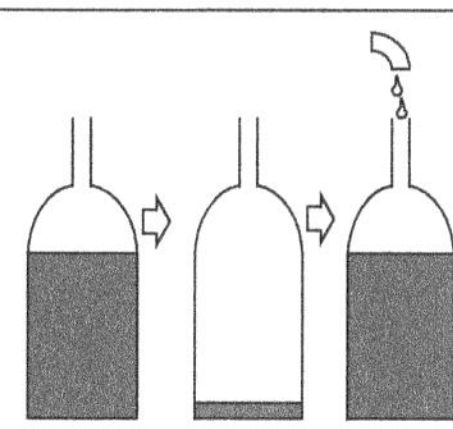	Making it possible to use products again rather than discarding them.

Figure 1.11.5 The six Rs of sustainability.

Recycle

It is essential that the recycling of materials used in products is maximised to maintain future resources and reduce levels of landfill. This is particularly the case for materials from non-renewable sources, such as most polymers, metals and glass. Local councils are now responding to this by supplying differentiated bins and collections for various types of waste. The suitability of materials being used in products is subject to scrutiny, since some are easy to **recycle** such as aluminium, but some, particularly composites like CFRP are very difficult or impossible to recycle. Recyclability also depends upon the extent to which the material can be maintained in its pure state. PET and PVC, for example, are very hard to differentiate in the recycling process, and it only takes one PVC bottle in a batch of 10,000 PET bottles for the resulting recycled PET to be potentially unusable. Most recycled PET containers end up being used for fibres when

Figure 1.11.6 Park bench made using mixed plastic lumber.

they are recycled, since making the residue food safe is very difficult due to the polymer's property of absorbing contaminants. Polyethylene and PP are often combined during recycling to create materials such as mixed plastic lumber, which is used mainly as a non-rotting timber substitute for external applications such as fences and benches.

Manufacturers place identification codes on polymer products to help consumers sort them more effectively, and therefore help to facilitate appropriate disposal. The codes do not indicate that the product is made of recycled polymer or that it is necessarily possible to recycle the product locally. The seven codes were introduced in the late 1980s, but there has been a considerable increase in the number of polymers that have become available since then, such as polyaryletherketone and liquid crystal polymers. This means that code 7 'other polymers' is being used more often than was first envisaged, so it is likely that at some stage there will need to be a reappraisal of the current codes. Although we usually associate identification codes with polymers, they are also starting to be used for other classifications of material such as glass and metal.

Recycling is irrevocably linked to the methods of manufacture that have been incorporated in products. Ease of **disassembly** is a critical factor that is being increasingly considered by designers of products such as cars and white goods. Legislative leverage from directives such as End of Life Vehicles (ELV) and Waste from Electrical and Electronic Equipment (WEEE) has been of crucial importance in this respect. Cars in particular pose many difficult problems with regard to recycling due to issues such as the depollution process required, and difficulties with disposing of automobile shredder residue (ASR). Also, the issues surrounding dealing with lithium batteries is becoming more significant due to the increasing number of electric and hybrid vehicles.

Dyson operates a free scheme to collect any of its vacuum cleaners that are no longer required when customers buy a new model, and will arrange for their collection and recycling.

The term 'cradle to grave' has been commonly used to mean that designers and manufacturers are considering the impact of the whole life cycle of the products they produce. This includes raw material extraction and production, manufacture, distribution, use and disposal. More recently, however, 'cradle to cradle' has become a preferable approach, since its key feature is that products are designed so that all of the materials utilised can be reused to manufacture new products. Accreditation has been introduced to promote this approach, and ensure that materials used in products always remain separate and are easily disassembled so they can be recycled, untainted, into the stock of raw material. The 'Think' chair was one of the first products developed using this philosophy, and is designed to be easily disassembled into individual sections to maintain material purity. Products that incorporate adhesives and other permanent methods to join dissimilar materials struggle to conform to this standard, as it is usually impossible to satisfactorily separate the various materials when the item is discarded.

An issue of increasing concern is the movement towards exporting waste for recycling to countries where the process is undertaken in highly unsafe conditions, and often by children. High levels of toxins often result from these processes, risking the health of workers and local people such as

Figure 1.11.7 Electronic scrap being recycled in China.

those living in Guiyu in China, where a significant proportion of electronic waste from western countries is sent for recycling.

Reduce

Is it possible to use less material and energy so that there is not as much depletion of resources? A major area requiring improvement in this respect is the wasteful amount of packaging that is often used for consumer products such as Easter eggs, which can be wrapped in aluminium foil set within additional layers of polymer and cardboard. Measures being taken by companies to **reduce** packaging include supplying concentrated forms of drinks, making re-fill packs available and supplying loose fruit and vegetables rather than pre-packaged.

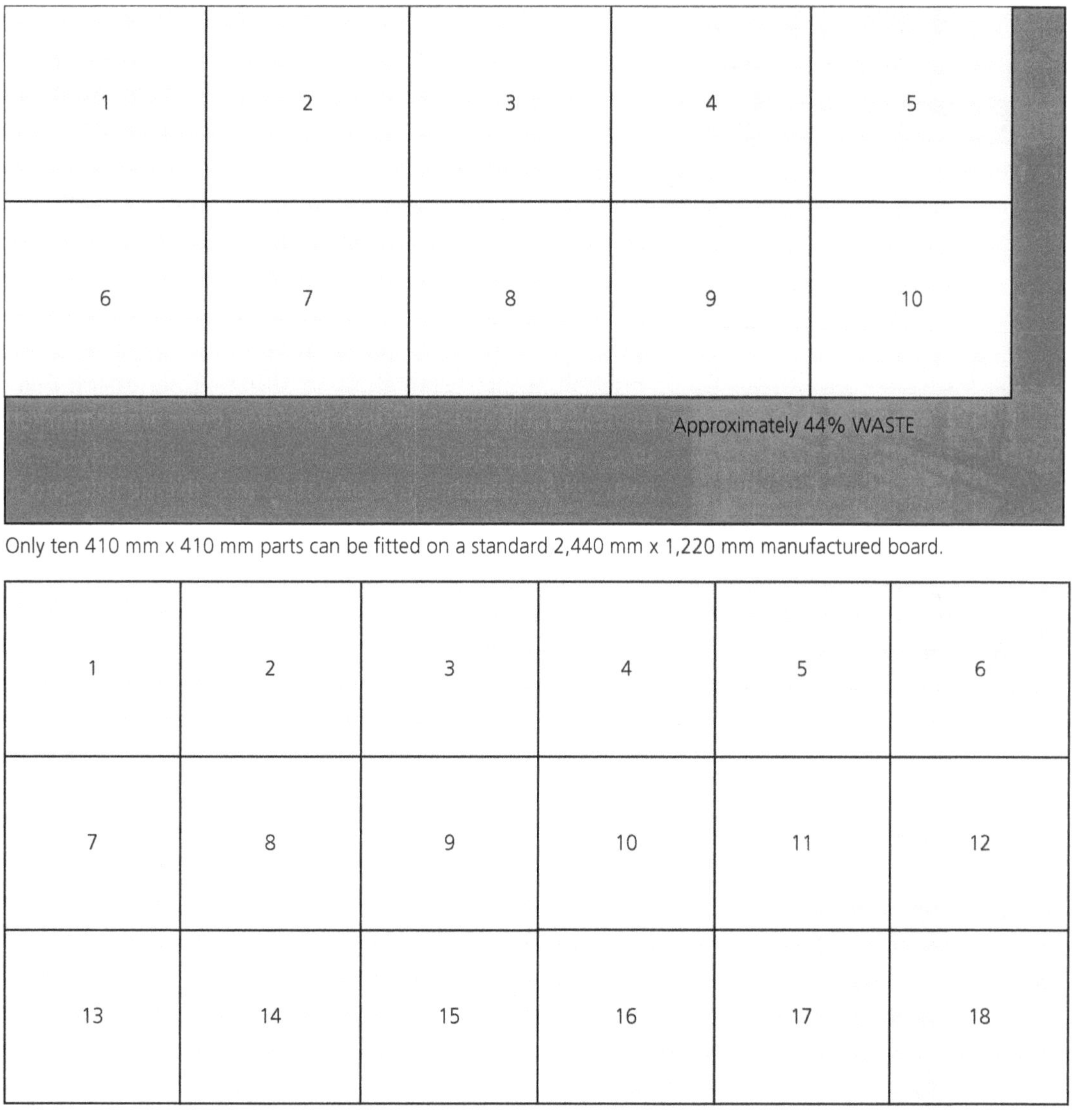

A standard 2,440 mm x 1,220 mm manufactured board will provide eighteen 405 mm x 405 mm parts with minimal wastage.

Figure 1.11.8 Reducing wastage when using manufactured boards.

ACTIVITY

Produce an annotated diagram showing how you minimised wastage when cutting out materials for one of your projects from a standard sized stock piece.

A company manufacturing shampoo bottles – Schwarzkopf – was able to make the bottles so that they were just as strong, but used 30 per cent less material, while also facilitating higher packing density, so that distribution was also more cost effective. CAD is often employed to achieve these reductions.

Ikea's sustainability strategy document pledges to make their distribution system more sustainable by, for example, increasing the level of loading for vehicles so that they are more efficient and by using more electric vehicles. Ikea also use a range of methods to reduce the amount of materials used in their furniture. Many of their table tops are hollow honeycomb structures, similar to that used in doors, which maintain reasonable strength while using a fraction of the amount of material that would be required for solid versions. They have also developed a method for constructing legs for tables that requires minimal materials.

Figure 1.11.9 Dining chairs constructed from multiple pieces of oak glued together to reduce wastage.

Some solid hardwood furniture manufacturers reduce the amount of timber wasted during production by using techniques involving gluing together numerous smaller pieces to create larger sections for chair legs and other parts. This makes it possible to considerably decrease the proportion of material that has to be discarded during processing.

Refuse

When choosing which particular product to buy, there may be a significant variation in the sustainability of different designs. If manufacturers make the relevant features of their products clear at the marketing and retailing stage, appropriate decisions can be made. As consumers become more aware of sustainability issues, they may decide that products such as ones made from sustainable resources and offering easy routes for recycling are preferable to those that make use of finite resources and are assembled using a combination of materials and adhesives that makes recycling difficult.

Figure 1.11.10 Wooden chairs that can be easily dismantled and recycled.

Consumers can make a difference by **refusing** to buy products that are harmful to the environment. Here are some examples of products that might be worth reconsidering:

- cars with poor fuel consumption or high levels of toxic emissions
- single use bottled water, when a refillable container could be used
- timber-based products from unsustainable sources
- snacks packaged in multi-layer foil that cannot currently be recycled
- electrical products that cannot be easily repaired
- inefficient light bulbs
- oil-based polymer products, when environmentally friendly bio-polymer alternatives are available.

Figure 1.11.11 A synthetic rattan, resin and metal chair that will be more difficult to recycle effectively.

Repair

Many products are not designed to be **repaired**, due to the greater cost of manufacturing with easily dismantled temporary fastening methods such as screws, nuts and bolts, compared to quicker, cheaper production methods such as adhesives and plastic welding. This issue also links to the aesthetic value of products that is highly sought after by many consumers, since designing products that are slim and attractive is much easier without potentially bulky temporary fixings.

It is now unusual for discrete components in circuit boards to be repaired, as it is much more likely that the whole circuit board will be replaced. This makes it easier to train service engineers, since diagnostic equipment can be used to identify faulty circuits that can then be unplugged and replaced with new units. It is not usually cost effective for the circuit board to be repaired, so they will very often be sent for recycling, with gold, silver and copper content in ICs and other components being significant targets for recyclers.

Figure 1.11.12 Modular component parts of a Fairphone.

Fairphone 2, a modular design in which reparability has been incorporated as a top priority, has addressed the problem of smartphones being discarded rather than repaired, but its development and adoption is still at a relatively early stage.

Repairs are often made to the fibre reinforced polymer composite components in cars, bicycles, aeroplane bodies and marine craft. The considerable initial cost of the moulding is accompanied by the difficulties of recycling, so a number of strategies have been developed for such repairs, including patches, scarf joints, plugs and resin infusion. This contrasts with conventional vehicle body repairs, where any significant damage will normally result in a full panel being replaced and the damaged sheet steel part being sent for recycling.

Community repair initiatives are becoming more common, such as the Restart project, which aims to facilitate local groups who help one another to repair products rather than discarding them. Another important part of their work is to try to counter the throw-away culture by opening people's eyes to the satisfaction gained from repairing rather than throwing items away. A large selection of self-help repair guides and videos are available on websites such as iFixit and YouTube, and this has led to a significant number of people undertaking repairs for themselves.

Rethink

This is the consideration of alternatives to traditional products, methods and materials. Being prepared to implement changes in lifestyle can also help to achieve sustainability goals.

Examples of **rethinking** include:

- taking advantage of a cycle to work scheme rather than purchasing a car
- having solar panels fitted to reduce domestic energy use
- keeping a phone for a longer period before upgrading
- installing a water butt system for garden watering
- using digital versions of documents such as tickets rather than printing them
- buying loose vegetables from a greengrocer rather than pre-packaged from a supermarket
- using solar or wind-up electrical products rather than battery powered
- composting food waste and using it to feed home-grown produce
- joining a car-share scheme.

Figure 1.11.13 Commuters taking advantage of a cycle to work scheme.

Reuse

Reusing products is not something that comes naturally to affluent communities, who are more than willing to purchase new items. These new items are often only affordable because they have been produced in countries where the workforce is poorly paid. This contrasts with the way that children in developing countries have to make do with toys that are made with reused materials such as wire and flip flop soles.

There are however some encouraging signs that reusing products and materials is gaining more popularity with examples such as these:

- Müller, the large dairy company, is planning to join those already offering household milk deliveries in returnable glass bottles. This may give further hope for the retention and promotion of this highly recyclable and hygienic container.
- The View Tube is a legacy project from the 2012 London Olympic Games – it provided a public viewpoint for watching the construction of the Olympic Park and also functions as a community hub. It is one example of many useful buildings constructed using old shipping containers.
- Matching bricks are often specified for extensions and other building projects. Companies specialise in supplying reused bricks so that clients can avoid using new ones, which may not blend in correctly.
- Children are now encouraged to use a refillable water bottle at school rather than purchasing a new bottle each day.
- Websites such as Freecycle and Freegle, where people advertise unwanted items, are becoming an increasingly popular way of passing on products to those who need them, rather than having to discard them.
- Companies such as Ikea, partnered with local charities, are now offering a furniture take-back service to give deserving families the chance to make use of unwanted items.

Figure 1.11.14 Milk being delivered in returnable glass bottles.

ACTIVITY

Choose a suitable product and analyse it by considering how well it fits in with the six Rs:

1. How easy is it to recycle?
2. Are there any aspects of it that could be reduced, for example, to use less material?
3. Are there any aspects of the design or choice of materials that we should refuse to accept?
4. How easy is it to repair?
5. How might you rethink the design to make it more sustainable?
6. Is it possible to reuse the product or any elements of it in other products?

Figure 1.11.15 Discarded tents at Glastonbury are upcycled by WiTHiNTENT to create clothing.

Upcycling

This is the creative reuse of unwanted products and materials, resulting in outcomes with a higher quality or environmental worth than the original. This type of activity is often undertaken at a domestic level, creating items such as:

- personalised designs made using Scrabble letters
- decorations from old jigsaw pieces
- pallets used to create furniture
- glass jars used as vases
- storage boxes from ice cream containers.

It is increasingly viable, however, for **upcycling** to be carried out on a commercial scale. WiTHiNTENT is an example of a company that has seized an opportunity to commercially develop discarded products. In their case, it is the tonnes of unwanted tents left behind after music festivals that they turn into a range of showerproof festival clothing and accessories. Companies have taken up many other opportunities, such as:

- lighting for bars and restaurants made from old bottles and redundant plumbing fittings
- guitars made from motor oil cans
- plectrums made from discarded credit cards
- 'folded book' decorations
- bags and belts made from discarded tyre inner tubes.

Maintenance

Some of the issues linked to repairing products are covered in the six Rs section, but here we will look in more detail at some of the strategies that can be used by designers and manufacturers to facilitate or control maintenance of their products.

An important consideration is the extent to which the designer wishes to make maintenance possible, and who is able to carry it out. Some products are difficult to repair due to the need for watertight seals and other elements that help to protect the internal workings. Technologically advanced products are sometimes designed in a way to actively deter non-specialists from attempting to repair or maintain them. Modern cars can be made difficult to maintain by anyone other than main dealers due to the need for specialised tools and diagnostic equipment, which can effectively bar many local garages and DIY enthusiasts.

Some products, such as mobile phones, tablets and laptops, can be difficult to repair because the requirement for making the products as slim as possible has led to the installation of elements such as non-removable solid state hard drives and batteries, which are held in place by adhesives. These design features make it more likely that the product will have to be returned to the manufacturer if parts fail, or it may have to be scrapped, possibly prematurely, due to the prohibitive cost of repairs.

A more consumer-friendly approach is to use temporary joining methods such as screws, to facilitate access to the internal parts and to incorporate easily accessed integral fittings such as those used to provide access for battery replacement in remote controls. Integral fittings and clips, which do not require the use of tools, are ideal for situations in which consumers need to remove covers on a regular basis. It should be noted that care is needed in their design to ensure that they are as accessible as possible, and if they are used on children's toys they will need additional safety features.

The waste and resources action programme (WRAP), a UK charity concerned with using resources in a more sustainable way, has published several 'buying specification guides for durability and repair' for electrical products. The laptop guide, for example, contains recommendations to manufacturers on points such as:

- availability of instruction booklet
- resistance to damage through avoiding weak points in the case
- vibration resistance

Figure 1.11.16 Modern SLR camera having its firmware updated.

- well placed ventilation and fans
- reasonable duration of parts availability
- prevention of water ingress
- standardisation of spare parts across a range of products
- ease of access to replace parts.

The increasingly ubiquitous nature of microprocessors to control a host of functions in modern electrical products such as cameras, games consoles, media players, TVs, washing machines and refrigerators, has made upgrades for firmware (control software) particularly important. These upgrades ensure that the products continue to work effectively and facilitate improved features. Computer and mobile phone operating systems (OS) are updated on a regular basis to deal with problems such as security issues that have arisen and to provide ongoing enhancements, although sometimes these updates can lead to hardware becoming prematurely obsolete, as it is incompatible with the latest OS.

Ease of manufacture

Manufacturers are keen to make their products as efficiently as possible, in order to reduce their costs and to minimise the chance of mistakes being made that could affect the quality of the finished items. Examples of some of the strategies that might be adopted include:

- using a modular approach in the design, so sub-systems are easier to track and fault-find
- using standardised parts and sizes across different versions/models to reduce the overall number of parts required
- using advanced soldering techniques such as wave or reflow, rather than manual soldering
- injection moulding parts that are required in large numbers rather than machining them from solid material
- employing self-finishing materials such as polymers to avoid the need for painting and other coating methods
- using redistribution techniques whenever possible, rather than fabrication or wastage techniques, as this will usually prove more efficient when large production runs are undertaken, because the number of manufacturing processes will be reduced
- using factory planning and layout simulation software to ensure optimal layout of the machines, robots and other elements of the manufacturing system
- using adhesives rather than mechanical fasteners
- reducing the thickness of cast and moulded parts by incorporating ribs and webbing to maintain appropriate strength in the parts that are produced
- using integral snap fittings in many polymer products, which can be designed to be either separable or fixed, because they remove the need for additional manufacturing processes such as the application of fixings or adhesives, and can be easily incorporated in the mould design
- using moulded posts into which self-tapping screws can be secured to hold parts as this can considerably improve efficiency of manufacture, by making it unnecessary to incorporate more complex fixings

Figure 1.11.17 Robots applying glue during the construction of an Audi A3.

- moulding a texture into the surface of a polymer as this can be simply incorporated into the design of the mould and can add tactile features, mask moulding blemishes, enhance aesthetics and simulate the appearance of other materials such as wood and leather
- buying in pre-made components from external suppliers, which can often prove to be more efficient than making them in-house.

You may notice that some of these techniques, such as the use of adhesives and non-separable integral fittings, can run counter to the sustainability goals set out in the six Rs. Therefore, manufacturers do face dilemmas in terms of balancing profitability in the efficiency of their manufacturing strategies and their commitment to the environment.

Disassembly

Disassembly means taking apart a product for repair or when it is at the end of its useful life and can, if this element of the design has been sufficiently prioritised, result in discrete parts made of pure materials. These can be returned for re-processing with no degradation, as outlined in the explanation of the 'cradle to cradle' approach (see page 160). All too often, unfortunately, cost implications put disassembly low down the list of priorities for some manufacturers, and this can result in products that are difficult to effectively dismantle into separate parts and materials.

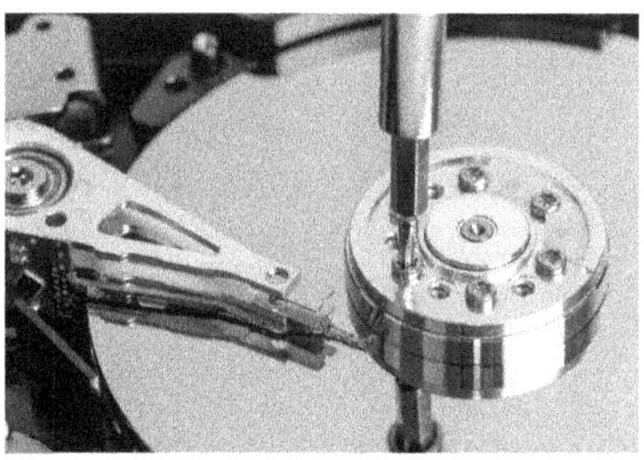

Figure 1.11.18 Specialised screws in a computer hard drive to deter untrained disassembly.

Integral fittings – as long as they are designed in such a way that they are easily released – can make disassembly very straightforward. Simple polymer clip-based fastenings are often used, such as those found in many vacuum cleaners, as well as screws with standard heads, such as Pozidriv, which can be used by anyone. More specialised forms of screw head, such as Torx, Pentalobe and Tri-Point, are used where manufacturers wish to deter untrained persons from dismantling their products, since specialised tools are needed rather than those commonly found in home tool kits.

BMW is one of the companies at the forefront in terms of implementing and pre-empting ELV (End of Life Vehicles) legislation and ensuring that their vehicles are designed for easy disassembly. In 1994, several years before the first ELV directive, they set up a recycling and dismantling centre in Unterschleissheim in Germany, to research the disassembly potential of new vehicles.

Smart materials may have the potential to help in the disassembly of products, and research has been undertaken in this field, including the use of SMA (shape memory alloy) and SMP (shape memory polymer) fastenings to hold together elements of products like mobile phones. At the end of their useful life, the phones would be put into disassembly ovens that cause the smart material to reach its transition temperature, resulting in automatic or semi-automatic disassembly of the product. This 'active disassembly' technology is at the development stage and has not yet been taken up by major manufacturers.

ACTIVITY

Obtain an unwanted, discarded electronic product such as a phone, camera or computer mouse and carry out a disassembly to see whether it comes apart easily. When you have disassembled the product, look for examples of how it has been designed for maintenance or ease of manufacture, and produce annotated diagrams to explain your findings.

Some pen manufacturers are now adopting an approach whereby users of their products can disassemble the pen for themselves, dispose of the parts that cannot be recycled and the PHA or starch based bio-polymer body of the pen can be composted. Some manufacturers claim that 98 per cent of the pen can be recycled in this way, and some pens incorporate plant seeds to further encourage sustainability.

KEY TERMS

Six Rs of sustainability: the key points that need to be considered in order to make products sustainable.

Sustainability: the management of resources to minimise depletion and pollution.

Recycle: recovering the parts and materials from unwanted products to be used again.

Disassembly: taking apart products for maintenance or to reclaim parts and materials.

Reduce: minimising wastage and excess materials or packaging.

Refuse: avoiding unsustainable materials or products.

Repair: designing products so that they can be put back in working order if something breaks or goes wrong.

Rethink: considering the adoption of more sustainable alternatives.

Reuse: making it possible to use products or parts again.

Upcycling: creative reuse of unwanted products and materials to manufacture higher value items.

KEY POINTS

- Using appropriate materials has a significant impact on the use, care and disposal of products.
- The six Rs of sustainability provide guidelines for environmentally friendly design and manufacture.
- Designs can be modified in a variety of ways to make them more efficient to manufacture and to reduce the number of manufacturing processes.
- Designers and manufacturers need to carefully consider how their products will be maintained and repaired with the help of appropriate features such as fixings, standardised parts and software upgrades.
- Some products or parts can be upcycled for use in higher value items when they are no longer required.
- Designers need to consider how their products will be disassembled at the end of their useful life, so that materials and parts can be used again, preferably without contamination or damage.

Check your knowledge and understanding

1 Choose four products – such as four chairs, four bags or four toys – which are designed for contrasting situations. For example, in the case of toys, one might be for a Christmas cracker, one to facilitate educational play for children with learning difficulties, one a construction kit for older children and one for energetic outdoor play at a nursery. For each of the four products, explain the designer's choice of materials and comment on their suitability in terms of manufacture, maintenance, repair and disposal.

2 Use a dimensioned diagram to show how a manufacturer might make economical use of half a standard manufactured board to create all the parts needed for a toy box with a base, four sides and a lid that has the largest volume possible.

3 List the six Rs of sustainability with an example of how each of them might be used to guide the design of a sustainable product.

4 Explain the term 'upcycling' and list four of your own examples of how this can be done.

5 What is the role of automatic machines in producing large numbers of parts very economically?

6 Make notes on features that you have seen incorporated in products by designers to ensure that they are easy to repair and maintain.

7 Make a list of product features designed to make manufacture as easy as possible.

8 Explain the importance of easy disassembly to enhance the sustainability of a product.

Further reading

Cradle to Cradle by William McDonough (Vintage). This book discusses how products can be designed so that, after their useful lives, they will provide something new – the 'cradle to cradle' model.

Sustainable Materials, Processes and Production by Rob Thompson (Thames and Hudson Ltd.). This book explores the environmental impact of materials, manufacturing processes and product lifecycles, and discusses how to select and use them in an intelligent way.

www.recyclenow.com

www.recycle-more.co.uk

These websites provide lots of useful information on recycling, including how popular household and workplace items are recycled and the location of local recycling facilities.

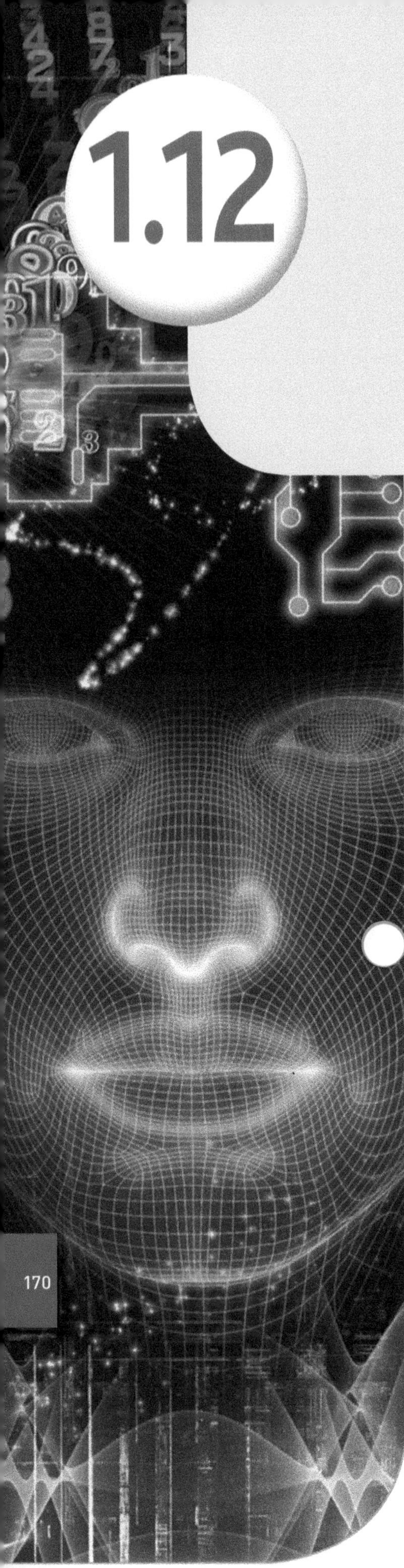

1.12 Feasibility studies

A level only

This topic is for A level only. If you are studying at AS level you do not need to cover the content in this section.

LEARNING OUTCOMES

By the end of this section you should have developed a knowledge and understanding of:

- the use of feasibility studies to assess the practicality for production of proposed designs, including the testing of prototypes with potential consumers.

A-level students need to understand how feasibility studies are used in the design process and in industrial contexts, and explain the reasons for their use.

Computer modelling in production planning

Modern manufacturing industries usually adopt '**lean manufacturing**' methods. This is an approach to efficient manufacturing that was first used by Toyota in the production of cars. Lean manufacturing concerns minimising waste in terms of time and materials. Computer modelling of production is an essential part of achieving lean manufacture.

Before manufacturers invest large amounts of finance into the production of new products, they may use computer modelling to assess the practicality of making the item. An example of this might be in the manufacture of cars. In the design of the production and assembly line, computer software will be used to plan out the production cells, the location of work stations, storage, tools, equipment and robots that operate along the assembly line. By using computer modelling, the manufacturers can test the efficiency of the production line and assess if there are going to be problems with the manufacture. It is possible to predict where there might be bottle necks on the production line due to some processes taking longer than others. These predictions can be used to plan where '**buffer zones**' might be needed. Buffer zones are temporary storage areas where partially completed work is stored while slower tasks or processes are completed.

Figure 1.12.1 A car assembly line.

Computer modelling can be used to work out the **cycle time** for a production process, for example a car manufacturing operation. This information is then used to plan what is known as the **takt time**, which is the rate at which an item needs to be made to meet customer demand. Other factors that influence production time include breaks, maintenance activities and cleaning time, which must also be considered in assessing the feasibility of production.

Below is an example of how takt time is calculated:

$$\text{takt time} = \frac{\text{available production time per day}}{\text{customer demand per day}}$$

$$\text{takt time} = \frac{\text{1,000 minutes per day}}{\text{500 pieces per day}} = \text{2 minutes}$$

Feasibility studies and costings

Simple software such as a spreadsheet can be used to calculate costs of production. These are split into direct costs such as materials and labour, and indirect costs such as factory lighting, rent and salaries. The costs can then be compared against the price that will be charged to the customer, to determine potential profits and whether it is feasible or practical to go into production.

More sophisticated computer modelling is available to calculate the costs of production of specific products such as injection mouldings. Software is available to calculate the costs of moulds including any cores for hollow products, the material and cycle time for the machine. This would be set against the number of units to be made given the cost of production.

Feasibility modelling in design

An example of how computer modelling might be used to test the feasibility of designs is in testing ergonomics. The layout of a kitchen is a very good example where computer software might be used to assess the efficiency of the arrangement of units and appliances. Computer modelling of the kitchen shows if there is a good working triangle, and if the spacing of appliances and accessibility of the units is efficient. Kitchen designers also use computer models to show potential clients what their kitchen might look like. This can help the client to make decisions on the layout of their kitchen before the units and worktops are built.

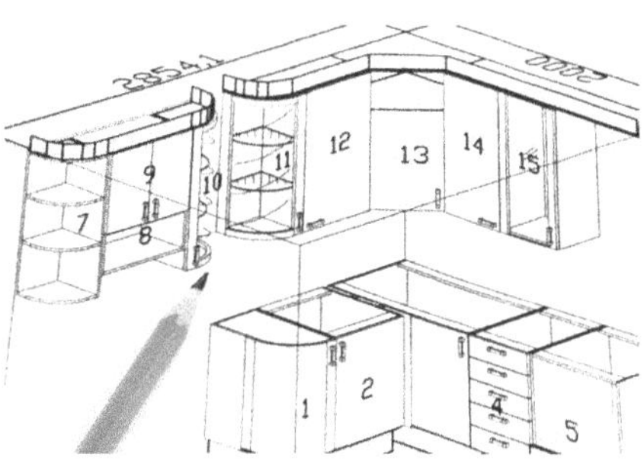

Figure 1.12.2 CAD drawing of kitchen units.

Another example of computer modelling to assess the feasibility of a design is in the development of aircraft engines. Aircraft engines for large jet aircraft such as a Boeing 787 are hugely complex assemblies of hundreds of components. Before such engines are put into production, computer modelling is used to test the sequence of how the parts will be assembled, and how they can be disassembled for maintenance.

Testing prototypes

An alternative to using computer modelling for testing feasibility is to build a **prototype** to test with the potential consumers. A prototype can be a functioning or non-functioning model of the product which, once developed, will go into production. This allows designers and manufacturers to get feedback about their product ideas, before putting the product into production.

Figure 1.12.3 3D rendering of a cardboard coffee holder.

A simple example of testing with prototypes is shown in Figure 1.12.3. Here, a designer has made a prototype of a coffee cup holder. This can be evaluated by potential consumers to test how easy it is to carry and to open to remove the cups. Manufacturers can also examine the carrier after use to evaluate the strength of material and its suitability, before going into production. The prototype can also be used by printers to discuss the type of graphics that are required by a client.

Prototypes can also be used in the evaluation of food packaging (See the examples of snack bar packaging in Figure 1.12.4). Designers and manufacturers can use such prototypes to test which shape and size is preferred by consumers. They could also use the prototypes to test and evaluate which type of card or box the snack bars might be sold in.

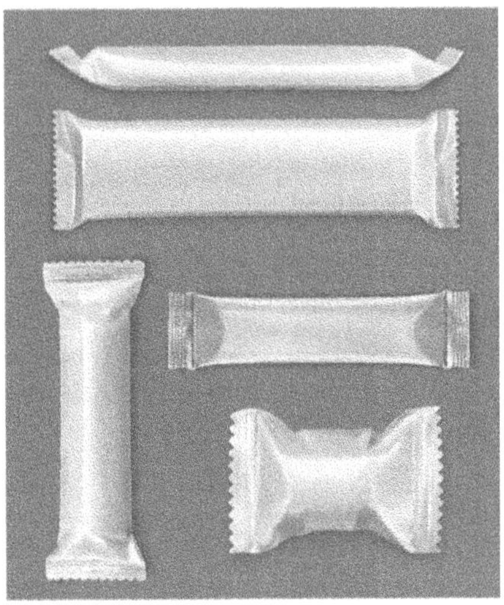

Figure 1.12.4 Prototype snack bars packaging.

Engineers may use large-scale prototypes to test and evaluate products, in order to improve and perfect them before going into production. Figure 1.12.5 shows a prototype of an aircraft that is being evaluated to assess the accessibility of critical parts which require maintenance. Engineers developing products as complex as an aircraft will make several scale and full-size prototypes. This will also include mock-ups of the interior cockpit to evaluate the layout, clarity and accessibility of instruments (see Figure 1.12.6). Testing will be done with a number of pilots to ensure the interior is suitable for the required anthropometric range.

Figure 1.12.5 Testing the prototype.

Figure 1.12.6 Inside the cockpit.

ACTIVITY

Use the internet to find a definition of the term 'feasibility study' and find at least one example relevant to design or manufacturing.

KEY POINTS

- Feasibility studies are used to assess the practicality of a project before investing in materials and manufacturing equipment.
- Feasibility studies can be used to assess ergonomics, especially of environments or interiors.
- Computer modelling can be used to carry out feasibility studies to test projects virtually.

KEY TERMS

Lean manufacturing: the concept of reducing waste in manufacture. This can be waste in time or materials, or idle machines.

Buffer zone: an area adjacent to a point on a production line for the temporary storage of partially completed products.

Cycle time: the amount of time it takes for a manufacturing process to be completed.

Takt time: a German word meaning 'beat or rhythm'. It is used in lean manufacturing to describe the maximum amount of time in which a product must be produced to meet consumer demand.

Prototype: a functioning or non-functioning pre-production model that can be used to evaluate with potential consumers.

Check your knowledge and understanding

1 What is a feasibility study?

2 What is the purpose of a feasibility study?

3 List three examples of feasibility studies that might be carried out in industry.

Further reading

You can read about the elements of a good feasibility study here: www.projectsmark.co.uk/elements-of-a-good-feasibility-study.php

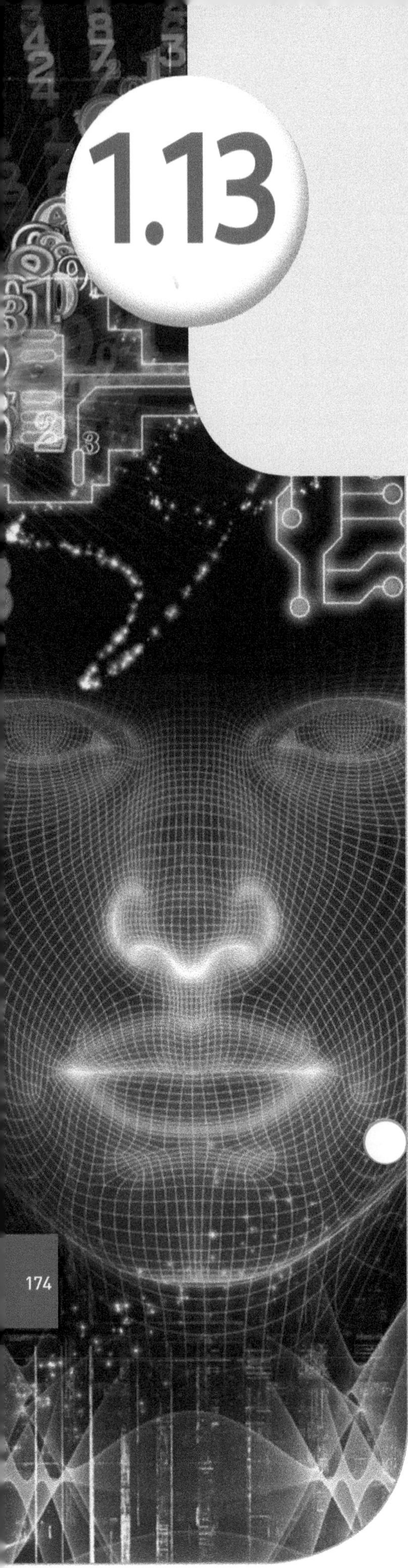

1.13 Enterprise and marketing in the development of products

A level only

This topic is for A level only. If you are studying at AS level you do not need to cover the content in this chapter.

LEARNING OUTCOMES

By the end of this section you should have developed a knowledge and understanding of:

- the importance of marketing and branding – including customer identification, labelling, packaging, corporate identification and the concept of global marketing
- the ways in which products are advertised and promoted, including the use of new technologies
- product costing, calculation and profit
- the role of entrepreneurs, marketing and collaborative working with the designer in the development of new and innovative products.

This section of the book will explain how designers work with entrepreneurs, and sales and marketing teams, and use research and customer information in developing products and packaging. It will also discuss how manufacturers promote their products in a global market.

Customer identification

Market research is used to identify customers. Designers developing products need to know as much as possible about the customer for their products. Key characteristics of customers include:

- age
- gender
- disposable income
- residential location
- recreational interests.

Once a designer knows what age group they are designing for, they may use certain aesthetic features that are relevant to that age group. Similarly, the aesthetics of a product might be influenced by gender. For example, a designer developing packaging for a deodorant for a male might use colours such as blacks and silvers, while deodorant packaging for the female market might include soft pastel colours.

The amount of disposable income of customers is a key consideration because designers will need to know whether to develop a 'budget' product or one that is aimed at a higher income, perhaps a luxury item,

The residential location of the typical customer may also influence the type of product that is developed. For example, if the customer typically lives in a one-bedroom flat, the designer might need to give more careful consideration to the size and ease of storage of the product they are developing.

Typical recreational activities can influence the design of products. For example, if the customer is interested in adventure sports or fitness products, the designer will try to use aesthetic and functional features, and choose materials that are in keeping with these themes.

Market research can also identify why the customers are motivated to buy certain products. Typical reasons include:

- social and emotional needs
- family needs
- budget pressures
- brand preferences.

Again, when the designer is equipped with this information, they can develop products that are matched more closely to the needs of the customer.

Corporate identity

Corporate identity is the branding that is used to present the image of a company to the public. Designers use colour or combinations of colours, logos and typeset (letter fonts) to create a 'brand' and brand name. This is vital in establishing and maintaining customer loyalty to a brand, and for a product or a range of products marketed under this brand name. When consumers try a product and like it, they will look for the same product again. The key to identifying their favourite product will be the distinctive corporate identity of the brand.

Manufacturers protect their **brand identity** using registered trademarks. This prevents rival companies copying their logos, colour schemes and specific letter font styles.

Packaging design

Packaging design is a key aspect of product marketing and branding. Many manufacturers use their packaging design as an integral part of brand identity. An example of this is Coca-Cola and their distinctive bottle shape. Over the years, the bottle has changed very little, and its iconic shape has become synonymous with the brand – so much so that the bottle is also used in the logo on all Coca-Cola packaging.

Labelling

In addition to promoting the brand, labelling might be used in marketing through the use of slogans or offers to attract customer attention. Labelling is also used to inform consumers about important information, for example dietary information is often printed on food packaging. Other important information such as allergens, 'use-by' date and – in the case of meat and dairy products – the country of origin is also printed.

Products such as children's toys might have age restriction labels to warn consumers of choking hazards and unsuitability for children under three years of age. Other products might have labelling to indicate that they meet British standards and have been tested by the BSI, or the CE mark to indicate they meet basic safety standards for sale in Europe.

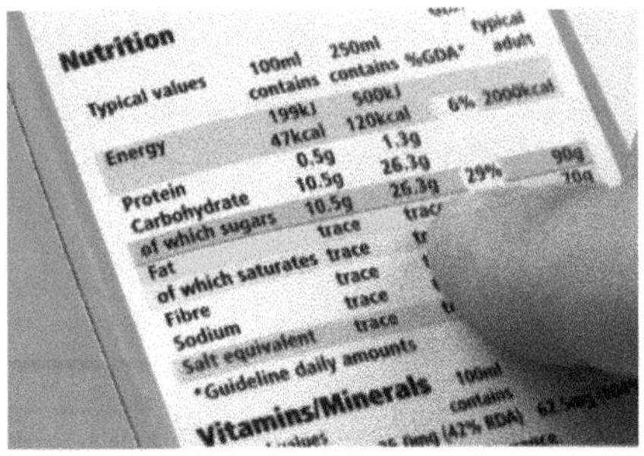

Figure 1.13.1 Typical food labelling.

Figure 1.13.2 BSI kitemark.

Global marketing

Global marketing is the process of promoting a final product or service worldwide. The purpose of this is to reach international markets, which has the potential to rapidly grow a business.

Different marketing strategies will be used depending upon the region in which the product is being promoted. For example, large fast food chains have different menus that are suitable for use in different countries. Such fast food companies focus on marketing foods that are popular in a particular country, and promote different foods in other countries. Global marketing requires designers to be sensitive to the religious and cultural beliefs in countries in which they are promoting products, so that customers are not offended and decency laws are not broken. It is important to understand the language and culture of customers when trading internationally. Another important factor is in the design of packaging. In many countries, food packaging is expected to show images of the contents. There is an urban myth that a baby food manufacturer used a cute picture of a baby as part of its branding, and when used on the front of jars, customers in some countries were offended.

Companies that are most successful with global marketing are those that develop products with universal appeal such as food, cars, mobile communications and electrical goods. Large corporations such as Apple, IBM and Microsoft, or international companies such as JCB will operate from offices in different countries. However, the rapid expansion of the internet has meant that even very small companies can trade internationally.

Advertising and promotion

The promotion and advertising of new products is crucial in several stages of the life cycle of a product. During the introduction phase, it is normal for marketing departments to use a wide range of media to advertise their products, so that interest and demand is created with consumers. There are many forms of advertising including:

- TV commercials
- radio commercials
- magazine advertisements and articles
- internet pop-up adverts
- billboards
- bus shelter posters
- social media
- in-store demonstrations.

The use of social media and the internet has transformed how products are advertised. Manufacturers using social media and the internet are now able to reach potential customers all over the world with relative ease and at a lower cost than other techniques such as TV commercials. In addition to this, 'cookies' are used. These are packets of data exchanged from the website server to the customer's browser, which allow websites to store certain information about the customer and their preferences. When you have visited a website using a cookie, it can be used to track what you were browsing. The next time you use the internet, pop-up display adverts will appear to promote products or services you were looking at. This is known is as 'retargeted marketing'.

A variation of this is the use of 'personalised video ads'. Companies are starting to use video adverts which target consumers who have previously shown an interest in a specific product. Personalised video ads are likely to contain images and campaign text, and a 'buy now' button to direct customers to a 'checkout'. This form of advertising is likely to expand as mobile video becomes widely available and has obvious advantages over static graphic videos.

Often, when signing up for free email accounts and social media sites, an individual enters information about their interests and preferences. This information is stored and the appearance of their page, including advertising, is customised to suit them. This means that marketing can be targeted towards customers who are more likely to have an interest in purchasing the product.

The use of social media has also seen a new phenomenon known as '**viral marketing**'. This technique relies on people passing a marketing message to their friends. If a large number of recipients pass the marketing message on, the overall growth snowballs and the message 'goes viral'. This type of marketing is often used in the initial launch of products, and in the growth stage.

During the growth stage of a **product life cycle**, marketing departments often use magazine articles and other techniques to promote the features of their product and its advantages over the products of their competitors. Advertising in specialist magazines allows marketing messages to reach the correct target market, for example motoring accessories might be advertised in automotive magazines.

Product costing, calculation and profit

The cost of a product is the total cost of its manufacture. Costs can be divided into direct and indirect costs.

Direct costs are those that are directly associated with making a specific product, and include direct labour (wages of people making the product), direct material costs (materials used to make the product), direct consumable production supplies and direct factory overheads (machine tools and equipment used to make the product).

Indirect costs are those that are not specifically associated with making a particular product, but are involved with running a business as a whole. This might include things like administration, personnel and security, maintenance and depreciation of equipment, and electricity.

To calculate the unit cost of production (the cost of making a single item), the calculation is as follows:

$$\text{total direct material costs} + \text{total direct labour costs} + \frac{\text{total manufacturing overhead costs}}{\text{the number of units made}} = \text{product cost per unit}$$

You should be familiar with how to calculate costs of making a product. An example is given below.

A **designer-maker** of furniture making handmade wooden dining chairs needs to calculate what they need to charge per chair. The furniture maker wants to earn a gross salary (before tax) of £22,000.

Step 1

Calculate the cost of annual overheads (studio rent, phone, mobile phone, insurance, business rates and marketing costs).

In the example here, the annual overhead cost is £12,500.

Step 2

Decide on how much time will be spent per week actually making the product.

In the example here, we will say 40 per cent of a typical 40-hour working week. We will assume the furniture maker will take four weeks off per year.

Step 3

Calculate the hourly overhead costs.

Hourly overhead costs:

48 weeks × 40 hours per weeks × 40% = 768 hours/year

Divide the yearly overhead cost by the number of hours per year:

£12,500 ÷ 768 = £16.28

Step 4

Calculate the required hourly wage:

£22,000 ÷ 768 = £28.65

Step 5

Calculate the total hourly rate (hourly wage + hourly overhead cost):

£16.28 + £28.65 = £44.93

Step 6

Work out the time it takes to make one chair.

In our example, this is 2.5 hours.

Step 7

Multiply the hourly rate by the time it takes to make one chair:

2.5 × £44.93 = £112.33

This shows the total direct cost to make one chair.

Step 8

Add the material cost to the hourly rate. In our example the material costs are £15:

£15 + £112.33 = £127.33

Therefore in our example, the furniture maker would charge £127.33 for the manufacture of a single chair.

It is important to note that retailers would typically double the cost price to arrive at their recommended retail price. Retailers have overheads such as premises, shop fittings, sales staff, business rates and energy costs. They also have to make a profit, which of course is the purpose of being in business.

Entrepreneurs and collaborative working with designers

Figure 1.13.3 James Dyson is an entrepreneur.

An **entrepreneur** is a person who demonstrates initiative and invests capital into an idea – perhaps setting up a business, project or other venture. Usually, the entrepreneur takes the risk by investing their own capital or by obtaining finance from banks and other investors.

Occasionally, entrepreneurs will have the skills to invent and design their products, for example Sir James Dyson who created the Dyson range of vacuum cleaners and a wide range of other products. More commonly, entrepreneurs need to work with designers and engineers to turn their ideas into reality.

Designers in industry do not usually work in isolation. The normal practice is to work collaboratively with other designers either face to face or virtually using video conferencing and other file sharing software. Video conferencing is done in real time, but cloud-based file sharing allows designers and engineers to work on the same project at different times. This can include working with people in different countries and even different continents with totally different time zones. Designers often work on highly complex projects which require skills and knowledge that a single person cannot be expected to have. Typically, designers might have particular skills and expertise in some areas, but will need to work with other designers and engineers who have the skills necessary to complete the project. Designers work collaboratively in large companies such as Dyson, Apple and major car manufacturers. In such companies, teams of designers – each with their own area of expertise – work alongside engineers, ergonomists, manufacturing specialists and project managers to develop their products.

Some designers work collaboratively in associations or studios that specialise in working with entrepreneurs to bring their ideas into reality. An example of this is London-based Seymour Powell, which is one of the world's leading design and innovation consultancies. Seymour Powell has developed hundreds of products in collaboration with many companies that are household brand names.

ACTIVITY

Using a wage of £15.50/hour and an hourly overhead cost of £16.75, calculate the cost of making your project at school or college.

KEY TERMS

Brand identity: the use of logos, colours, typeset/letter fonts, and slogans to promote customer loyalty to a product.

Viral marketing: the process of passing a marketing message from one person to many others using social media.

Product life cycle: the stages of a product from introduction, growth and maturity to decline.

Designer-maker: someone who both designs and makes products.

Entrepreneur: a person with initiative who invests capital in a project or idea.

KEY POINTS

- Designers often work collaboratively with entrepreneurs, marketing departments, engineers and other designers when developing products.
- Packaging has several functions: to protect, preserve and promote products.
- Advertising and marketing are key to the success of products, and different forms can be used at different stages in the life cycle of a product.
- Due to legal requirements, packaging may need to include labelling such as dietary information and safety information.

Check your knowledge and understanding

1 What is the role of an entrepreneur in the development of products?

2 Why do designers often work collaboratively with other designers?

3 Why is packaging important in the marketing of products?

4 What is the purpose of brand identity?

5 How is the cost of a product calculated?

Further reading

www.seymourpowell.com

Packaging the Brand: The relationship between packaging design and brand identity by Gavin Ambrose.

The Creative Entrepreneur by Isa Maria Seminega.

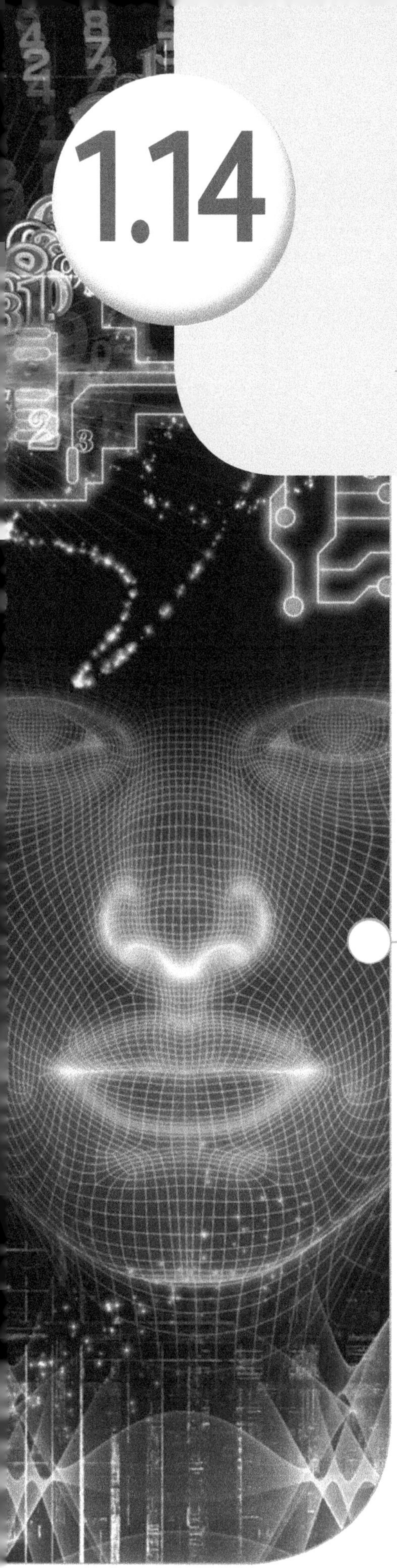

1.14 Design communication

LEARNING OUTCOMES

By the end of this section you should have developed a knowledge and understanding of:

- how to use a range of communication and presentation techniques for conveying design proposals, including:
  - report writing
  - graphs
  - tables and charts
  - 2D and 3D sketching
  - mixed media and rendering to enhance drawings and to show dimensions and details for manufacture.

Throughout your product design course, you will use a wide range of design communication methods in your project work. This will allow you to develop a good understanding of how they are used in the design process. This chapter will explain each of the communication methods and their uses.

Report writing

In industry, designers and engineers, often working in a team, document their work in a design report. This is a record of the project that remains after the project has been completed and the design team has disbanded or commenced a different project.

Design reports may start with a section known as the 'executive summary'. This is a concise summary of the design itself, the reasons behind the design and a summary of how effective it is.

The next section is the introduction, which describes the design problem that was being tackled and explains background details such as the context, client and possible customers. This section can also include the alternative design ideas and the selection process of the idea that was developed.

The third section of the report is known as the conclusion and describes the selected design, the problems encountered in the project and how these were overcome. The third section evaluates how well the design works in comparison to the original specification criteria. It also explains how the design was tested. If the design did not meet the requirements of the specification, the conclusion will analyse why and how the design could be modified to make it a success.

From time to time, designers may also have to write technical reports. For example, they may have to write up the results of carrying out practical tests, targeted market research or a review of products that might influence the development of a product specification and designs. It is important that reports are written in clear English and use the correct technical terms, such as the names of specific materials. Reports may be illustrated with tables of data, graphs and charts. You may use report writing in your project work, either in project research or evaluation.

Graphs, tables and charts

It is important that you are able to identify the different types of graphs and charts, and explain their use. You should also be able to read and interpret data that is presented in a range of different formats.

Bar charts

Bar charts are the simplest way of representing data. They are used in many different reports and in the media, so they are widely recognised.

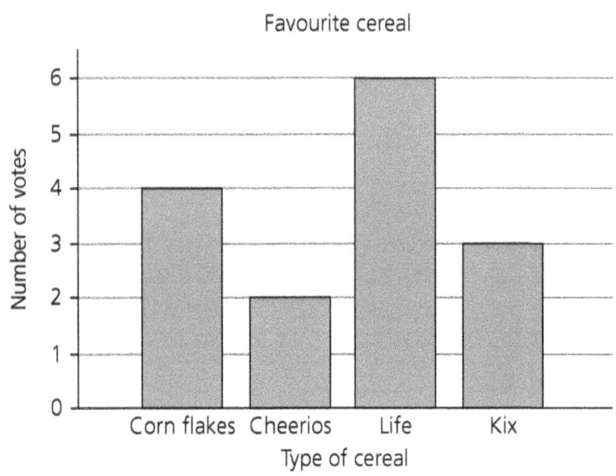

Figure 1.14.1 Bar chart showing favourite cereals of a sample of sixth form students.

Pictographs

Pictographs show data in a way that is easy to understand. They are quick to read and therefore suitable for a presentation, but they are not very accurate.

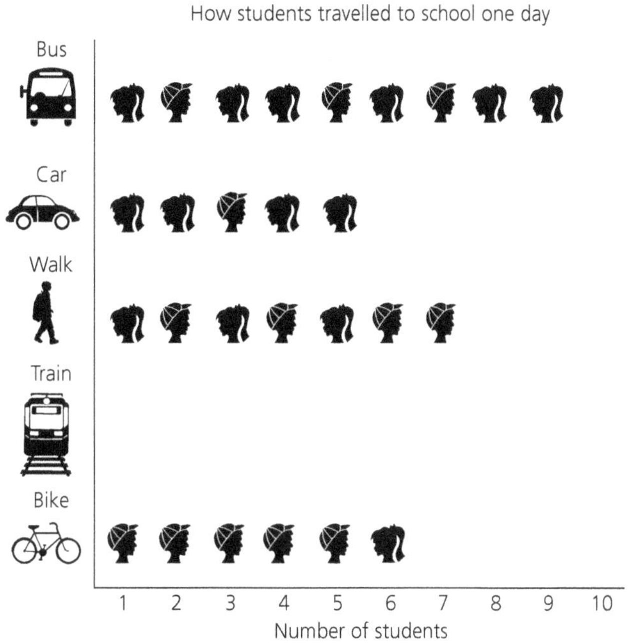

Figure 1.14.2 Pictograph showing how students travelled to school.

Histograms

Histograms are similar to bar charts, but the data is shown in ranges. They are useful to show the frequency distribution of data, and are particularly useful in representing the results of research.

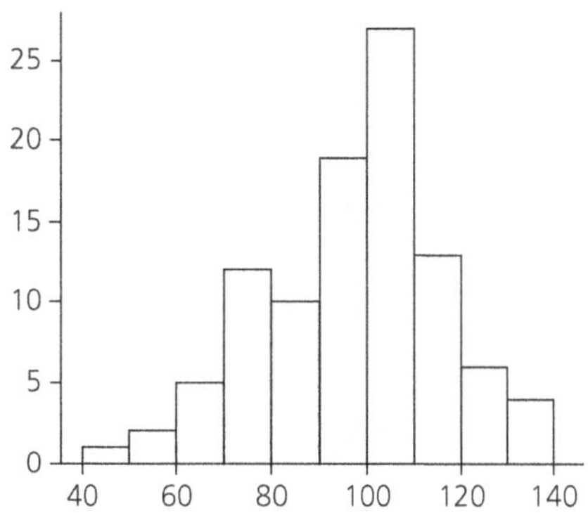

Figure 1.14.3 Example histogram.

Line graphs

Line graphs or line charts show data changes over time. They are relatively easy to understand, and the data can be shown accurately because it is plotted against a scale.

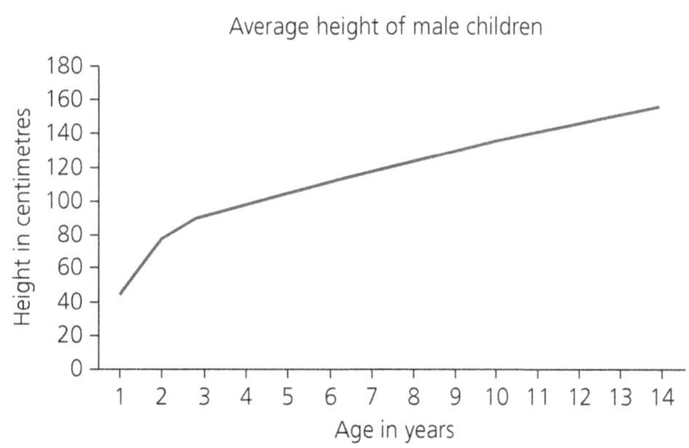

Figure 1.14.4 Line graph showing the average height from a sample of 180 children.

Pie charts

Pie charts show the distribution of data represented by segments of a circle. The size of the segment represents the relative proportion of data compared to the whole. Pie charts are easy to read and understand.

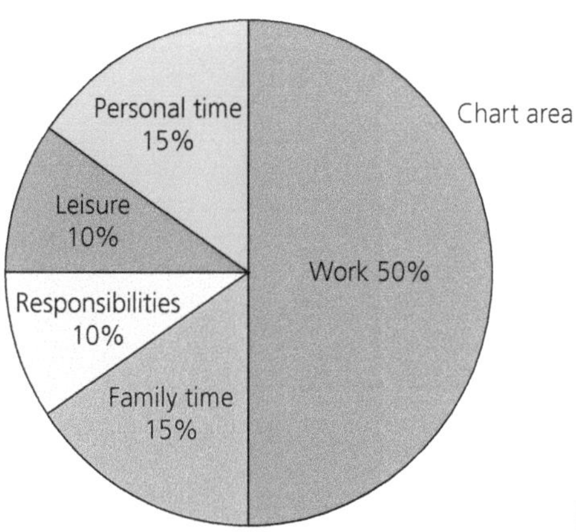

Figure 1.14.5 Pie chart showing the proportion of time spent on different activities in a day by a typical working parent.

Data tables

You will be familiar with tables and their use in presenting data. You will need to be able to develop the ability to interpret data and perhaps identify patterns, spot incorrect data or add missing data. Where possible, you should practise using tables in your project work including anthropometric data (human body measurements), cutting lists and costings.

Age bracket, years	Anthropometric data*			
	Age, years	Weight, kg	Height, m	BMI, kg/m^2
Males				
20–29	23.20 ± 3.27	71.44 ± 10.74	1.76 ± 0.07	23.15 ± 2.13
30–39	32.70 ± 2.26	72.29 ± 7.63	1.75 ± 0.08	23.55 ± 1.42
40–49	44.27 ± 3.10	76.85 ± 13.73	1.69 ± 0.07	26.63 ± 3.12
50–59	54.80 ± 3.39	81.65 ± 8.63	1.76 ± 0.05	26.50 ± 3.04
60–69	63.70 ± 2.63	73.18 ± 3.91	1.70 ± 0.08	25.52 ± 2.84
70–80	77.00 ± 2.79	65.62 ± 11.89	1.64 ± 0.05	24.31 ± 3.90
Females				
20–29	22.40 ± 2.55	59.18 ± 8.80	1.62 ± 0.06	22.47 ± 3.52
30–39	33.30 ± 2.91	65.51 ± 8.37	1.66 ± 0.07	23.79 ± 2.96
40–49	44.80 ± 2.78	63.40 ± 9.92	1.62 ± 0.09	23.96 ± 2.29
50–59	54.20 ± 3.29	61.69 ± 5.96	1.58 ± 0.06	24.88 ± 1.94
60–69	64.70 ± 3.83	65.19 ± 7.95	1.59 ± 0.05	25.87 ± 2.11
70–80	75.70 ± 3.16	61.30 ± 6.62	1.56 ± 0.07	25.14 ± 2.61

*Values expressed as mean ± SD; each age bracket comprised 10 subjects.

Figure 1.14.6 Typical table of anthropometric data: Table shows weight, height and BMI of adult males and females.

2D drawing

Designers may use a range of 2D drawing techniques to quickly sketch design ideas or to produce more technical drawings, which might be used to aid the manufacture of the product. Here, we will focus on the common technical drawing techniques.

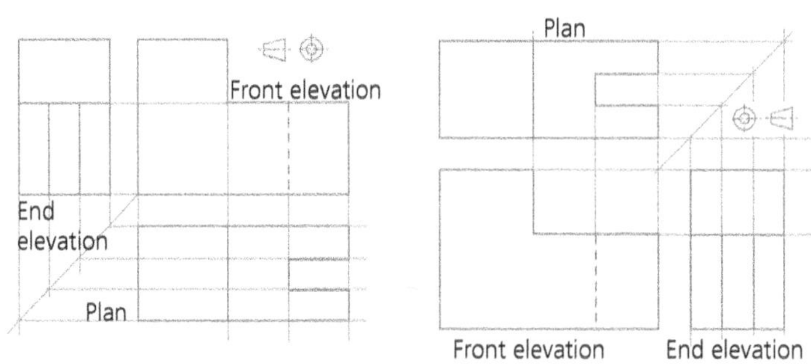

Figure 1.14.7 First and third angle drawings.

Orthographic projection

Orthographic projection is used to draw the front, plan and end 'elevations' of an object or building. Such drawings are drawn in a standardised way so that they can be understood by engineers and manufacturers worldwide. This standardisation includes the position of the elevations relative to each other, how the object is dimensioned or drawn to scale and the line types used.

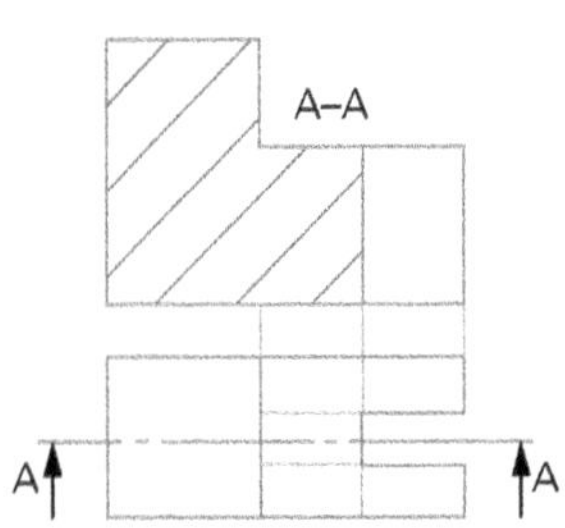

Figure 1.14.8 An example of a sectional drawing.

Sectional drawing

Sectional drawings are used to show internal, hidden details of an object. In a simple section drawing, typically the object would be 'cut' in half along a 'cutting plane', which might be labelled A–A as in Figure 1.14.8.

3D drawing

Isometric

One of the main 3D sketching methods used by designers is **isometric**. It is one of the simplest 3D drawing methods and is typically used to sketch design ideas. Isometric drawings can be drawn freehand or using a 30/60 set square as they use 30-degree lines. Typically, designers use an isometric '**crate**' to construct drawings, with all lines parallel to the edges of the crate.

Figure 1.14.9 shows an isometric drawing with a set square.

One-point perspective

One-point **perspective drawing** is another simple 3D drawing method, which is typically used by architects and interior designers to sketch buildings or rooms. Perspective gives the drawing depth and realism – when we look at objects in real life, they appear smaller the further away from them you are standing. With perspective drawing, objects drawn in the foreground are large, and those that are closer to the vanishing point are smaller.

One-point perspective can also be used to draw design ideas in 3D and, again, these can be constructed within a 'crate' if required. One of the advantages of using this technique is that the designs can be viewed from different angles to look at the top of the object, the left, right or underneath. This is determined by whether the drawing is started above or below the horizon line, or to one side or the other of the vanishing point.

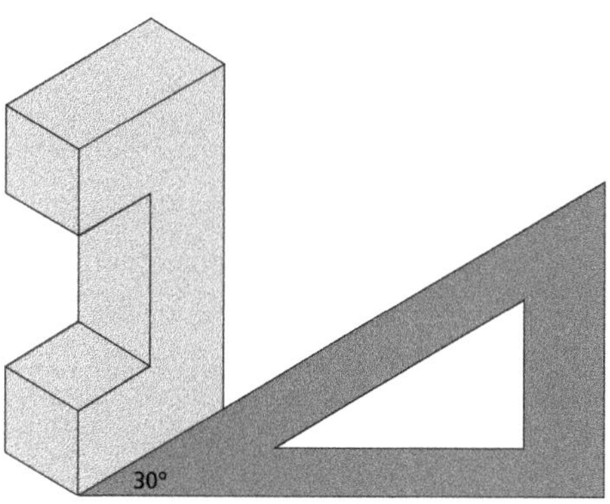

Figure 1.14.9 An isometric drawing. A 30-degree set square shows the angle used in isometric.

Figure 1.14.10 Examples of one-point perspective.

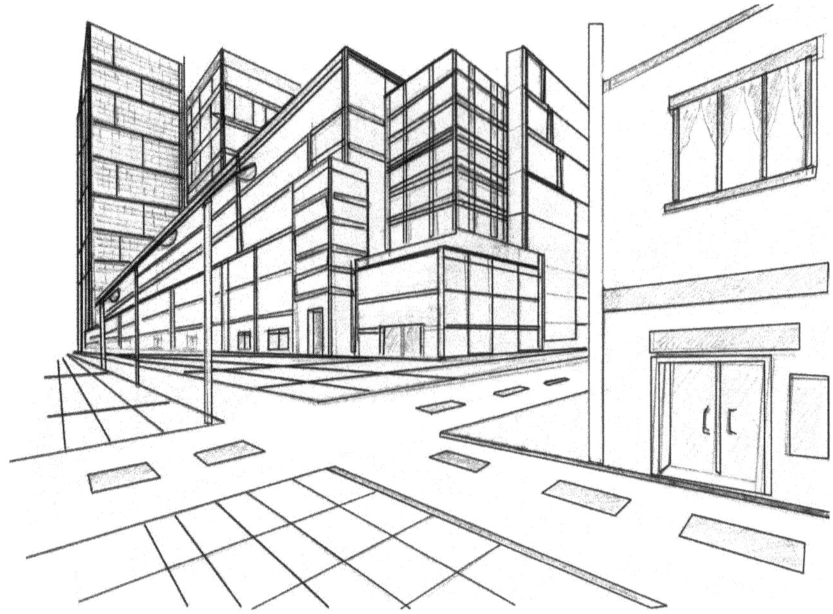

Figure 1.14.11 Examples of two-point perspective.

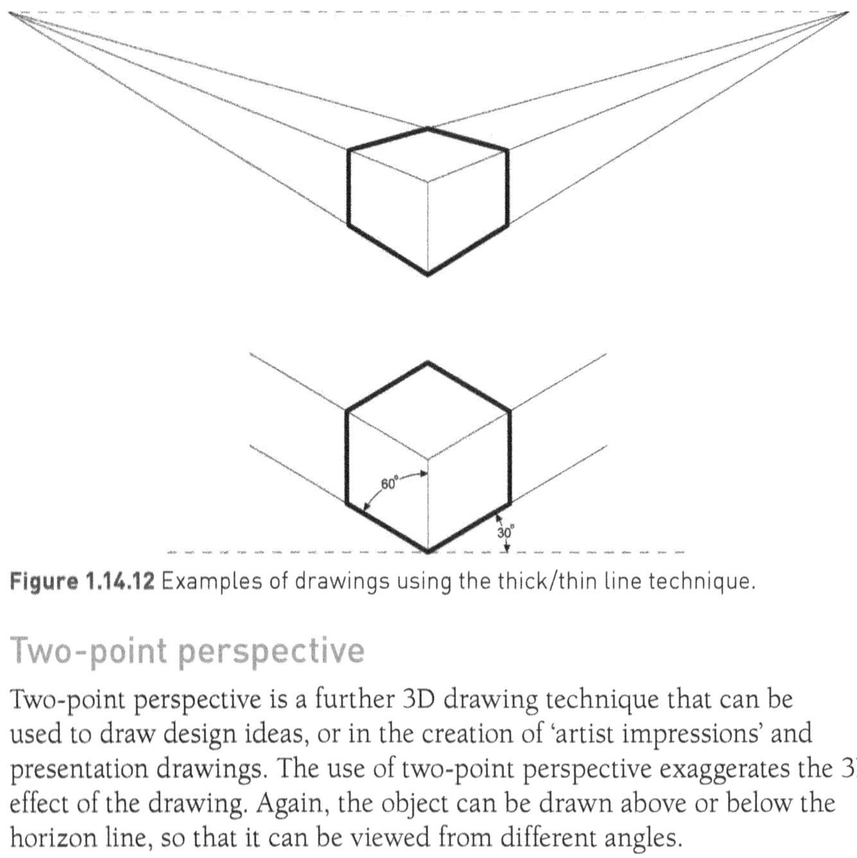

Figure 1.14.12 Examples of drawings using the thick/thin line technique.

Two-point perspective

Two-point perspective is a further 3D drawing technique that can be used to draw design ideas, or in the creation of 'artist impressions' and presentation drawings. The use of two-point perspective exaggerates the 3D effect of the drawing. Again, the object can be drawn above or below the horizon line, so that it can be viewed from different angles.

Thick/thin line technique

Thick/thin line is a technique used to make 3D drawings stand out. Usually (but not always) the external edges are bold and thick, and the internal lines are thin.

Mixed media and rendering

Marker rendering

Spirit-based markers are widely used by designers to enhance their drawings. They can be used to provide a base colour on which additional layers of colour or tone can be added.

Figure 1.14.13 shows how 2D drawings can be made to look 3D with the use of markers and a fine line pen. White reflection lines have been left to give an additional 3D effect.

Figure 1.14.13 Marker rendered ideas drawing.

Figure 1.14.14 Mixed media rendering of a kitchen interior.

Mixed media drawings

Mixed media drawings make use of pencil, marker pen, watercolour and fine line pen.

Watercolour is useful to cover large areas quickly. Pencil and fine line pen is used to pick out the detail of the drawing. Using mixed media can add realistic textures to drawings.

Texture rendering

Texture can be added to drawings using a fine line pen. By adding textures, drawings can be made to look like specific materials. Typical fine line pen textures include wood grain, cross hatching and stippling.

Figure 1.14.15 Examples of textures created with a fine line pen.

ACTIVITY

Sketch three cubes using isometric, one-point perspective and two-point perspective. Then render each cube with a fine line pen to represent different materials, for example wood, concrete and reflective glass.

KEY TERMS

Orthographic projection: a technique used to draw the front, plan and end elevation of an object.

Sectional drawings: a technique used to draw the hidden details and internal features of an object.

Isometric: a 3D drawing technique using 30-degree lines – typically used in sketching ideas.

Crating: using a 3D box and constructing drawings inside them, parallel to the outer lines of the box.

Perspective drawings: 3D drawings that use one or two vanishing points. They are often used for design ideas or presentation drawings.

KEY POINTS

- Graphs, charts and tables are an important communication method, and you should be able to produce them, read them and interpret the data found in them.
- You need to be familiar with a range of 2D and 3D drawing techniques – how to construct them, their uses and advantages.

Check your knowledge and understanding

1 Name a 3D drawing technique and explain how it is used in the design process.

2 What are the differences between first and third angle projection?

3 What are the advantages and disadvantages of using a pictograph in presenting data?

4 What would you use marker rendering for?

5 How can you represent materials and textures in drawings?

Further reading

Presentation Techniques: A guide to drawing and presenting design ideas by Dick Powell.

AQA GCSE Design and Technology: Graphic Products by Eamonn Durkan, David Dunlop and Geoff Westell. Pages 40–46 and 55–60 give a good overview of different communication methods. Pages 70–71 give an introduction to information graphics.

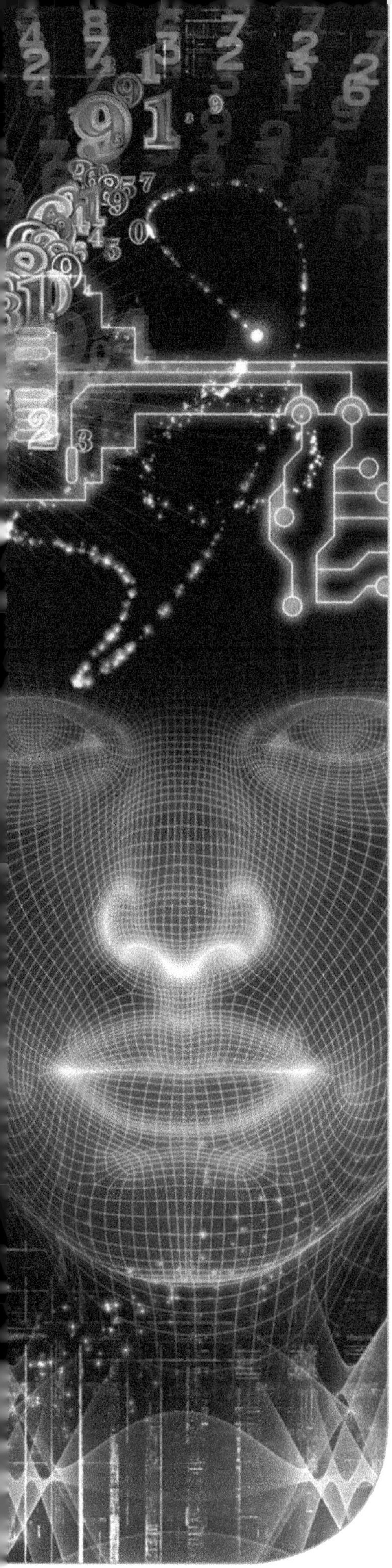

Section 2 Designing and making principles

In this section you will learn about the following:

2.1 Design methods and processes

2.2 Design theory

2.3 Technology and cultural changes

2.4 Design processes

2.5 Critical analysis and evaluation

2.6 Selecting appropriate tools, techniques and processes

2.7 Accuracy in design and manufacture

2.8 Responsible design

2.9 Design for manufacturing and project management

2.10 National and international standards in product design

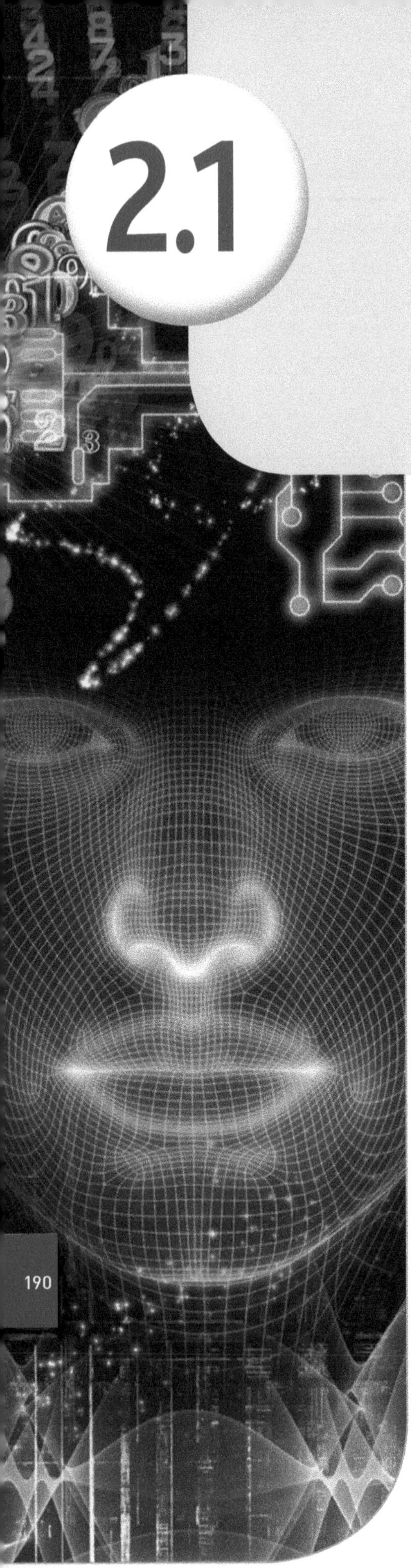

2.1 Design methods and processes

LEARNING OUTCOMES

By the end of this section you should have developed a knowledge and understanding of:

- UCD as a method of approaching a design challenge
- designing to meet needs, wants and values
- investigations to inform the use of primary and secondary data: market research, interviews, human factors, focus groups, product analysis and evaluation, the use of anthropometric data and percentiles and the use of ergonomic data
- the development of a design proposal
- the planning and manufacture of a prototype solution
- the evaluation of a prototype solution to inform further development.

There are many theories about the process of designing. Traditionally, this was considered to be a linear process that started with a design problem and ended with the evaluation of a prototype. Other theories about design thinking describe the process of developing products as an **iterative** process.

Iterative design process

The iterative design process is a circular set of steps that a designer will go through to solve a problem. The designer or team of designers and engineers may not go through the process one stage at a time, but may go backwards and forwards between steps. For example, it is quite normal for designers to sketch first design ideas, then pause to carry out research, then continue designing and pause again to find out further information. At any point in the design process, they may carry out evaluations – for example, to analyse research or evaluate ideas and models.

User-centred design

User-centred design (UCD) is an approach to design that ensures your product will be easy to use. Designers who adopt this approach normally work to the principles set down in the international usability standard ISO 9241-210:2010. In a competitive, global market made up of discerning consumers, the user experience is very important. If designers can say that what they are developing conforms to international standards, they are likely

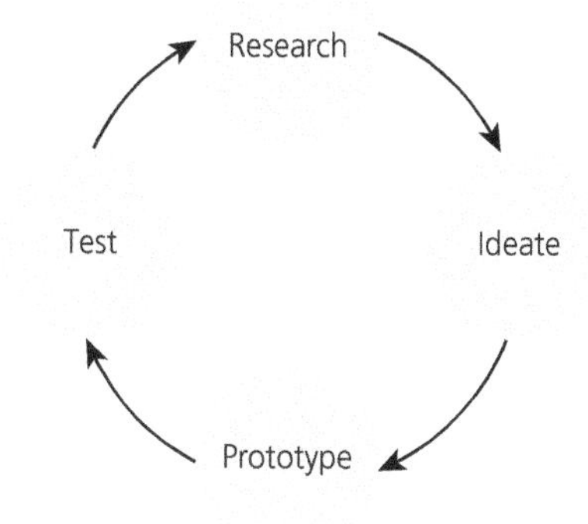

Figure 2.1.1 Circular design process.

to obtain more support from investors or entrepreneurs. The principles of ISO 9241-210:2010 are:

- The design takes full account of users, the task they perform with the product and the environment in which it is used.
- Users are involved throughout the design and development process.
- The design is refined by user-centred evaluation.
- The design process is iterative.
- The design considers the whole user experience.
- The design is developed by a multi-skilled, multi-disciplinary team, for example product designers, engineers, industrial designers, ergonomists and sales and marketing.

Ergonomics are a key consideration in products that have been developed using UCD methods. Products should be both easy and comfortable to use. In order to achieve this, designers use **anthropometric** data (body sizes) to ensure a product 'fits' their intended user. In addition to this, they observe people using products, and employ consumer focus groups to identify the problems with existing products. Designers can only develop successful products if they fully understand how users interact with products and how the user experience can be improved.

Designing to meet needs, wants or values

When developing products, it is important to consider the needs, wants and values of consumers. People are more likely to purchase a product if it meets their needs, wants and values. In order to establish what the needs, wants and values are, designers must conduct and use research.

Conducting research enables designers to identify consumers' physical needs. For example, the physical needs of children will be very different from the physical needs of the elderly, wheelchair-bound, disabled, blind or deaf. If products can be closely matched with the physical needs of consumers, they are more likely to be successful.

Identifying the emotional needs of consumers can make a significant difference to the success or failure of a product. Finding out what people like or dislike, their aspirations and ambitions, can help the designer match the product to the consumer. Many consumers are very brand conscious and want to be seen using top brand products. Designers and manufacturers exploit this emotional need and develop products that are perceived as top brands or high quality.

Research can identify the intellectual needs of consumers. Products have to be matched with the intellectual level of the user. For example, small children might be attracted to brightly coloured, simple toys but teenagers will want something that is more stimulating and with a higher level of challenge to keep them interested.

Values and sociological needs should also to be taken into account when developing products. For example, eco-conscious consumers want to live a sustainable lifestyle and therefore demand products that are not damaging to the environment.

Investigative methods

Designers make use of a wide range of investigative techniques when developing products. One such technique is 'immersion', which involves designers putting themselves into the environment or situation that the customer or user is in. For example, a designer developing products for use by pregnant ladies might wear a 'pregnancy suit', simulating the size, weight and restricted movement of someone who is pregnant. Another example might be using a thick glove to simulate the experience of someone who has arthritis or a loss of dexterity, or frosted glasses to simulate the experience of someone who is visually impaired. This type of immersive investigation enables designers to gain a real understanding of the problems users have with products.

Other primary investigation techniques

Investigative techniques from primary sources are the most useful when developing new products. Examples include:

- Interviewing potential users – this investigates their needs or wants with regard to new products.
- Focus groups – these are panels of potential customers who represent the target market. They are used to obtain qualitative data such as opinions about existing products, or to obtain feedback on prototypes.
- Market research – this is usually carried out using questionnaires and surveys. Quantitative data can be obtained which can influence design thinking. Examples of data that can be obtained include income level, disposable income, age range, hobbies and interests, typical shopping habits and recent purchases. Market research can identify the needs of consumers. It can also explain why consumers choose one product over another. Information such as this can help designers to make more informed decisions about how they will develop products.
- Analysing, disassembling and evaluating existing products – this identifies useful materials, components, mechanisms, structures and other useful features that can be adapted in a new product.
- Practical testing of materials, components, construction methods and finishes – this identifies what might be used in creating a new product, and how well the product works/lasts (and appears) in usage.
- Human factors such as ergonomic issues can be considered by observing potential customers using a product – this can identify whether the consumer has difficulties in using a product and potential improvements can be considered. Watching customers use a product and discussing its usability with them can also identify whether there are any issues in terms of comfort. An example might be observing someone using a computer workstation, and assessing whether they can sit comfortably and use the computer without any strain issues. Anthropometric data (human body measurements) can be taken directly from the client or volunteers in a test group or be obtained from secondary sources.

Secondary investigation

Carrying out investigations from secondary sources can also be useful when developing new products. Examples include internet or book research to:

- explore historical and contemporary designers or design movements to identify styling influences which might be used in new products
- identify suitable materials, components and construction methods
- obtain anthropometric data that is relevant to the age group of the target market. Anthropometric data is expressed in percentiles – a percentile represents a percentage position in a range of data. Usually designers focus their designs on a specific percentile range of anthropometrics. Typically, for the adult market, designers use data around the 50th percentile which is the average for male and female consumers. Designers use this information to ensure that products are designed to 'fit' the body size of the average male and female in the target market.

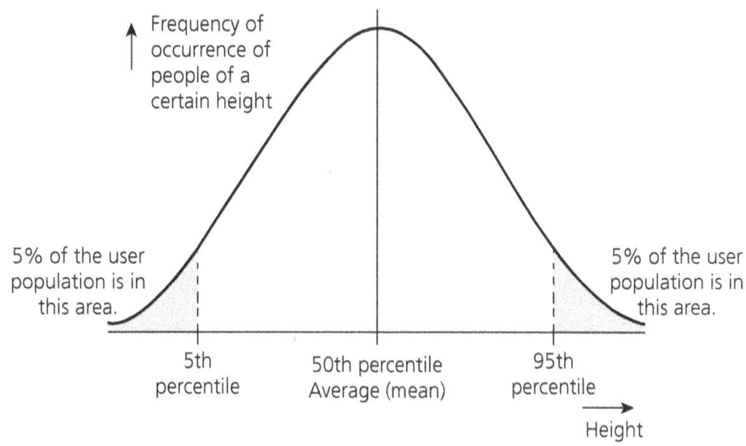

Figure 2.1.2 Graph showing the typical distribution of anthropometric data for height.

The development of a design proposal

Developing a design proposal concerns sketching and modelling ideas in order to work out a solution to a design problem. Design ideas are usually annotated with critical and evaluative comments that explain how ideas meet the specification. The designs may be evaluated by the client or potential customers to select a single design to develop.

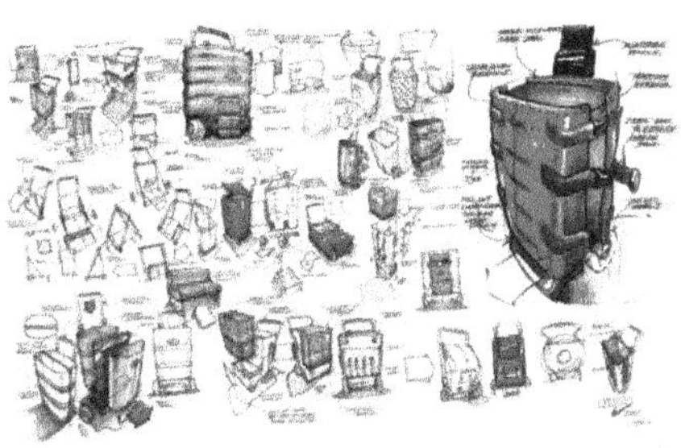

Figure 2.1.3 Example of design idea sketches.

Once a design idea is chosen, the designer develops it into a final design proposal. This includes developing the aesthetics of the product, ensuring it is the correct size and proportion, and considering alternative materials and construction methods to select which is the most appropriate. This might involve making test pieces to try out different materials and making methods. During development, scale and full-size models may be used to evaluate with clients, and if the product is to be manufactured in industry, models might be evaluated with production engineers.

At this stage, working drawings with dimensions are created. These are elevation drawings showing the front, plan and end view of the product and include sufficient detail that the product could be made.

The planning, manufacture and evaluation of a prototype

Once a design has been finalised, its manufacture is planned. This is a step-by-step plan detailing all the stages of manufacture. It includes all of the tools and equipment that would be used, and a quality control (QC) check that would be carried out in its manufacture. A cutting list is prepared, which details all the materials and components that are required for the manufacture.

Prototype manufacture can use any material or combination of materials. It The prototype can be made by hand or a combination of hand and machine processes. It could also be made by rapid prototyping.

When the prototype has been made, it is tested and evaluated. This might be done with clients and potential customers, and enables improvements to be identified before the final product is put into production.

ACTIVITY

Evaluate a product such as a computer chair. Test the chair and evaluate its function and ergonomics. Test how easy it is to adjust the chair. How accessible are the control knobs that adjust the chair? How comfortable is the seat pad and back rest?

Produce an annotated diagram of the chair and describe possible improvements that could be made.

KEY TERMS

Iterative: describes the process of repeatedly going through the circular design process. An iteration would be one cycle of the process.

User-centred design (UCD): developing products with the end user fully in mind. Such products are easy to use.

Ergonomics: concerns the design of products that are easy or comfortable to use. Ergonomics is the study of the interaction between products and humans.

Anthropometrics: body measurements.

Primary investigation: research carried out first hand (not using the internet, books or magazines).

Secondary investigation: research information that is gathered from books and the internet.

KEY POINTS

- The design process is not a linear process that starts with a design problem and finishes with a practical product, but is a continual cycle that seeks to constantly improve products and the user experience.
- UCD employs techniques that investigates the needs and wants of the user, and aims to make products more usable.
- Primary investigation is essential for finding out information that will be most useful in developing a design specification and generating design ideas.
- Anthropometric data is an important area to research in order to ensure prototypes are the correct size for people to use.
- Ergonomics are a key consideration when investigating how a product can be developed.
- When developing design ideas and a final design proposal, it is crucial to involve clients and potential customers to obtain feedback.
- Market research is used to obtain information on people's values, needs and wants.

Check your knowledge and understanding

1 Describe how focus groups might be used in the early stages of a design project.

2 How might a designer carry out primary research?

3 How might designers use anthropometric data when designing a prototype?

4 Explain why designers might disassemble products in the investigation stage of the design process.

5 Why do designers use immersive techniques when carrying out research?

Further reading

The Design of Everyday Things by Donald A. Norman

How Designers Think: The Design Process Demystified by Bryan Lawson

2.2 Design theory

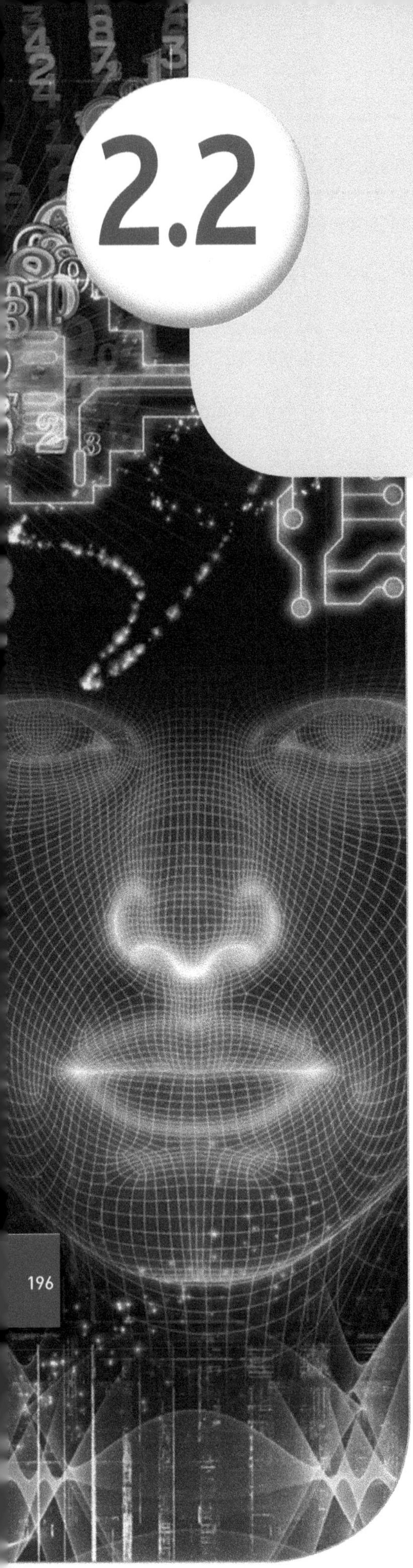

LEARNING OUTCOMES

By the end of this section you should have developed a knowledge and understanding of:

- key historical design styles, design movements and influential designers that have helped to shape product design and manufacture
- key design styles and movements
- examples of designers and their work, and how their designs were influenced by design principles.

Successful product design relies on a combination of functional performance and aesthetic appeal. The view of Donald Norman – a design expert and Director of The Design Lab in California – is that usability is key in product design. This is true if we consider the hundreds of designed products we interact with on a daily basis such as lamps and mugs. All of these small products are used without thought as to who designed them, and this is the aim of most designers – not to be noticed.

When we talk about particular products, it tends to be due to negative interactions, and experiences that cause us discomfort or annoyance. Donald Norman, however, also recognises the need for emotional attachment to products. By this, he means the relationship we have with products that are not purely functional, different or just slightly harder to use than the norm. These products are selected for sentimental or aesthetic reasons, when we want to display our personality, to stand out from the crowd or to bring back fond memories as we use the product.

In this chapter, we will look at key design movements, and designers and their theories with regards to design.

ACTIVITY

For a product you own, prepare a two-minute presentation explaining your emotional attachment to it.

Design influences

When we look at the work of past designers and design movements, we must consider them in context, comparing their work with the work of others of the same time period. It is impossible to gauge the impact or innovation of these designers and movements unless we compare them with their peers.

ACTIVITY

Research the work of Raymond Loewy, and produce a poster showing his major designs and the cultural, social and technological influences on his work.

We should consider a range of aspects when analysing designs from the past:

- cultural and social influences from the time
- major technological developments of the time
- key aspects associated with the movement/designer
- influences on design today.

Design styles and movements

Arts and Crafts movement

After the Great Exhibition of 1851, which was hosted in London to promote and celebrate the development of modern industrial technology, there was a significant reaction to some of the perceived changes brought about by the industrial revolution. The **Arts and Crafts movement** was born from a concern held by prominent designers and social activists of the time – that the use of machinery and factory-based production created products with unnecessarily ornate decoration, and meant that appreciation of the materials used in their construction was lost. They drew inspiration from medieval craftsmanship, reflecting natural forms in textures and surface design such as wallpapers and focused on the natural beauty of timber in the production of handcrafted furniture.

The industrial revolution in Great Britain during the latter half of the eighteenth century and the first half of the nineteenth century caused huge social change, reducing the reliance on craftsmanship and increasing the use of machine tools. This was combined with the division of labour within manufacturing, reducing the skills of workers who were expected to repeat individual tasks within the production process rather than complete the whole process.

Despite their concern over the developing machine age, members of the Arts and Crafts movement could see the benefits of modern manufacturing techniques to enhance honest designs which retained the qualities of the materials used in the manufacture.

Major contributors to the Arts and Crafts movement were William Morris, C.F.A Voysey and Richard Norman Shaw.

Figures 2.2.1 and 2.2.2 show key examples of Arts and Crafts design.

Figure 2.2.1 'The strawberry thief', textile designed by William Morris, 1883.

Figure 2.2.2 The passage at Red House, Bexleyheath, Kent.

Key features of the Arts and Crafts movement:

- Appreciation of the beauty of materials: the Arts and Crafts movement was keen to highlight the unique nature of materials, such as the grain and figure visible in solid oak. They resented the ornamentation of machine-produced products, which often hid this aesthetic beauty.
- Hand produced using craft skills: concerned that the machine-produced objects of the industrial revolution heralded the end of craftsmanship, the Arts and Crafts movement took inspiration from the hand-produced natural forms of medieval Europe.

Art Deco

Following the Art Nouveau movement (a style of art inspired by natural forms and structures), the Art Deco movement of the 1920s and 1930s was named after the Paris Exhibition in 1925 – the *Exposition Internationale des Arts Décoratifs et Industriels Modernes*. Before the exhibition, the movement was known as the modern style.

The discovery of Tutankhamun's tomb by Howard Carter in 1922 was a huge worldwide event, leading to increased interest in the art of the ancient world. This was a major influence on the design period, and is reflected in the use of simple geometric forms and stepped pyramid structures (ziggurats) in the architecture of the time. The end of the First World War and changing social class systems beckoned a new age of modern living, often symbolised in design by sunburst motifs.

Major contributors of the Art Deco movement were Clarice Cliff, Eileen Gray, Alvar Aalto and Walter Dorwin Teague.

Figures 2.2.3 – 2.2.5 show key examples of Art Deco design.

Figure 2.2.3 The Chrysler building.

Figure 2.2.4 Art Deco, South Beach, Miami.

Figure 2.2.5 1930s Art Deco wooden coffee table.

Key features of Art Deco design:

- Sunburst motifs: these rays or segments radiating from a central point were commonly seen in architecture and surface patterns.
- Ziggurat (stepped pyramids): the ziggurat is an element often seen in Art Deco architecture, especially the 'skyscraper' designs of New York, where buildings were restricted by the 1916 zoning resolution to reduce

the impact on light visible on the streets below. The zoning resolution required regular 'setbacks' as the height of the building increased, thus reflecting the ziggurat structures of ancient civilisations.

- Simple geometric forms: these were a distinct change from the natural realism associated with Art Nouveau.

Modernism

Figure 2.2.6 'Red and blue chair' by Gerrit Rietveld, 1917.

Evolving from the worldwide Art Deco style, Modernism is symbolised by key design schools such as the **Bauhaus** and De Stijl. As the First World War ended, both of these design schools were formed. The impact of the war on the infrastructure of Europe and the resulting need to rebuild presented them with the ideal opportunity to make changes and modernise.

Based in the Netherlands, the De Stijl movement focused on the use of basic rectilinear forms and primary colour schemes to produce abstract artistic pieces. Designs were largely asymmetrical. These forms and colour schemes were evident in furniture, interiors and architecture associated with the movement.

Major contributors of the Modernism movement were J.J.P. Oud, Piet Mondrian, Gerrit Rietveld and Robert van't Hoff.

Figures 2.2.6–2.2.8 show key examples of Modernist design.

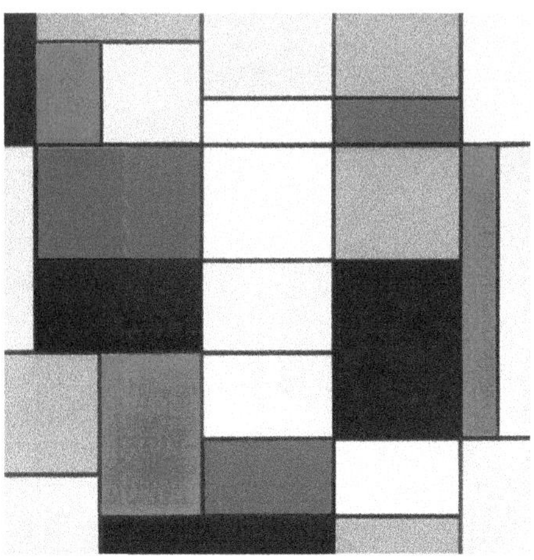

Figure 2.2.7 'Composition A' by Piet Mondrian.

Figure 2.2.8 'Schroder house' by Gerrit Rietveld, 1924.

The Bauhaus was a design school founded in Germany in 1919 by the architect Walter Gropius. Students at the design school undertook a programme of study giving them an appreciation of materials, manufacturing and form, before specialising in areas such as metalwork, furniture architecture and graphics. They aimed to be true to materials and relinquish ornamentation, focusing on the aesthetic associated with the manufacturing process.

Key features of Bauhaus products:

- **Form follows function**: the aesthetic appearance of a product is dictated by the way it works. This can be seen in the furniture of Marcel Breuer, who developed the use of tubular steel in furniture design, having been inspired by bicycle handlebars.

- Embracing the machine age: although the Bauhaus – like the Arts and Crafts movement – rebelled against ornamentation of designs, it differed in that it was keen to use modern machine-based manufacturing processes, seeing beauty in machined finishes and fabrication methods, as well as the natural appearance of the materials.
- Geometrically pure forms: inspired by the Art Deco movement, Bauhaus designs also used simple geometric forms, as seen in the MT49 tea infuser by Marianne Brandt.
- Everyday products for everyday people: in the aftermath of the First World War, Walter Gropius was keen to embrace modern manufacturing techniques as a method of providing affordable products for the whole population. This was another reason why ornamentation was avoided – because products were designed for easy mass production.

Major contributors of the Bauhaus design movement were Walter Gropius, Marcel Breuer, Mies Van Der Rohe and Marianne Brandt.

Figures 2.2.9 and 2.2.10 show key examples of Bauhaus design.

Figure 2.2.9 Bauhaus B3 Wassily chair, 1925.

Figure 2.2.10 Barcelona chairs in the Absolute Hotel, Limerick.

The work of the Bauhaus and Modernism was seen as extremely controversial at the time, with very few Bauhaus products making it to mass production during the lifetime of the design school. The introduction of tubular steel to furniture was such a huge change from the largely wooden interiors of the time, that it was often satirised – as seen in work by William Heath Robinson and K.R.G Browne in *How to Live in a Flat* from 1936:

'Whereas formerly the best furniture was made by carpenters, cabinet-makers, and similar skilled craftsmen, nowadays the trade is almost entirely in the hands of plumbers, riveters, blow-pipers and metal-workers of all sorts. As a result, the ultra-modern living-room resembles a cross between an operating-theatre, a dipsomaniac's nightmare, and a new kind of knitting.'

This can be likened to the reaction in the early 1980s to the designs of the **Memphis design** group.

Along with the use of tubular steel, the introduction of bent plywood furniture – epitomised by the work of Marcel Breuer when working with Isokon in the 1930s, and Charles and Ray Eames with the LCW (lounge chair wood) and DCW (dining chair wood) models – was synonymous with the movement. This functional and minimal material was formed to give continuous unbroken curves, also seen in the work of Scandinavian designers Alvar Aalto and Arne Jacobsen.

Figure 2.2.11 Chair by Marcel Breuer, London, UK, 1936.

Figure 2.2.12 DCW (dining chair wood) by Charles Eames.

Figure 2.2.13 Birch wood 400 armchair by Alvar Aalto.

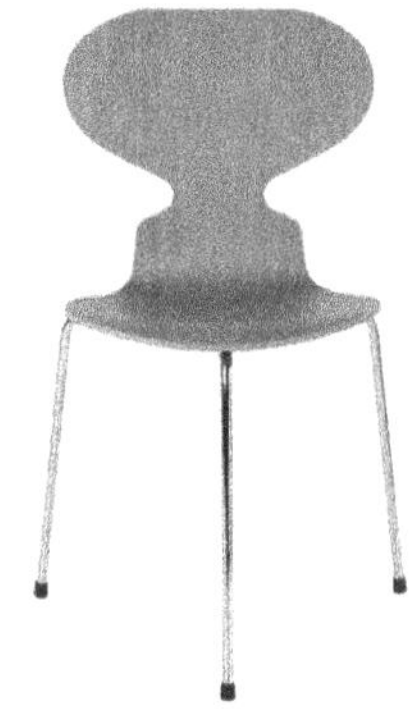

Figure 2.2.14 Laminated veneer Ant chair by Arne Jacobsen.

Streamlining

As Art Deco developed, the evolution of **streamlining** saw the increased use in architecture of flowing curves and smooth exteriors, bisected with chrome detailing. This approach was seen in car design as early as the 1920s, when aerodynamics began to affect developments in car body design. The application of streamlining to household objects such as clocks and fridges was seen as a sign of modernity, and developments in materials such as Bakelite enabled replication of these flowing curves.

Figures 2.2.15–2.2.17 show key examples of streamlining.

Figure 2.2.15 Tesla Talisman 308 U Bakelite streamline radio.

Figure 2.2.16 The iconic Midland Hotel, Morecambe, Lancashire, England.

Figure 2.2.17 1934 Chrysler Imperial Airflow.

Major contributors of streamlining design were Raymond Loewy, Norman Bel Geddes and Henry Dreyfuss.

Post-Modernism

The Modernist theory of design rebelled against the excessive ornamentation of design. Post-Modernism is seen as a counter rebellion against the simplicity of form and purely functional nature of Modernism.

During the 1970s and 1980s, the Memphis design group epitomised this challenge towards Modernism, and produced a range of playful products designed more as sculptures to be admired, with compromises made regarding practicality. The use of anthropomorphic (giving human appearances or traits to objects) and zoomorphic (giving animal appearances or traits to objects) characteristics within their designs added quirky associations for the user.

Key features of Memphis design:

- Bold and colourful playful designs: Memphis designs challenged the simplicity of Modernism, and often used a range of bright colours, including those directly opposite on the colour wheel.
- Simplistic juxtapositions of geometric forms: similar to Art Deco designs, Memphis designs used geometric forms, but the position of these could seem quite random, reflecting a desire to produce abstract sculptural products. This can be seen in the design of the Tahiti Lamp shown in Figure 2.2.18.
- Challenging forms that often compromised on function: the desire to produce sculptural designs often took precedence over functionality as seen in the Carlton dresser shown in Figure 2.2.19, where the angled shelving reduced the storage surface.

Figures 2.2.18 and 2.2.19 show key examples of post-Modernism design.

Figure 2.2.18 Carlton dresser, Ettore Sottsass, 1981.

Figure 2.2.19 Tahiti lamp, Ettore Sottsass, 1981.

Furniture designs by designers such as Marc Newson and Danny Lane challenge perceptions of how certain products should look; they were seen more as catwalk pieces than main stream high street products.

Marc Newson stated that the Lockheed Lounge designed in 1986 needed 'only to be more comfortable than a bus stop'. The aim of the project was to produce a chaise longue from what appeared to be a single 'blob' of mercury.

Figure 2.2.20 Marc Newson's Lockheed Lounge recliner.

ACTIVITY

Compare and contrast the Lockheed Lounge by Marc Newson with the Cesca chair by Marcel Breuer.

You should refer to:

- materials
- function
- aesthetics.

Designers and their work

Philippe Starck

The portfolio of work undertaken by the French product designer Philippe Starck is eclectic to say the least. He has famously worked on kitchenware for Alessi, and some of his designs can be regarded as postmodern sculptural pieces, focusing on aesthetics before function. However, his work in architecture and interior design is aimed at 'democratic design' for the population rather than the elite.

Starck challenges perceptions and aims to enhance the experience of the user through playful and thought-provoking design. His desire to experiment with modern materials and manufacturing processes make him a highly influential designer.

Look up the Juicy Salif product online. It is a single-piece aluminium casting, designed after Starck squeezed lemon juice over squid in a restaurant. The juicer functions successfully, although there are flaws. Key features of the design:

- The design has a very high centre of mass and becomes unstable during juicing.
- The head of the juicer is large enough for a lemon, but when juicing an orange, much of the fruit is missed due to the small diameter.
- The form of the product is striking and lends itself to being displayed in a kitchen, rather than stored in a drawer as with most juicers.
- The tripod legs exit the body of the juicer at an angle to prevent juice running down the legs. This also widens the base of the juicer to increase stability and allow a glass to be placed under the juicer.

James Dyson

James Dyson is a British design engineer who has developed a wide range of highly successful household products through the application of innovative technology to existing products. His developments of wheelbarrows, vacuum cleaners and hand dryers have revolutionised the way we see functional and essential household items.

With the Dyson vacuum cleaner DC01, he introduced 'bagless' cleaning to homes, utilising technology used in dust extraction systems to separate the collected particles from clean air. Through detailed product analysis of existing vacuum cleaner technology, he identified problems with the bag collection system, which blocked with the collected dust and reduced suction from the cleaner.

Key aspects of the DC01 vacuum cleaner:

- By changing the dust collection system, Dyson was able to separate the dust without reducing suction.
- Contrasting colour schemes mean that key features are highlighted, aiding the ease of use by the consumer.
- The 3D form of the product is produced largely from injection moulded ABS components, apart from the clear dust bin, which is polycarbonate (PC).
- A large number of integral fixings allow disassembly of the product and aid maintenance.
- The 3D form is sculpted around the technology within, which means the user can follow the path of the dust after collection. This may increase their confidence in maintaining the product.
- The clear dust bin can be seen as a positive feature, giving the user a sense of achievement as they see the dust they have collected. Also, this gives visual feedback as to when the container needs emptying.
- The negative aspect of the clear dust bin is that, once collected, the dust within the bin is unsightly if not emptied.

Margaret Calvert

Margaret Calvert is a highly influential graphic designer who, working with Jock Kinneir, developed the Transport font and many of the standard pictograms used on UK road signs. The development of stylised silhouette forms for the signage produced simple and clear communication with motorists. Their work in the development of signage introduced a mix of upper and lower case lettering to road signs, after testing proved that this improved readability at high speeds.

Figures 2.2.21 and 2.2.22 show key examples of Margaret Calvert design.

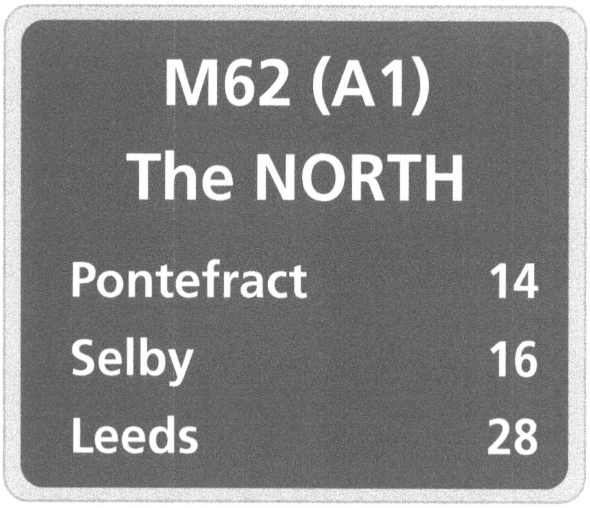

Figure 2.2.21 UK motorway destination distance sign.

Figure 2.2.22 Roadwork warning sign.

Dieter Rams

The German functionalist designer Dieter Rams has had a huge impact on product design in the latter half of the twentieth century. His simple approach to design and application of ten key design principles has been referenced by many contemporary designers, who have been influenced by his work with German manufacturer Braun. During his early work for Braun, he worked with Hans Gugelot, a product designer and lecturer working at the Ulm school in Germany.

Dieter Rams and Braun did for consumer electronics what the Bauhaus did for furniture – taking ornamental wooden casings and replacing them with functional minimal designs in white and grey, as demonstrated in the SK4 radio set in Figure 2.2.23 and the subsequent iterations of the product. These developments, as with the Bauhaus, relied heavily on technological developments – specifically the transistor and thermoplastics.

Dieter Rams came up with some key principles of good design:

- Good design is innovative: it is a product that uses modern materials, technologies and approaches to solve a design problem with innovation.
- Good design makes a product useful: a product must be functional, and successfully solve the design problem it set out to address.
- Good design is aesthetic: as a functionalist, Dieter Rams believes in the phrase 'form follows function', although the aesthetic success of a product depends on the market it is designed for.
- Good design makes a product understandable: the use of a product should be intuitive and require minimal guidance. Any guidance required should be evident through the form of the product.
- Good design is unobtrusive – this refers to the idea that form follows function, and all aspects of the product should be designed for a purpose, with no feature added purely for aesthetics.
- Good design is honest: this sounds peculiar, but we are often led to believe a product is something that it is not, such as a table with a wood veneer pretending to be a solid timber product.
- Good design has longevity: a product has longevity if it stays relevant over a long period of time.
- Good design is thorough down to the last detail: all aspect of a product must be designed, down to the smallest of details, such as the ridges on the battery cover of a remote control.
- Good design is environmentally friendly: the environmental impact of a product is key to its success. This could be down to the use of materials or processes during production, use or disposal.
- Good design is as little design as possible: it concentrates on the essentials – a true minimalist statement.

Figure 2.2.23 Dieter Rams and Braun SK4 radio set and record player.

Charles and Ray Eames

Charles and Ray Eames were an American husband and wife design partnership, famous for their work on moulded furniture both in plywood and polymers. Their modernist house made from used parts found in catalogues (the Eames house) reflected the cubist architecture of the De Stijl movement. Charles and Ray's work on the LCW (lounge chair wood) evolved from their work with the USA Navy, developing leg splints from laminated plywood forms. The application of this technology to furniture design paved the way for single form seating both in plywood, and later in polymers.

Figures 2.2.24 and 2.2.25 show key products of Charles and Ray Eames.

Figure 2.2.24 Eames lounger by Charles and Ray Eames, 1955.

Figure 2.2.25 Upholstered fibreglass shell office chair by Charles and Ray Eames.

Marianne Brandt

Marianne Brandt was a student at the Bauhaus design school, and became the head of the metalwork department in 1928. She developed a range of geometrically pure kitchenware products, which were successfully marketed and sold at a time when most Bauhaus products were still regarded as too controversial for the mass market. The simplicity of form used in her designs has ensured their longevity and relevance in modern design.

ACTIVITY

Use the principles of good design written by Dieter Rams to compare the Tahiti lamp by Ettore Sottsass with the Anglepoise lamp by George Cawardine.

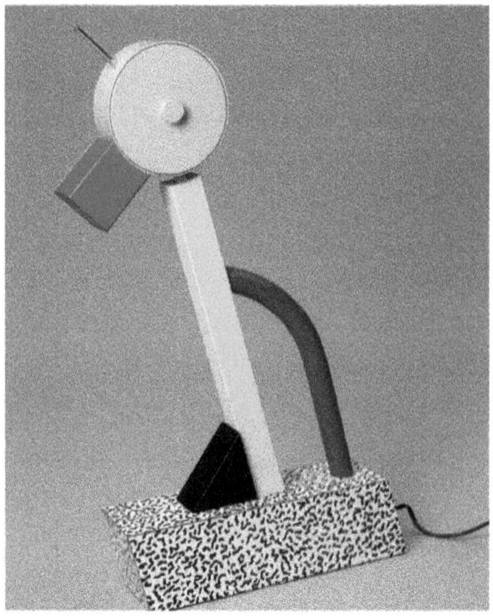

Figure 2.2.26 Tahiti lamp, Ettore Sottsass, 1981.

Figure 2.2.27 Anglepoise lamp, originally by George Cawardine.

KEY TERMS

Arts and Crafts movement: a nineteenth-century design movement that rebelled against the use of machines in design.

Bauhaus: an early-twentieth century German design school started by Walter Gropius.

Form follows function: the aesthetic appearance of a product is dictated by the way it works.

Memphis design: a late-twentieth-century design group who challenged modernist design views.

Streamlining: the development of products using flowing curves and chrome detailing inspired by the increased study into aerodynamics in the early twentieth century.

KEY POINTS

- The Arts and Crafts movement drew inspiration from medieval craftsmanship, reflecting natural forms to produce handcrafted furniture.
- The Art Deco movement drew inspiration from ancient Egypt to produce simple, stylised products, using geometric forms that reflected a modern approach to design.
- The Bauhaus design school embraced the machine aesthetic to produce functional products designed for the mass market.
- Post-Modernist design challenged the minimal designs of Modernism, adding bold colours and sculptural aesthetics to products.
- Phillipe Starck challenges perceptions and aims to enhance the experience of the user through playful and thought-provoking design.
- Dieter Rams developed a range of functional principles for good design.

Check your knowledge and understanding

1 Name four key aesthetic characteristics associated with the Art Deco design style.

2 What was the major social change that affected the Bauhaus design school?

3 Why was the Arts and Crafts movement concerned about the introduction of the factory system to product manufacture?

4 Which female designer developed the British transport font?

5 From where did James Dyson gain inspiration for the cyclone?

Further reading

Great Designs by DK books

Design: The Definitive Visual History by DK books

Scandinavian Design by Charlotte and Peter Fiell

Less but Better by Dieter Rams

Eames by Gloria Koenig

www.eamesoffice.com

www.vitra.com/en-gb/corporation/designer/details/charles-ray-eames

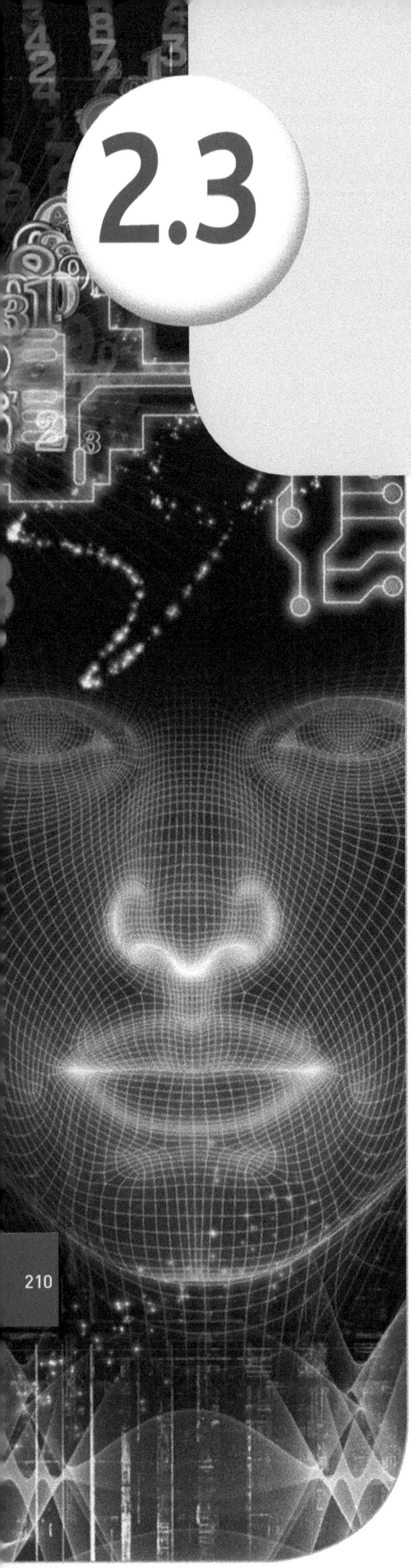

2.3 Technology and cultural changes

LEARNING OUTCOMES

By the end of this section you should have developed a knowledge and understanding of:

- how socio-economic influences have helped to shape product design and manufacture
- major developments in technology and how they are shaping product design and manufacture
- the social, moral and ethical responsibilities of designers and manufacturers.

If you are studying at A-level you should also have developed a knowledge and understanding of:

- the **product life cycle** and how designers refine and re-develop products during their life cycle.

Designers' activities are inevitably shaped by society, since they are endeavouring to design a variety of products that need to be appropriate for different groups of people with a range of backgrounds. Sensitivity in their approach is required to ensure that their designs are acceptable for the target consumers. They will also have constraints and opportunities afforded by social trends, as well as developments such as those affecting materials and technology, which they can incorporate in products to improve their appeal and profitability. These products will be likely to go through many stages of evolutionary improvement and augmentation as they proceed through their life cycle, before they are ultimately replaced by newer versions.

Socio-economic influences

Socio-economics provides understanding of how society and the economy interact. This enables designers to recognise more effectively the opportunities that exist for the successful development and marketing of products. This might range from technology companies marketing expensive devices for consumers who regard spending money on the latest smartphone or tablet as a priority, to the increasing popularity of furniture stores selling high volume, budget furniture to cater for the changing needs of the population. Demographic trends, such as couples often starting families later in life than they did formerly, larger numbers of young people progressing to university and an ageing population, contribute significantly to the factors that affect demand for certain products.

Figure 2.3.1 Fokker D.VII Triplane with tubular steel construction.

Post-First World War, the Bauhaus and development of furniture for mass production

The post-First World War Bauhaus design school (1919–32) believed in uniting art and industrial design, and had a revolutionary approach to designing, which dispensed with the normal barriers surrounding different crafts and skills, and encouraged experimentation. A major influence on their furniture designs was the large-scale wartime adoption of metal tubing as a replacement for wood, which was sometimes weak due to its irregular grain structure. One such example is the Fokker D.VII Triplane which, although it was not regarded as the best performing warplane, nevertheless proved to be outstandingly robust and reliable with its pioneering steel tubing construction.

Marcel Breuer is the Bauhaus designer most associated with experimentation with the use of steel tubing for furniture. His model B3 chair (later to become more commonly known as the Wassily chair), which was also influenced by Breuer's interest in bicycle design, has become an iconic representation of how the inherent strength and elasticity in tubular steel can improve on traditional furniture manufacturing materials and techniques, and facilitate mass production through the utilisation of industrial machinery and modern materials. The simple, easy to mass-produce tubular steel forms pioneered by the Bauhaus transformed the way that furniture was designed and made. It meant that machines could be used to create an entire frame for a chair or a table without needing to rely on traditional handcrafting skills, and this paved the way for many of the industrial furniture-making techniques that are still in use today.

ACTIVITY

Use wire and cardboard to create a model of a Bauhaus chair or table, to investigate how simply it can be done.

Figure 2.3.2 Marcel Breuer's B3 Wassily chair.

Figure 2.3.3 Mass production through the utilisation of industrial machinery.

The Second World War, rationing and the development of 'utility' products

The Second World War (1939–45) brought about a severe shortage of many goods. **Rationing** was therefore introduced in Britain to provide fair distribution of essentials such as food, clothing, petrol, timber and furniture. The shortage of timber for furniture manufacture was further exacerbated by the destruction of many houses and their contents during bombing raids, and the switching of a great deal of manufacturing capacity towards the war effort. Furniture rationing was accompanied by the setting up of a committee, led by the furniture designer Gordon Russell. His remit was to ensure that scarce resources were used in an optimal way by designing and manufacturing a range of appropriate items. The **utility**

Figure 2.3.4 Simple, plain, undecorated utility furniture.

furniture scheme was one of many utility schemes, including footwear and clothing, designed to ensure that basic, simply designed products were available in response to the shortages being experienced. Utility furniture was initially only available to newly married couples and those who had lost their homes in bombing raids. The furniture itself was designed to be strong, with no superfluous decoration, borrowing from the traditions of the Arts and Crafts movement from the beginning of the century. Much of the furniture used traditional, vernacular methods, involving locally sourced materials, such as those embraced by traditional Windsor chair makers, who used beech grown in local woods in Buckinghamshire. As a designer, Gordon Russell was influenced by William Morris and the Arts and Crafts movement. He welcomed the opportunity afforded by state control of furniture and furnishing standards to bring to fruition some of the 'fitness to purpose' ideas that he had admired in the designs promoted by Morris forty years earlier. An official CC41 logo (originating from the one used by the utility clothing scheme's 'civilian clothing 1941') was used to identify approved pieces of utility furniture.

Contemporary times, fashion and the demand for mass produced furniture and decorative products

Products designed in the aftermath of the Second World War bore the hallmarks of the utility programme, in that they were well made but lacked the excitement and design 'touches' that people craved following a period of enforced drabness. However, designs were slowly beginning to change. In the USA, designers were beginning to incorporate new ideas into the design of their products, to appeal to an eager public. For example, in 1934, Raymond Loewy – one of the USA's foremost industrial designers – redesigned the Coldspot refrigerator to make it appear more 'streamlined', and this resulted in sales increasing fourfold.

The Government was concerned that competition from other countries and the legacy of wartime utility design would result in Britain being left behind as a designing and manufacturing nation, so the **Council of Industrial Design (COID)** was set up in 1944, and was successful in introducing a programme of training, assessment and information to improve standards of design. One of the COID's main achievements was to hold two influential exhibitions: the 'Britain Can Make It' exhibition in 1946 and the 'Festival of Britain' in 1951. In 1972, the Design Council took over from the COID, and they function today as the Government's advisory body relating to design, with a particular remit to 'address all aspects of design including product, service, user experience and design in the built environment'.

British consumers were not particularly welcoming to the 'streamlined' ideas of American designers such as Raymond Loewy and Henry Dreyfuss, as they were wary of the consumerism that was gaining momentum in the USA, and there still existed a desire to get back to pre-war normality. A significant opportunity arose, however, with the potential market created by teenagers' insatiable appetite for novel products and their willingness to embrace anything new. This included products such as miniaturised radios and record players which were now entering the market, facilitated by the development of the transistor.

Figure 2.3.5 The 'Britain can make it' exhibition in 1946.

The 1940s and 1950s also coincided with the rapid development of polymers and related moulding techniques. This gave designers new opportunities for product ideas that can be seen being adopted later in the output of designers such as Verner Panton and Robin Day. It is the latter's 1963 PP chair which has sold millions, and paved the way for an explosion in the variety of furniture design available to the public. The influence of Scandinavian design at this time was also very significant, with designers such as Arne Jacobsen and Alvar Aalto's plywood furniture creating a foundation for Gillis Lundgren's 'Lövet' table in 1956. This table, with simple removable legs, was revolutionary as it was the first example of furniture that could be dismantled for ease of transportation and storage. Lundgren designed the 'Lövet' table for Ikea. He was one of their key designers in the development of KD (knock-down) furniture, going on to design the 'Billy' bookcase, which has reached world sales exceeding 40 million. This revolution in the furniture industry made it possible for consumers to buy reasonably priced, fashionable furniture without worrying about what they will do about replacing it when fashions change. Prior to this, for most people, furniture purchase was regarded as virtually a 'once in a lifetime' experience and the furniture was, therefore, expected to be highly robust and capable of outliving the customer. This new approach to furniture and furnishing offered consumers a more enjoyable and affordable experience as well as the opportunity to make changes as fashions came and went.

Figure 2.3.6 Stacked Robin Day PP chairs.

Figure 2.3.7 Ikea's 'Lövet' table.

ACTIVITY

Study one of the Robin Day PP chairs shown in Figure 2.3.7, and produce an annotated diagram highlighting its main features, advantages, disadvantages and manufacturing methods.

Do the same with a conventional wooden dining chair, particularly comparing the manufacturing methods with those of the Robin Day design.

Major developments in technology

Microelectronics

Basic electrical batteries and circuits had been in existence from the start of the nineteenth century, but it was not until the invention of the vacuum tube (thermionic valve) at the beginning of the twentieth century that the era of electronics arrived. These devices could control and amplify the flow of an electric current, and their invention facilitated the development of early electronic products such as radios, TVs and computers. Unfortunately their large size, high level of energy consumption and unreliability meant that there was a limit to the range and effectiveness of products that could be designed.

Figure 2.3.8 Two thermionic valves with the transistors that replaced them.

The invention of semi-conducting devices, with the first transistor being developed in the 1940s by William Shockley and his team at Bell Laboratories in the USA, was pivotal to the development of the information technology on which we now rely. They discovered that the semi-conducting properties of materials such as silicon could be employed to create electronic switches and amplifiers, and this opened up a new world of miniaturisation that was given a major impetus by the development of integrated circuits (ICs) in the 1950s and their adoption from the 1960s onwards. ICs consist of multiple transistors and other components formed on a single silicon chip. The miniaturisation that they facilitate introduced the term **microelectronics**, since they made massive reductions to the size of components, circuits and products possible. Refinement of these techniques has brought us to a point now where over 10 billion transistors can be incorporated in just one device. This is referred to as 'ultra-large-scale integration', and has made it possible for computers and mobile devices to proliferate. This has had multiple effects on society – both positive and negative – ranging from the benefits of powerful medical scanning capability to the negative aspects of a more connected, technological world, such as cyber bullying via social media.

The increasing number of transistors makes devices more powerful in terms of the processing that they are capable of and the amount of data that can be stored. Moore's Law, named after Gordon Moore – one of the founders of IC manufacturer Intel – correctly predicted in 1965 that every year, ICs would contain twice as many transistors as they had the year before. He changed his prediction in 1975 to a doubling of transistors every two years. Recently however, a saturation point has started to become apparent. This may need to be considered by designers in planning future products, although developments in nanotechnology could become a significant factor. The following table shows how ICs, which are the key element in microelectronics, have become more powerful since their commercial introduction in the 1960s.

Table 2.3.1 How ICs have become more powerful since the 1960s.

Integration level	Year	Number of transistors
Small-scale integration (SSI)	1964	1–10
Medium-scale integration (MSI)	1968	10–500
Large-scale integration (LSI)	1971	500–20,000
Very large-scale integration (VLSI)	1980	20,000–1,000,000
Ultra-large-scale integration (ULSI)	1984	1,000,000 +
The most powerful IC currently available has 10,000,000,000 transistors.		

The technological developments that have taken place during the second half of the twentieth century and beyond, particularly those in the field of microelectronics, have been inextricably linked with the development of computers and the establishment and refinement of the now ubiquitous internet (interconnected networks) and World Wide Web (collection of web pages). This has facilitated fast communication via fibre-optic cables and wireless networks, along with vast resources of web pages and interactive content that contribute significantly to the 'information age' that we now inhabit.

ACTIVITY

Find out more about how multiple ICs are made using silicon 'wafers'.

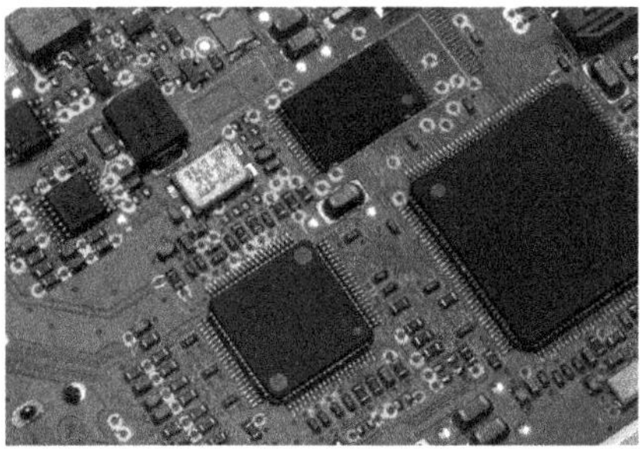

Figure 2.3.9 ICs on a computer circuit board.

Figure 2.3.10 Fibre-optic cables.

The impact of microelectronics on designing and manufacturing practice

Microelectronics-based technological developments have made a massive impact on designing and manufacturing techniques because they have made it possible to incorporate powerful microprocessor ICs (the main element in a computer) in CNC (computer numerically controlled) machines. They have also made the computers themselves capable of carrying out, at high speed, the millions of calculations required for manipulating designs in a CAD (computer aided design) program. Table 2.3.2 shows some contrasting examples of designing and manufacturing practice in the 1950s and now, to highlight the scale of the innovations and improvements that have been made possible by technological developments related to microelectronics.

Table 2.3.2 The innovations and improvements that have been made possible by technological developments related to microelectronics.

	1950s pre-microelectronics examples	Current examples
Research	Information from books and journals Personal interviews with target consumer groups Film-based photography dependent on laboratories, chemicals and darkrooms Parts manufacturers' paper catalogues Communication by letter and landline telephone	Internet searches for a range of online resources Web-based market research Materials databases Digital photographs available for instant access Parts manufacturers' PDF data sheets Mobile phone, text and social media
Generating and refining design ideas	Pencil sketching Working drawings hand produced, using instruments and drawing boards, by teams of 'draughtsmen' Mistakes have to be erased and drawings may need to be started again Similar parts have to be drawn separately Limited ability to reproduce copies (blueprints) Inefficient physical filing and storage system for drawings Hand-produced colour renderings requiring great artistic skill Widespread use of slide-rule calculations Limited access to early computers for solving engineering problems	Graphics tablet with 'pen' Highly sophisticated 2D and 3D CAD facilitating parametric (linked) designs Easily modified CAD details CAD parts libraries A range of additional CAD 'tools' to aid the designer Highly realistic CAD renderings automatically generated with pre-set materials, finishes and backgrounds Decals and other details easily added within rendering software Existing products can be scanned to input data for CAD designs Stages of design development can be saved for reference and recall
Design collaboration and communication	Face-to-face meetings Landline telephone calls between offices Drawings sent by postal service or courier (fax machines did not become common until the 1960s) Generally an 'over the wall', compartmentalised, approach to communication, leading to delays and more potential for misunderstandings and errors	Web conferencing Email Texting Social media File sharing on intranet and internet (e.g. Dropbox, Google Docs) Working from home is facilitated Photo and video sharing websites Several designers can collaborate on one CAD design 'Concurrent engineering' is facilitated, meaning that different processes can be carried out at the same time
Modelling and testing ideas	Handmade models, using traditional materials, for clients to see Destructive testing of full-size prototypes Calculation-based predictions of the behaviour of mechanisms and structures, possibly with the limited help of early electronic calculators and – in very few cases – early computers	Virtual models in simulated scenarios can be seen by clients 3D printed models and prototypes for testing and checking FEA (finite element analysis) to check stress effects Mould flow analysis (MFA) to check possible production problems Complex computer calculations and predictions
Manufacturing	Large number of skilled operatives required for lathes, milling machines, etc. Many companies operated large-scale 'apprentice' schemes to maintain skill levels Limited use of numerically controlled (NC) machines reliant on punched paper tape programs Parts stocked in large numbers and monitored by paper-based recording systems Material and parts handling is usually manual and labour intensive	Smaller number of skilled operatives needed for a vast range of CNC automatic machines, including lathes, milling machines, routers, lasers and robots Robotic devices and AGVs (automatic guided vehicles) handle materials and parts JiT (Just in Time) manufacturing is facilitated by computerised monitoring systems, which reduces stock levels and cuts wastage 3D printing is increasingly widespread and encompasses a wider range of materials, including metals

	1950s pre-microelectronics examples	Current examples
QC (quality control) and testing	Mainly manual inspection and measurement of samples using micrometers, verniers and gauges Visual checks Use of relatively basic laboratory equipment	Automated scanning of products for faults such as cracks and imperfections Rejection of parts based on digital imaging and code recognition Probe-based computer measuring systems used to check accuracy Emphasis on QA (quality assurance) facilitated by access to 24/7 process monitoring by computer systems

Figure 2.3.11 1950s drawing office.

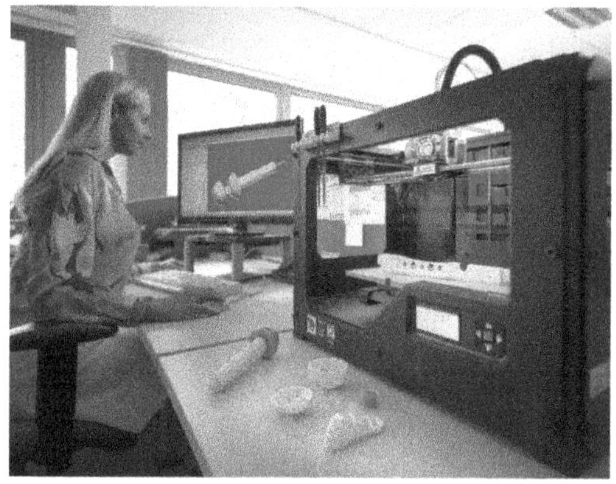

Figure 2.3.12 Designer using CAD system and 3D printer.

Figure 2.3.13 1950s telephone: telephone calls being directly connected at an exchange.

The impact of the use of microelectronics in products

Everything we have seen in Table 2.3.2 relates to how microelectronics has advanced designing and manufacturing. However, it has also provided the essential components at the heart of a new generation of products that had either never previously existed or had been very inconvenient, bulky and energy intensive.

The following images demonstrate the scope of the miniaturisation, and increase in capability and user convenience that has advanced product design since the 1950s. This has all been made possible by the introduction of microelectronics, and the advances that this also facilitated in the development of wireless communication.

Figure 2.3.14 Current telephone: wireless cellular networks facilitate the use of mobile phones.

Figure 2.3.15 1950s radio: showing the thermionic valves in an early radio.

Figure 2.3.16 Current radio: portable digital audio broadcasting (DAB) radio capable of receiving clear signals and additional broadcast information.

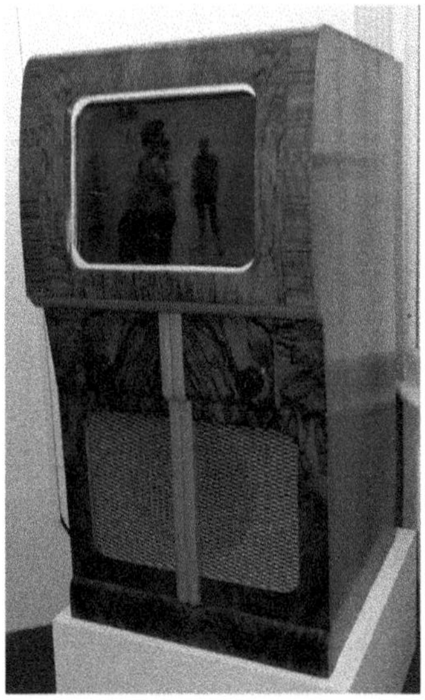

Figure 2.3.17 1950s television: bulky, furniture-size television with small, low-quality monochrome screen.

Figure 2.3.18 Current television: slim, large screen, ultra-high definition, light emitting diode (LED), 3D, smart, colour television.

Figure 2.3.19 1950s computers: thermionic valve circuitry meant that computers needed to be room-size and had limited capacity.

Figure 2.3.20 Current computers: ULSI ICs make it possible for powerful computers to be very slim and portable.

Figure 2.3.21 1950s music player: analogue, fragile and scratch-susceptible records were the only way to listen to a limited choice of music.

Figure 2.3.22 Current music player: it is now possible to digitally 'stream' virtually any music into your home with studio-like audio quality.

ACTIVITY

Choose one of the microelectronics-based products shown in the images that you feel has made the most impact on society, and list the positive and negative points that you have identified.

The impact on product design of microelectronics and other technological advancements

Below are some of the developments in personal music players in the space of just 40 years, from the 1950s to the 1990s. You can compare these models to current ones that you own to see how developments have continued.

Date: 1954

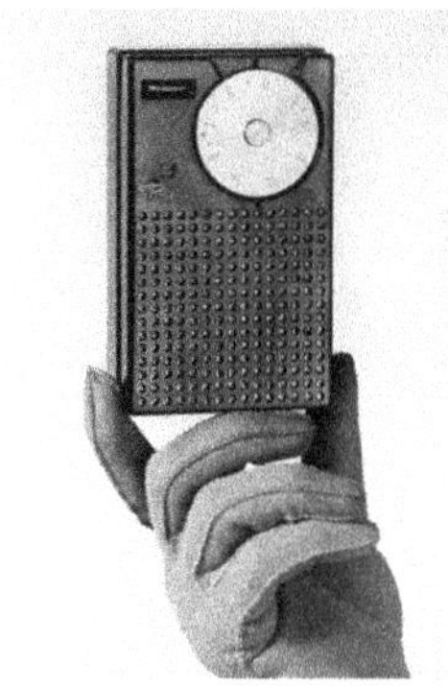

Figure 2.3.23 Regency TR1 portable radio.

Examples of some significant developments:

- radio with four transistors
- carbon zinc battery.

Date: 1963/4

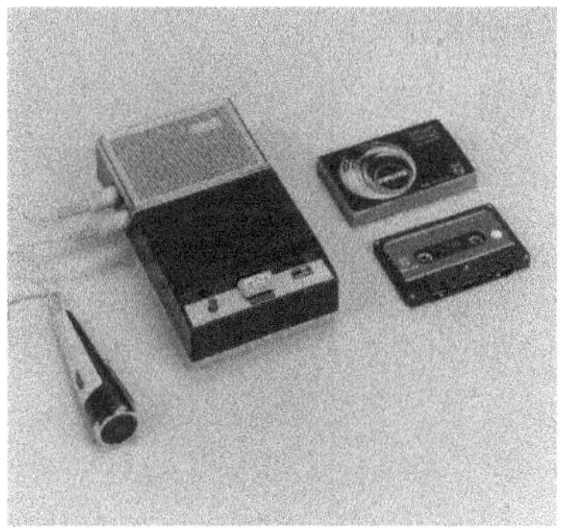

Figure 2.3.24 Philips EL3300 compact cassette.

Examples of some significant developments:

- ten transistors
- electro-mechanical speed control
- first successful, standardised compact cassette tape.

Date: 1979

Figure 2.3.25 Sony Walkman.

Examples of some significant developments:

- stereo
- cassette and LSI IC control and amplification facilitates revolutionary level of portability
- AA batteries (standard or rechargeable).

Date: 1984

Figure 2.3.26 Sony Discman (CD Walkman).

Examples of some significant developments:

- digital
- compact disc
- laser
- VLSI IC control and amplification
- rechargeable Ni-Mh batteries
- LCD display.

Date: 1992

Figure 2.3.27 Sony MZ1 MiniDisc.

Examples of some significant developments:

- recordable magneto-optical disc (works like a hard drive)
- compressed sound files (ATRAC).

ACTIVITY

Produce your own table like Table 2.3.2, tracing the major stages of technological development of a product of your choice since the 1950s, such as a:

- camera
- printer
- light bulb
- video recorder/player
- cooking hob.

New materials

Advances are constantly being made in the development of new materials for use in product design, and the last 70 years has seen a massive change from almost total reliance on metals, timber, ceramics and glass to the much more polymer and composite-based range of materials currently available. Here are some examples of developments in materials, some of which are only just being introduced and others that are more established.

Glulam

Figure 2.3.28 Glulam roof structure.

Glulam is the use of several pieces of timber that are glued together to create strong, composite components for use in buildings, bridges and other structures. It is stronger than using solid timber because the careful choice of laminates enables the natural defects found in timber, such as knots and shakes, to be almost eliminated. Although this composite material has been in use for many years, it is currently undergoing a revival in popularity due to a number of factors:

- it has better strength-to-weight ratio than steel due to minimisation of defects
- ease of forming and shaping with lower energy costs
- variants such as cross laminated timber (CLT) are more sheet-based and have strength in both directions and give great potential for imaginative designs
- it is a **sustainable** resource.

Kevlar®

Kevlar® is a form of aromatic polyamide (a polymer with ring-like molecules connected in long chains) artificial fibre that has tremendous toughness and tensile strength due to the density and chemical bonds present in its molecular structure. It is woven into a variety of materials that are strong and light and these woven materials are often combined with various resins into advanced composite materials. The uses of Kevlar® range from bulletproof vests and puncture resistant tyres to sophisticated aeronautical applications. The Boeing 787 Dreamliner, for example, utilises Kevlar® honeycomb panels as part of its 50 per cent composite structure.

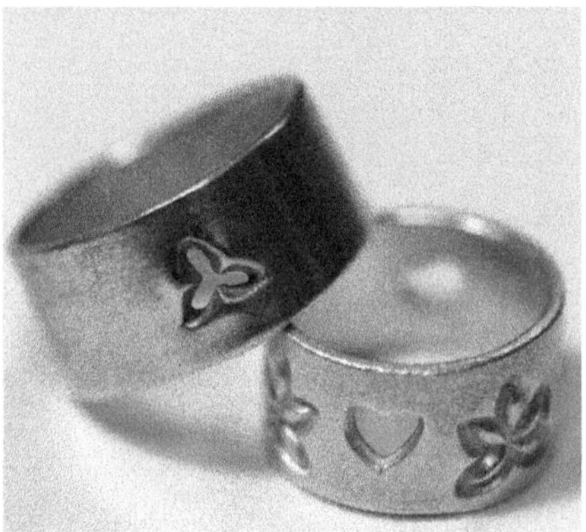

Figure 2.3.29 Jewellery made using PMC.

Precious metal clay

Precious metal clay (PMC) is a craft material that consists of microscopic particles of gold, silver and other metals that are bound together in a pliable medium so that they can be shaped into jewellery and other products. When they have been shaped, a sintering process is carried out, which means raising the temperature of the clay until the particles fuse and the binding medium is burned off. The latest PMCs only need ten minutes of heating at approximately 700 °C for this to happen, and will shrink in size by about 10 per cent in the process. Therefore, this needs to be carefully factored into the design of the product, particularly if it is a ring.

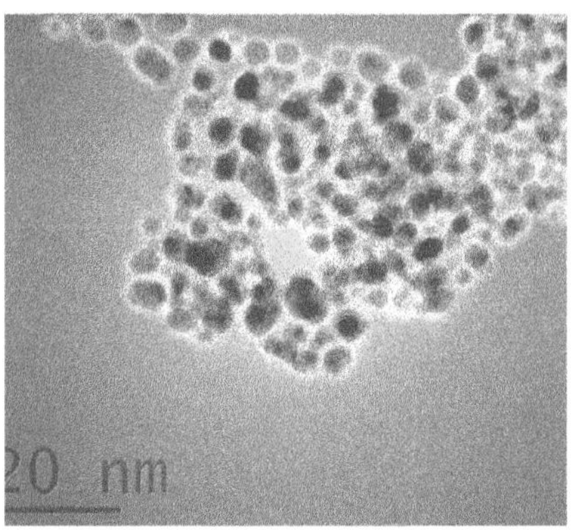

Figure 2.3.30 Gold nanoparticles for use in coatings and IC manufacture.

Nanomaterials

Nanotechnology involves the precise manipulation of nanomaterials, which are created using particles in the atomic and molecular size range of 1 to 100 nanometres (0.000001 mm to 0.0001 mm). It is already being used in the form of additives in sunscreen and cosmetics, and is also being investigated as a way of making more advanced 'nanoelectronic' devices, which would be far smaller and have greater capacity than current microelectronic devices. It also offers ways of coating materials and products, as well as the potential for revolutionary medical and energy applications, such as Samsung's exploration of the possibility of using graphene-coated anodes to extend battery life. Graphene is a nanomaterial consisting of a two-dimensional form of carbon with a honeycomb-like atomic structure. If it is rolled into tubular form, it displays highly desirable levels of tensile strength and hardness, combined with heat resistance and high electrical conductivity. These carbon nanotubes are incredibly thin as well as hollow, and therefore they have the ability to be used in a range of medical applications, such as delivering drugs.

Although they have great potential for the advancement and improvement of product design, the lack of recyclability and potential toxicity of some nanomaterials mean that caution will need to be exercised in their adoption.

ACTIVITY

Try to find out about other new materials that are becoming more widely used.

New methods of manufacture

As we have seen earlier in this chapter, the twentieth century saw the introduction of massive improvements in our ability to manufacture products. One of the significant consequences of this has been to change the nature of the workforce; we have moved from needing large numbers of highly skilled manual machine operators, relying on years of apprentice training and experience to manufacture components, to a much smaller number of highly skilled technicians designing, building, programming and maintaining computer-based manufacturing devices. The monitoring of these machines only requires a relatively small number of semi-skilled operators. These machines are often set up as a manufacturing cell so that multiple types of machining operations can be executed, usually facilitated by robotic devices and AGVs to transfer materials and components.

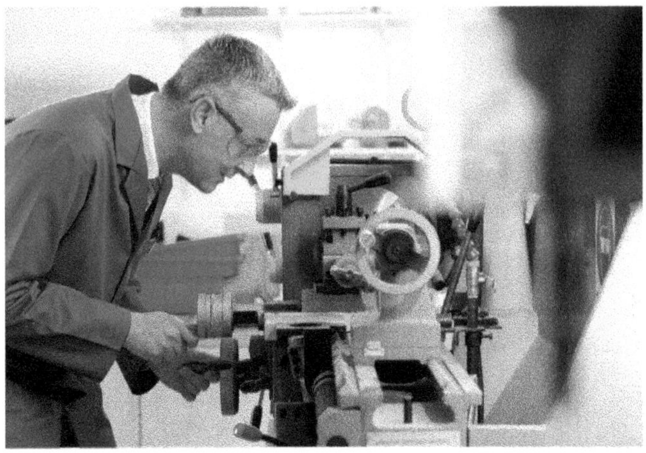

Figure 2.3.31 Manually controlled lathe.

As well as the ongoing impact and development of CAM (computer aided manufacturing), robotics and related manufacturing technology, there are many other innovative techniques that are starting to become more prominent.

Electrohydraulic forming

Electrohydraulic forming is the single stage forming of complex sheet metal parts using a single sided former, by the action of a shockwave generated by an electrical spark in a tank of water. Similar processes are used which require the use of explosives.

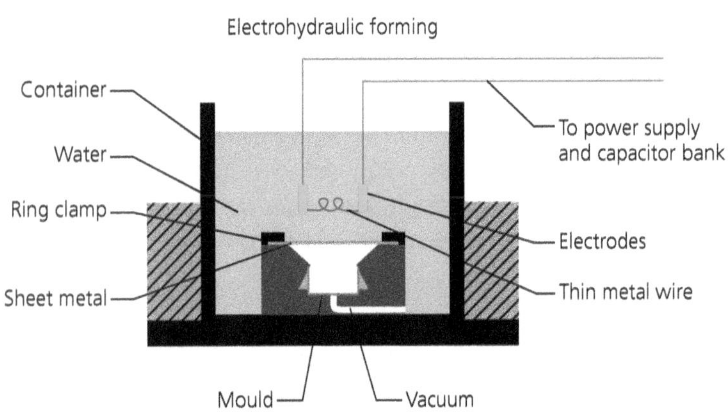

Figure 2.3.32 Electrohydraulic forming.

Although this technology was first introduced in the mid-twentieth century, it is likely to become increasingly adopted due to a number of factors:

- It only needs a one sided former rather than the two required for a conventional press.
- It can produce deep, complex and fine detailed shapes.
- It can deal with a range of materials and thicknesses.
- It is a single stage process.
- It is very fast.
- Material is evenly distributed, avoiding potential weak points.

Advanced 3D printing of metals

An important method of printing metal products is to use a form of selective laser sintering (SLS) called direct metal laser sintering (DMLS). In this process, a laser is used to sinter (fuse) metal particles layer by layer, to build up the required form. With this method, it is possible to create parts that are strong and lightweight. It is also possible to achieve complex internal features like undercuts and internal voids that are incompatible with conventional manufacturing techniques such as casting and machining. It is particularly appropriate for making one-off prototypes and test parts. This manufacturing technology has similarities to the principles used in the early 'powder bed' 3D printers, which selectively add a binding fluid to a starch or plaster powder to create a solid model.

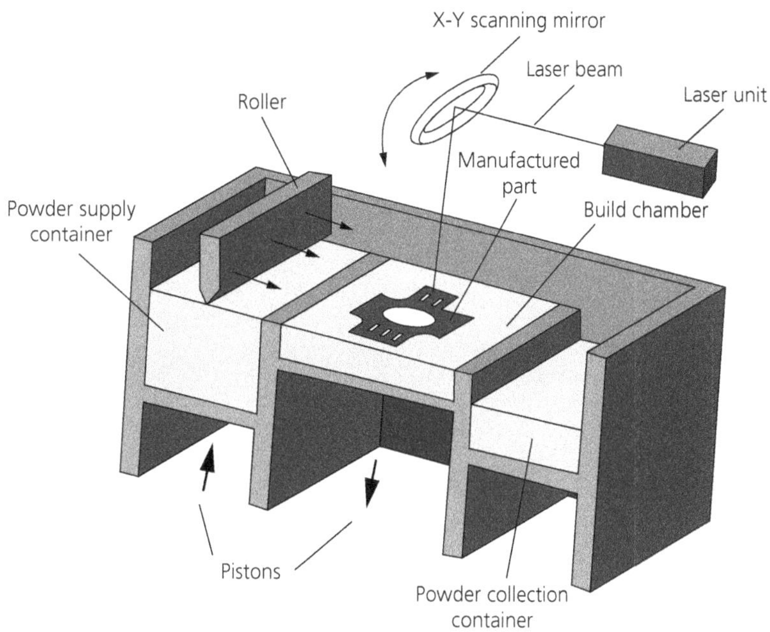

Figure 2.3.33 DMLS.

Fibre injection moulding

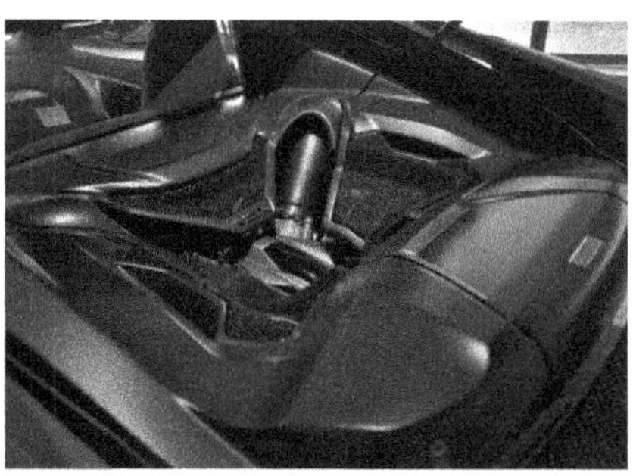

Figure 2.3.34 Carbon-fibre reinforced polymer automotive components.

Fibre injection moulding is a recent augmentation of the standard injection moulding process, in which pellets of glass or carbon fibre-filled polymers such as polyamide (nylon) are used. In the long fibre injection moulding variation of this process, reinforcing fibre roving (a number of slightly twisted strands on a continuous reel) is incorporated directly into the polymer being moulded. These processes are becoming increasingly popular in the automotive industry, because they are capable of producing parts that are very strong, stiff, lightweight and economical to mould. Additionally, some polyamides can be electroplated in the same way as ABS and metals, to give the quality finish that is often required. Fibre injection moulding techniques can also benefit by offering greater sustainability, because they facilitate the reuse of carbon-fibre fabric off-cuts and waste left over from conventional carbon-fibre reinforced resin manufacturing.

Laser beam welding

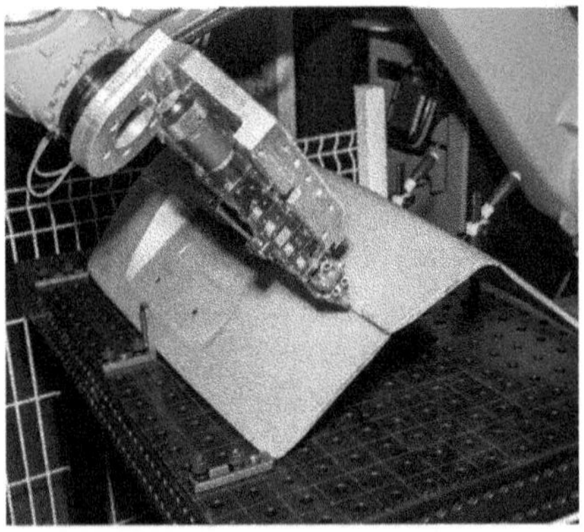

Figure 2.3.35 Laser beam welding using a robot in the automotive industry.

Laser beam welding (LBW) is a technique that is increasingly adopted for applications such as automotive manufacture in which the intense heat of a laser beam is used to join multiple pieces of metal. The process is fast and is capable of producing narrower and deeper welds than conventional techniques. Sometimes twin laser beams are used, and it can also be combined with other types of welding such as MIG and TIG, to increase their efficiency and speed.

Additional advantages of LBW include:

- It welds a wide range of metals and is capable of welding dissimilar metals.
- The area affected by heat from the laser is narrow, therefore minimising distortion.
- The weld is smooth enough to require no further finishing processes.
- It is more accurate than other methods.
- It is unnecessary to use filler rods.
- Small, thin components can be welded with less likelihood of damage.

Disadvantages include:

- high capital cost
- the need for a clean environment to protect the optics
- additional health and safety considerations.

ACTIVITY

Produce a simple table to compare the advantages and disadvantages of a wide range of welding methods, including LBW.

Physical vapour deposition

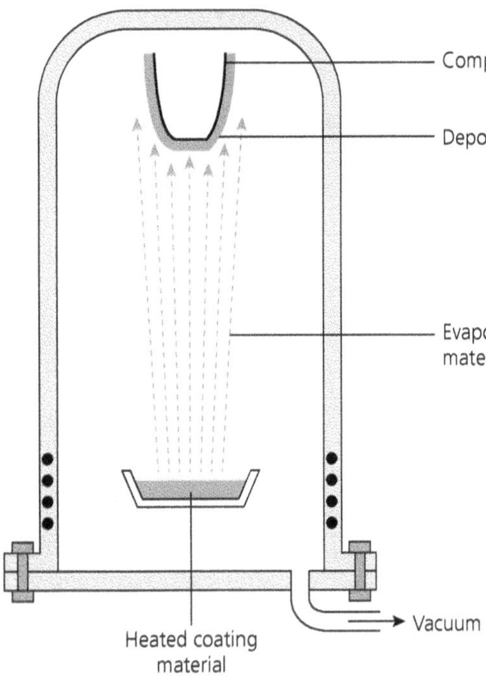

Figure 2.3.36 Physical vapour deposition.

Physical vapour deposition is a method of producing thin films of material or coating products with a finishing surface, as an alternative to methods such as electroplating. It is used for the production of semi-conductor components, food packaging, machine tool cutting tips and decorative products. It works by heating the base material so that it vaporises, which then passes through a vacuum to condense on the target material, depositing a thin layer of the required material.

A related process called chemical vapour deposition is also used for similar purposes, but relies on chemical reactions rather than evaporation.

The internet of things

The internet of things (IoT) is the connection of a range of devices to one another over networks such as wi-fi and the internet. It has massive potential for improving the way that manufacturing works, as well as more mundane domestic tasks. For example, smart fridges use cameras and radio frequency identification (RFID) scanners to recognise whether there is any margarine, salad, eggs and so on left, and automatically orders some more without you needing to intervene.

If you scale up this idea, it is possible to visualise the machines that are used in a JiT manufacturing system being set up to automatically ensure that there is a prompt supply of parts when they are needed. An additional advantage of such a system is that it facilitates a dynamic response to anything that changes in the system, such as a machine fault resulting in the need to reorganise schedules. It also facilitates 'predictive maintenance', whereby sensors constantly monitor the condition of elements of a machine; the data collected would indicate when servicing, repair or replacement of various parts is needed, and this would be automatically scheduled by the system.

Advancements in CAD/CAM

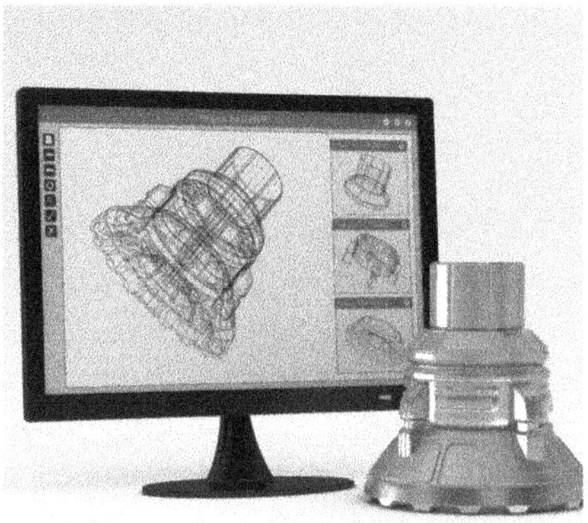

Figure 2.3.37 CAD design manufactured using CNC machines.

The growth in the use of standardised file formats such as Drawing eXchange Format (DXF) and STereoLithography (STL) to connect CAD and CAM continues to be a key factor in the increasingly influential manufacturing role of this technology. This is because the design files that are produced when products are designed in CAD systems also facilitate the generation of manufacturing files, which can then be used on a wide range of different machines. The need for skill in the operation of machines is reduced, since the codes generated by the manufacturing files automatically operate the CNC machines to make the part required.

This trend towards computer-based design and manufacture received an enormous boost with the development of 3D printing capability, as mentioned in the section on rapid prototyping in Chapter 1.8 (see page 129). This facilitates the manufacture of components that would previously have been impossible using conventional techniques.

Figure 2.3.38 Designer using a VR headset.

A number of strategies are now under consideration or in active development to further enhance the compatibility and capability of CAD/CAM systems. For example, the use of extensible markup language (XML) offers the ability to provide improved compatibility for files generated in different software packages. This is becoming increasingly important as the number of different CAD/CAM packages continues to increase.

Another development is integrated realisation – the dynamic use of design tools such as Finite Element Analysis (FEA) or Computational fluid dynamics (CFD) during the early stages of designing parts and assemblies, to calculate the impact of decisions being made at the design stage. This is preferable to having to wait until the end of designing to carry out simulations because it considerably speeds up the whole process, and problems can be seen much earlier therefore modifications can be put in place to rectify these.

ACTIVITY

Use a car manufacturer's website to find out how you can customise the design of a vehicle. Make a simple chart with some of the options that can be chosen for features such as:

- engine
- gearbox and transmission
- in-car entertainment and navigation
- bodywork and trim
- wheels.

At the moment, most CAD/CAM software used by companies is limited to specific computers or workplace networks, but development of Cloud-based (internet-based) CAD/CAM software packages is taking place, which will make it possible for designers and engineers to access CAD/CAM software wherever they are and thus improve productivity. They will not need to have the software installed on their computers or tablets; instead it will be accessed through the internet. Software available in the Cloud is currently mainly used for word processors and spreadsheets such as those provided by Google Docs and Microsoft OneDrive.

Mass customisation – which is covered in the section on scales of production in Chapter 1.6 (see page 106) – is already offering customers options to personalise the products that they buy online, such as trainers sold by Nike. The next step in the development of this is likely to be towards a greater CAD input by customers, to provide more options for the personalisation of products.

The use of virtual reality (VR), is becoming more important in CAD, since it provides an environment where much more realism can be experienced through the use of devices such as gloves with haptic (sensory) feedback and 3D headsets. It is now being harnessed so that designers will, for example, be able to 'shape' pieces of virtual material with their hands and see the resulting form in a three-dimensional environment, rather than relying on the traditional keyboard and mouse to input their ideas and viewing the results on a flat, 2D screen.

Social, moral and ethical issues

Social issues are those such as environment, health, poverty, discrimination and unemployment that affect a significant number of people. Moral and ethical issues are related to people's beliefs, such as what they believe is right and wrong. Morality was a key feature of the thinking of William Morris, the Arts and Crafts champion and reformer. He saw the growth of machine-based manufacture evidenced by the Great Exhibition of 1851 and the consequent decline in craft skills as being detrimental to the structure of society. Despite his efforts, the demand for mass-produced products grew, and this has placed a great deal of power to shape society into the hands of designers and manufacturers. The power gained by large manufacturing companies became a target for the designer Victor Papanek, who was very critical of many modern products and alarmed by the way designers mainly concentrated on society's 'wants' rather than their 'needs'.

The need for designers to consider their responsibilities in relation to these issues, and how they apply to various products and systems, will vary considerably for different societies and particular groups within them. However, they have a duty to ensure that their practices and designs are appropriate for consumers.

Some companies are very much aware of their social responsibilities, and go to considerable lengths to ensure that they maintain an appropriate balance. Corporate social responsibility (CSR) is a self-regulatory scheme which gives companies a framework for ensuring that their level of social responsibility and sustainability is optimised. Lego Group is an example of a company that has made ambitious pledges in this respect. It has, for example, set a target of 100 per cent renewable energy capacity, and has also made a commitment to reach a target of moving to sustainable raw materials as an alternative to those that are oil-based by 2030. It is also in partnership with the World Wildlife Fund (WWF), with a commitment to spearheading change in the search for methods of reducing emissions. Disney is another example of a company that has a good reputation for CSR. It encourages workers to volunteer for charity work and has provided significant support in natural disaster situations such as earthquakes. It has also been involved in planting trees in rainforests, using funds raised from natural history films.

Products designed and manufactured for military use are, understandably, often criticised as having major detrimental social effects due to the negative aspects of conflict. It is, however, worth considering that military requirements not only advanced the study and application of anthropometrics and ergonomics, but also led to the development of a wide range of products that are universally regarded as having positive benefits, such as:

Figure 2.3.39 Radar.

- PillCam® – an internal medical imaging system for early detection of cancer was developed from missile guidance technology.
- Radar – the system that facilitates the safety of today's essential air travel by providing an accurate image of the position of aircraft, was developed and refined for military purposes during the Second World War. Additionally, microwave ovens came about following further radar-based research.
- The EpiPen is a type of automatically injecting syringe that is used extensively by diabetics and people who suffer from allergic reactions. They were first developed in order to enable soldiers to protect themselves from chemical weapons and nerve agents.

There are a number of other examples of products with similar origins in military applications, such as:

- global positioning satellite (GPS) navigation systems (1970 US defence systems)
- Penicillin (first put into wide-scale use in the Second World War)
- drones (initially developed during the First World War)
- jet engines (initially developed during the Second World War)
- nylon and other synthetic materials (initially developed during the Second World War).

The contradictory nature of these examples highlights the need for making a balanced judgement about the negative and positive effects of products and systems; what is detrimental for one group of people can be beneficial for others.

Sustainable materials and ethical production methods

Our standard of living is currently dependent on the ready availability of a wide range of materials, many of which are obtained from finite sources. As we have seen, companies are looking towards sustainable materials that can be substituted for the oil-based polymers currently being used. Starch-based materials such as PLA are growing in popularity, and have the additional advantage of being biodegradable. Timber is another example of a sustainable material, which is protected by schemes such as the one operated by the Forest Stewardship Council (FSC).

Figure 2.3.40 Aftermath of the Minas Gerais dam burst.

Metal production relies on the extraction and processing of ores, and this can be very damaging to the environment due to the large scale of mining and quarrying operations. The work involved in obtaining these ores can be very dangerous, and some companies have been accused of having a disregard for their workers' and local people's welfare. An example of this occurred in the Minas Gerais region of Brazil in 2015, when a dam burst in an iron ore mine, killing several workers and local people as well as engulfing the nearby town in mud. Accusations have been made that the appropriate safety standards had not been implemented.

Another disaster, on a far larger scale, occurred in connection with the processing of chemicals for pesticide manufacture. In 1984, the Union Carbide plant in Bhopal, India suffered a leak of methyl isocyanate, which resulted in the death of thousands of people in the plant and local towns, as well as serious health effects for thousands more. The cause of this disaster has been subject to much controversy and argument, but the siting of such a plant near to a large centre of population clearly exacerbated the problem, and health and safety standards were subject to much criticism.

In the last two examples, we have considered methods of extraction and preparation of materials that could be regarded as being questionable, but conditions for workers manufacturing and assembling products can also be subject to criticism. An example of this is the Foxconn iPhone factory in China, where several workers committed suicide. The cause was allegedly the pressures of working in a stressful environment.

How can companies avoid these pitfalls and work towards an **ethical** approach? Here are some examples of strategies that they might adopt:

- Carry out a sustainability review prior to starting a project.
- Check that materials are being sourced as locally as possible along the lines of the 'Grown in Britain' campaign.
- Use responsibly managed resources overseen by schemes such as the FSC.
- Ensure that the workforce operates fairly and safely under the Ethical Trading Initiative promoted by the International Labour Organisation (ILO), particularly in connection with issues such as child and slave labour.
- Ensure that products are labelled and marketed honestly to aid consumers.
- Adhere to appropriate compliance schemes such as BS EN 60335 for electrical appliances.
- Use **Fairtrade** certified products whenever possible, and seek certification for their own practices.

ACTIVITY

Carry out your own research to see if you can find out about products that have allegedly been manufactured in unethical ways.

Culturally acceptable and non-offensive products

The global nature of trade means that product designers need to be aware of the culturally diverse nature of their markets. One of the major areas to be aware of is the different connotation of colours of products, clothing and decorations. Here are some generalised examples

Table 2.3.3 Examples of different connotations of colours.

	Black	White	Red
UK/USA	Death, mourning, funeral	Bride, wedding, purity	Energy, warning, love
Eastern	Wealth, health, prosperity	Death, mourning	Bride, wedding, prosperity
China	Colour for young boys	Death, mourning	Good luck

Figure 2.3.41 White funeral flowers in Hong Kong.

An example of a company using an inappropriate colour is United Airlines, decision to hand out white carnations to passengers when it started a new first class route serving Hong Kong. They had not considered that white symbolises misfortune and death in that part of the world, and once they realised their mistake, they changed to red flowers.

The traditional, three-stone cooking stove used by millions of people in rural African communities are only about 10 per cent efficient. Unfortunately, efforts made by well-meaning designers to 'improve' the design of these stoves and reduce the vast amount of effort and time that goes into collecting firewood, as well as reducing the danger to health caused by smoke inhalation, have often been unsuccessful. The reason for this is the long-standing cultural tradition of using the version incorporating three stones, since the stones can have a symbolic meaning.

When a toothpaste company tried to market its product in some areas of southeast Asia with a slogan emphasising that it 'whitens your teeth', their advertising campaign was not successful. This was due to the local custom in some communities of blackening their teeth to make them more attractive, as they believe that only demons and wild animals have white teeth.

Designers need to undertake thorough research to ensure that these situations do not occur, in order to avoid alienating potential consumers and wasting money.

Products that are inclusive

The British standards Institution's (BSI) definition of **inclusive** design is 'the design of mainstream products and/or services that are accessible to, and usable by, as many people as reasonably possible ... without the need for special adaptation or specialised design', as set out in their BS 700 Part 6, 'Managing inclusive design' standards document. One of the key aspects of the approach to design that they promote is to ensure that the needs of a diverse range of people can be accommodated without them being stigmatised or limited in some way. The goal of designing products for the widest possible range of people has led to the term 'universal design' being used.

Figure 2.3.42 Specialist clothes store catering for a wider anthropometric range of customers.

The term 'accessible to, and usable by, as many people as reasonably possible' has obvious links with the dimensions of people, since there is a wide range of sizes of people who need to be able to use products. Anthropometric data gives designers a useful picture of the size, shape and strength of the population, and the 5th to 95th percentile is often used as a basis for deciding on the size and range of adjustability of products. The extremes are sometimes more difficult to cater for, and the fact that there are fewer people in these categories means that there is less profit to be made, so they are often regarded as a niche market.

Figure 2.3.43 Bright yellow handrails on a Nottingham tram.

Assumptions are often made that the number of people in society who require particular consideration in the design of products is relatively small. Research shows, however, that there is a much more complex picture, with a large proportion of the population sometimes finding that their requirements are not met. Therefore, designers need to bear this in mind when designing products. Examples of this already happening include:

- the adoption of updated standards for wider doors for wheelchairs in the design of new houses
- the production of a highly successful, attractive range of 'good grips' kitchen utensils, by Oxo, that can be used by a wide range of people and have won a number of design awards
- many accessibility improvements on public transport, including buses with floors that lower for pushchair and wheelchair users, and the incorporation of bright yellow handrails that are more easily seen by people with a visual impairment
- measures to enable people with a range of disabilities to safely negotiate pedestrian road crossings, such as raised pavement bumps, ramps and visual, audible and tactile feedback
- the installation of hearing induction loops in theatres for hearing impaired members of the audience
- a range of products such as kettle pourers, simplified remote controls with larger buttons, key and door handle adapters and modified gardening tools that can be purchased from a growing number of specialist retail outlets.

Although the Disability Discrimination Act (DDA) of 1995 does not specifically dictate requirements for product design, it was instrumental in bringing about many of the accessibility improvements that we have seen, and has strongly influenced design in areas such as public transport.

It can sometimes be a problem that designers are unaware of the difficulties encountered by disabled consumers of their products. For this reason, it is common practice for 'empathic research' to be carried out to give them an idea of what it might be like for the disabled to use their products. Examples of this include the simulation of arthritis using specially adapted gloves, and wearing obscured goggles to replicate levels of sight loss.

> **ACTIVITY**
>
> The next time you use public transport, try to identify as many inclusive features as you can, such as the yellow handrails shown in Figure 2.3.52.

The growth in participation and interest in the Paralympic Games has resulted in a wide range of specialised equipment for disabled athletes, and has also been effective in raising awareness of the potential that inclusive design has for improving the lives of the broader population.

Products that could assist with social problems

Figure 2.3.44 Litterbin to encourage good habits in young children.

Social problems cover a wide range of issues, from what might seem relatively trivial, to difficulties that seriously affect the lives and wellbeing of significant numbers of people.

An example of a social problem that does not affect lives seriously but does create ill feeling is litter. The problem of changing attitudes to dropping litter is a difficult one to deal with, but there is awareness that attitudes can be formed at a young age. A number of innovative designs for litterbins have therefore been conceived to encourage children to adopt good habits.

A significant social issue is the large number of accidents involving young drivers and, as a result of this, insurance companies are actively seeking ways of reducing the high number of casualties in that group of motorists. The increasingly ubiquitous nature of satellite navigation and vehicle tracking has facilitated the development and use of 'black box' devices that actively monitor driving data such as location, time, speed, acceleration and braking. This enables a constant check to be made on the standard of driving and thus reduce risk.

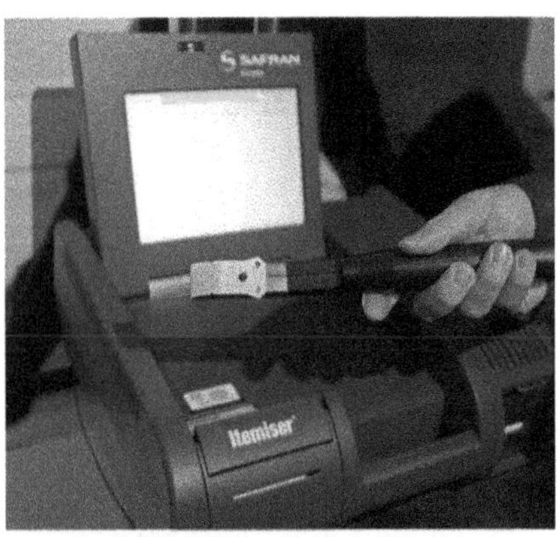

Figure 2.3.45 'Itemiser' scanning device being used by police.

Another example of a product responding to a significant social problem is the 'itemiser' scanning device being used by police to check for illegal drugs being taken into pubs and clubs or being used by motorists. This device simply needs a swab from the suspect's hand to check for recent interaction with illicit substances. Versions of this device are also employed in the detection of explosives at airports and other locations, to counter potential security threats.

Poverty, health and wellbeing

The declaration of human rights, first published in 1948, is a United Nations document setting out the rights of people all around the world, and has been published in 500 languages. Article 25 in the document says, 'Everyone has the right to a standard of living adequate for the health and well-being of himself and of his family, including food, clothing, housing and medical care'. This declaration offers a guide to what should be available, but unfortunately some of the most basic requirements remain elusive to many.

ACTIVITY

Sketch some of your own ideas for designs that could help disadvantaged people such as the homeless and poor.

Designers in developed countries understandably concentrate on catering for the market-led demands of consumers who can afford what they are offering, but there are examples of designers and organisations that have risen to some of the challenges set out in the declaration.

Trevor Baylis became aware of the unaffordability and unavailability of batteries to power radios in parts of Africa, and the consequential lack of information being received from broadcasts on issues such as Aids. As a result of this, he developed a wind-up clockwork radio which does not require batteries. He has sold millions around the world, and has been the inspiration for a number of other wind-up products such as torches and chargers.

'Crowd funding', where individuals can pledge money online to projects that they would like to support, can provide an alternative source of finance for altruistic product design projects. An example of this is 'Gravity light', which makes use of the stored energy in a simple weight that can be raised by the user. This is a project targeted towards poor families in Kenya, who often have to spend 30 per cent of their income on kerosene fuel for lamps. The Gravity light gives out five times more light than a typical kerosene wick, is much safer because it does not involve a flammable liquid and is free to use, providing light for about half an hour each time the weight is raised.

Figure 2.3.46 Homemade concrete compost bin.

Yanko, an online design blog and magazine, organised a competition to encourage designers to work on solutions to problems faced by the poor. They received a number of entries, including:

- the winning idea for a simple rainwater harvesting system, which makes use of normally discarded plastic bottles attached to existing gutters
- shelters that could be used by the homeless
- mobile storage units to facilitate educational projects in developing countries.

Intellectual property (IP) can sometimes be a hindrance to the development of affordable products, since the priority for owners of the IP for a design will normally be financial rewards rather than altruism. Organisations such as Practical Action, an international NGO (non-governmental organisation) that is challenging poverty in developing countries, prefers an 'open-design' approach that benefits the poor, as explained in the section dealing with IP in Chapter 1.10 (see page 154).

An example of Practical Action's initiatives to alleviate poverty in developing countries is in Sri Lanka, where it introduced a design for a three-ring concrete compost bin that can be made by local people. The bin is used to produce compost from organic waste that will enrich their soil, improve the crops that they grow and thus provide financial benefits, as well as improve their health and wellbeing. Its 'pathways from poverty' initiative in Bangladesh also facilitated improvements in health and wellbeing by introducing pumps, sludge carts and safety equipment, which enabled the very poorest members of society to carry out their unpleasant sewage pit emptying duties more hygienically, and also improved their social status.

The improvement of health and wellbeing is an ongoing area of major development in terms of products and technological advances, as well as medications and how they are administered. Recent health-related products that have been developed include the following:

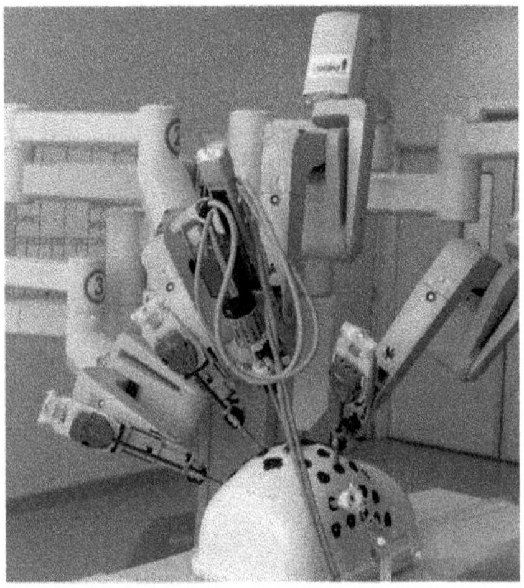

Figure 2.3.47 The 'da Vinci' surgery robot.

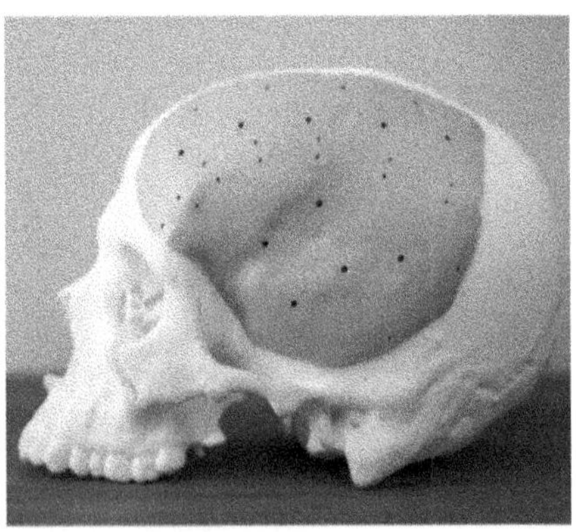

Figure 2.3.48 3D printed polymer skull repair implant.

- Prostheses (artificial limbs) have become much more sophisticated due to the incorporation of new materials, products and technology such as advanced composites, lightweight motors and nerve-linked microelectronic control.
- Minimally invasive robotic surgeries use high precision devices such as the 'da Vinci' robot, which leaves virtually no scarring and massively reduces both the amount of time it takes to recover and the possibility of infections. The surgeon is able to carry out highly complex surgery by remote control using a 3D high definition monitor and input device, and the robot's actuators, which move the surgical instruments as directed by the surgeon, are capable of moving in a much more flexible way than a human hand.
- Telehealth is the use of technological systems to improve healthcare by strategies such as facilitating remote monitoring of a patient's health status by their doctor. This is often used for patients suffering from chronic conditions, where data is transmitted via telecommunication systems such as the internet, and is particularly useful in rural and remote areas. Telehealth also encompasses aspects such as the sharing of patient notes, transmission of reminders and holding web-conferenced meetings between health professionals.
- Medical equipment, such as magnetic resonance imaging (MRI) and computerised tomography (CT) scanners have facilitated tremendous improvement in doctors' abilities to investigate patients' state of health. These highly sophisticated and very expensive pieces of equipment facilitate detailed, non-invasive investigation of possible irregularities by producing a series of 'sliced' views of internal organs, and assembling them as a 3D virtual model of a person's body or a section of it.
- Artificial organs are man-made devices that are used to replace the function of organs such as the heart and ear (cochlear implant). Artificial hearts have been in use since 1982, but a great deal of research is currently taking place to investigate a range of other organs that might be replaced artificially, such as the lungs, eyes and liver. Many of the current investigations involve the use of stem cells, which are cells that divide and 'grow' more cells of various types. This medical technology has, however, been patented, and is also the subject of controversy concerning ethical considerations.
- 3D printing offers a number of avenues for use in medicine. Experimental, investigative and actual treatment applications are under way to explore 3D printing examples such as:
  - ceramic-based scaffolds to support bone-growth
  - synthetic skin
  - medical equipment such as umbilical cord clamps in poorer countries
  - polymer skull repair implants to replace damaged sections.

Migration

Many people around the world become migrants and refugees due to conflict, poverty and victimisation in their home country. This results in massive temporary communities springing up all over the world in areas where they are seeking refuge. The infrastructure to support large numbers of displaced people sometimes does not exist in these places, and governments often turn to humanitarian organisations to provide support. The main requirements, as we saw earlier in the UN document, are food, clothing, housing and medical care.

Here are some examples of how designers, companies and NGOs have responded to these needs.

Food and water

The United Nations High Commissioner for Refugees (UNHCR) organisation has encouraged the adoption of trough-style solar cookers for use by refugees in Burkina Faso, where the climate makes it one of the best parts of the world for using solar energy.

SunDwater, an Israeli company, has developed water distillation equipment that can convert water that is contaminated, unsafe or salty into perfectly drinkable water which can be consumed safely. This system also uses solar power, so it can operate independently of electrical supplies.

Clothing

Donating discarded clothing for reuse by refugees from other countries is an excellent way of providing humanitarian aid at the same time as improving sustainability. However, sometimes the needs of people in crisis can be addressed in a more focused way. One such example is the work of Angela Luna, a young award-winning American designer who has created a fashion collection called 'Adiff' which is intended to be used by refugees. One of the garments is a cape that can also be used as a small tent. Her company motto is 'design intervention for global issues'.

Figure 2.3.49 Ikea Foundation 'better shelter'.

Housing

Shelter for refugees, particularly in hostile climates, is an urgent need that has been addressed in various ways. One such response is the Ikea Foundation's design and production of a flat-pack 'better shelter' which is being used by refugees in locations such as Macedonia, Iraq and Ethiopia. It can be erected without tools, sleeps five people, is well insulated and is fitted with a solar panel for charging lighting systems and phones.

Figure 2.3.50 'Hospitainer' on its way to a refugee crisis location.

Medical care

Access to hospitals, medical professionals and appropriate medicines is very difficult or impossible in refugee camps, and it often falls to NGOs to step in and do what they can to provide appropriate care. The United Nations Population Fund (UNFPA), for example, has provided 'hospitainers' and 'maternitainers', which are shipping containers with adapted internal designs for use as mobile hospitals and maternity units. These have been dispatched to a variety of refugee situations such as those in Iraq and Syria. The containers have been designed to provide life-saving care, with facilities for surgery and other medical procedures. Using adapted containers means that they are easily transported to where they are needed using standard lorries and ships.

Solar energy finds another application in providing the power required to run refrigeration systems at refugee camps, which not only help to maintain food for healthy eating, but also facilitates the essential temperature regulation of medicines.

ACTIVITY

Make a scale model of a shipping container with its roof removed, so you can add further internal dividers and other key features to show how it might be adapted for use as an emergency hospital.

Fairtrade

Consumers are now becoming more aware of the way that growers and handicraft producers in developing countries are often exploited and

Figure 2.3.51 Pushpanjali Fairtrade worker creating a decorative box in Agra, India.

ACTIVITY

Research the rights afforded to Fairtrade workers, and use a table to contrast these with the conditions that have to be endured by those who are less fortunate.

dealt with unfairly when trading their products. The Fairtrade movement seeks to give a better deal to such people in many countries, mainly in South America, Africa and Asia. Supermarkets have enormous buying power and often dictate very strict terms when dealing with suppliers, which sometimes leads to them and their families living in poverty, even when they are a vital element in the supply of food, such as the five billion bananas consumed in the UK every year.

The Fairtrade organisation enters into negotiations with supermarkets and other buyers to ensure that a fair minimum price is set for products, and also agrees on a premium to be added to the price of goods, which is used to make investments to support developmental projects. Its products display a certification label called the Fairtrade mark, which indicates that its standards have been attained. Traidcraft is an example of a Fairtrade organisation whose wide range of ethically sourced products are retailed through a website and other outlets.

Many organisations, including schools, colleges and local authorities seek Fairtrade mark status to support sustainable development, and benefit producers while encouraging active citizenship. This has resulted in an increased demand for food and craft products with Fairtrade accreditation. An example of a craft product producer operating as part of the Fairtrade movement is Pushpanjali, in the Agra area of India. It provides sustainable work, fair wages and a number of other benefits for local craftsmen and women, including equal rights, healthcare, education and training.

Product life cycle

A level only

The stages of the product life cycle

Products rarely stay on the market forever in their original form. The life cycle of a product normally progresses through four stages – set out below – and it is important for companies to plan the release of their products carefully. The time that a product spends in the 'maturity' phase of highest sales needs to be maximised, since the product will not become profitable until the money invested in research, development and new tooling has been recouped. If manufacturers fail to effectively plan a product's life cycle, they may have no new modifications or products ready for introduction to rectify the fall in revenue when sales start to decline. Smart phones are the most obvious manifestation of this type of approach, but it also applies to cars, computers, domestic appliances and a range of other products. A good example of a company that has been successful in maintaining the sales of its products by constantly making design improvements is Toyota. The Toyota Corolla has been redesigned 11 times in its 50-year production run, in an effort to ensure that it continues to be the world's bestselling car.

Introduction

This stage sees the product being launched, usually preceded and surrounded by a good deal of publicity, particularly regarding how effectively it supersedes previous models.

Sometimes products have evolved in response to the 'demand pull' of consumers for particularly desirable features, such as improved battery life or larger memory in a smartphone.

Manufacturers sometimes seek to capitalise on large sums spent on research and development by utilising the 'technology push' approach. This involves trying to convince consumers that they need to buy their latest products with the most up-to-date features. A dramatic example of this going wrong was Google's failed attempt to win consumers over to their concept of 'Google Glass'. This was a head-mounted computer device with camera, display and touchpad, which failed to capture the public's imagination, partly because it was so expensive and also because of safety and privacy concerns.

Growth

Once the product is on the market, it is expected that its sales will grow as consumers seek to replace older models and purchase the latest one. Advertising plays a key role for some products, to ensure that consumers are fully aware of the benefits of the new version.

Maturity

Eventually, the growth in sales will peak as the product is perceived to be desirable. It is important for companies to ensure that this stage is maintained for as long as possible to reap the maximum sales from their investment in the product.

Decline and replacement

Sales eventually begin to diminish as most interested consumers will by now have bought the product and, because it has been on the market for a significant period, they will be starting to anticipate replacement models. Eventually the product is withdrawn from sale, but by this time the replacement model should be growing its sales.

The need for consumers to replace products due to factors such as unavailability of spare parts, being unable to run the latest software or premature failure, is sometimes regarded as 'planned obsolescence'. This is the concept of companies designing and manufacturing their products in such a way that they have to be replaced on a regular basis. The Fairphone (see page 154, Chapter 1.10) is an attempt to move away from this approach by offering a product that has upgradability at the heart of its design. This type of design acknowledges the increasingly urgent need for products to be more sustainable.

Evolution of products

Evolution means the gradual change that occurs in products as designers and manufacturers progressively incorporate new ideas, materials, technologies, manufacturing methods and other aspects that offer scope for improvement and increased sales. Large companies have research and development (R&D) departments, whose role it is to explore and develop new ideas for products. This can be very expensive for companies, but they rely on the work that they do to generate future sales. Apple, for example, spent $2.5 billion on research and development in 2016.

Figure 2.3.52 Jet engine research and development at Rolls Royce.

Here are some examples of how products have evolved with the creation of 'desirable' new features, to maintain customers' demand and thus ensure that the profitability of their life cycle is maximised:

- Smart phone: higher definition cameras, more powerful processors, larger screens, latest connectors, subtle changes in shape and form, alternative finishes, larger memory and so on.

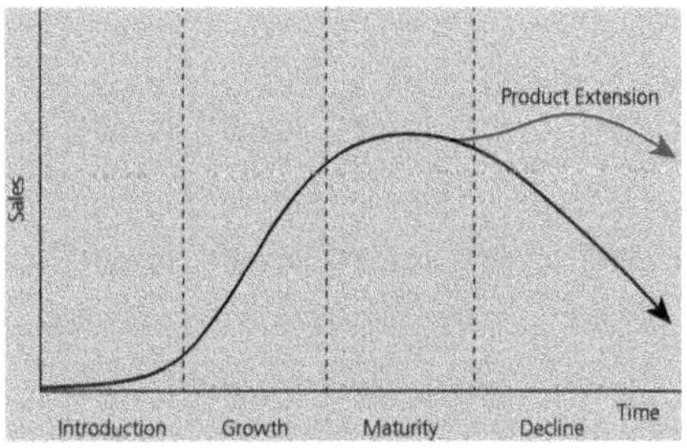

Figure 2.3.53 Simplified product life cycle graph showing the need to update products in order to maintain sales at the highest possible level.

- Television: slim design, LED, organic light emitting diode (OLED), thinner screen, 'smart' capability, high definition (HD), ultra-high definition (UHD), curved screen, high dynamic range (HDR).
- High end digital camera: increased pixel count, faster speed for bursts of shots, articulated screens, electronic viewfinder, HD and UHD movie capability, water resistance, multiple card capacity.

Progressive feature upgrades incorporated in new models of products result in many consumers eventually feeling that they need to replace their device to keep up to date. There is currently a trend for mobile phone owners to keep their phones for between two and three years, whereas a few years ago, it was not uncommon for them to be changed after only eighteen months. This means that companies have to work even harder to ensure that they are offering new features which consumers are eager to obtain. Other strategies to maintain sales and revenue include the development of complementary products such as smart watches and the offer of upgrade finance schemes.

ACTIVITY

Produce your own product life cycle graph for a series of related products from a specific manufacturer.

KEY TERMS

Product life cycle: the stages a product goes through from its introduction, growth and maturity to its eventual decline and replacement.

Socio-economics: how society and the economy interact to create particular circumstances.

Rationing: the limitation of the availability of certain goods, usually in response to a shortage created by a war or other crisis.

Utility products: post-Second World War basic products that were often rationed.

Council of Industrial Design (COID): a post-Second World War British organisation set up to improve design standards and competitiveness.

Microelectronics: miniature electronic devices and systems facilitated by the development of the IC.

Sustainable: something that has the minimum possible impact on the environment.

Ethical: something considered by society to be morally fair.

Fairtrade: a social movement to encourage the ethical treatment of farmers and workers in developing countries.

Culture: the combined ways of living developed by a group of people, that is passed from generation to generation.

Inclusivity: consideration of the needs of the widest possible range of people.

KEY POINTS

- The free thinking, modernist principles of the Bauhaus designers embraced First World War industrial manufacturing methods and materials, such as tubular steel, to revolutionise the design and mass production potential of furniture.
- Utility products, which were basic and used minimum amounts of material, were produced after the Second World War in response to material and energy shortages.
- The post-war period saw an increase in the range of decorative and fashionable products due to the impact of targeted design exhibitions, developments in materials, manufacturing and technology, and the growing influence of ideas from abroad.
- The processing power and small size of semi-conductor microelectronic components such as ICs have made a significant impact on design and designing.
- New materials such as graphene are being developed, which continue to extend the potential of product design.
- Innovative methods of manufacture, such as DMLS, offer designers and manufacturers increasing scope for the realisation of their ideas.
- Advancements in CAD/CAM, such as the introduction of Cloud-based software and VR, have increased the potential for designing and making products with the help of computers.
- Designers need to ensure that their products are as sustainable as possible and that the negative impact on society of their manufacture is minimised.
- The culture and ethical beliefs of consumers need to be considered by designers in order to ensure that their products are appropriate.
- Designers should take into consideration the broadest possible spectrum of people's needs when developing new products.
- Designers need to consider the requirements of all members of society, including those who are poor, vulnerable, ill or deprived of basic needs.
- The Fairtrade movement seeks better treatment for farmers and workers in developing countries. This is implemented by organisations such as the Fairtrade Foundation, which licenses the use of the Fairtrade mark on UK products that meet internationally agreed standards.
- The life cycle of a product must be carefully considered in order to maximise its lifespan, and ensure that upgrades and replacements are ready when sales decline.

Check your knowledge and understanding

1 How did the materials and manufacturing legacy of the First World War influence the Bauhaus designers?

2 What is 'utility furniture' and why was it introduced?

3 What was the role of the COID and how did it set about influencing British designers?

4 Name some of the other influences on British design after the Second World War.

5 What was the key development in materials and manufacturing that made it possible for Robin Day to design and manufacture his PP chair in 1963?

6 What was the important invention in the late 1940s that facilitated smaller electronic products such as the Regency TR1 portable radio?

7 What did the 1940s device mentioned in the previous question replace, and why was it such a significant improvement on its predecessor?

8 What was the next major breakthrough using semi-conductors, which became commercially available in the 1960s and started the era of microelectronics?

9 How much have the microelectronic devices mentioned in the previous question improved from the 1960s to the present time?

10 Make five direct comparisons between designing and manufacturing practice in the 1950s and the present day, to highlight some of the key advances brought about by microelectronics.

11 Name five products that marked major advances in the development of portable music players, and describe some of the significant technological developments that they incorporated.

12 Name three new materials and describe their applications in product design.

13 How does the IoT benefit manufacturers?

14 Name and explain two recent advancements in CAD/CAM systems.

15 Describe two examples where technology developed for military products has brought about significant benefit to the population as a whole.

16 Describe an example where the extraction or processing of materials for manufacturing products has had a seriously detrimental effect on society.

17 Why was it a mistake for an airline to hand out white flowers to its passengers in Hong Kong?

18 Describe two examples of product adaptations designed to make them more accessible.

19 Describe an example of a product specially designed to help poor people in developing countries improve their lives.

20 Describe two examples of technologically advanced products that have been able to improve people's health and wellbeing.

21 Describe two examples of products that have been developed by designers to solve some of the problems created by mass migration in areas suffering from conflict.

22 Explain the term 'fairtrade' and how it can bring about benefits to workers in developing countries.

23 List the stages of the product life cycle, and explain why it is important for designers and manufacturers to be conscious of how it applies to their products.

Further reading

Utility furniture of the Second World War by Jon Mills (Sabrestorm Publishing)

How computers work (How it works) by Ron White (QUE)

Made to measure: New materials for the twenty-first century by Phillip Ball (Princeton Paperbacks)

Design for the real world: human ecology and social change by Victor Papanek (Academy Chicago Publishers)

Design meets disability by Graham Pullin (MIT Press)

Consumer product innovation and sustainable design by Robin Roy (Routledge)

Product life cycle: The fundamental stages of every product by 50Minutes.com (50Minutes.com)

www.designcouncil.org.uk

www.yankodesign.com

These two websites include many case studies of products developed in response to different socio-economic influences and developments in materials and technology.

www.fairtrade.org.uk. This website provides information on the work of the Fairtrade Foundation.

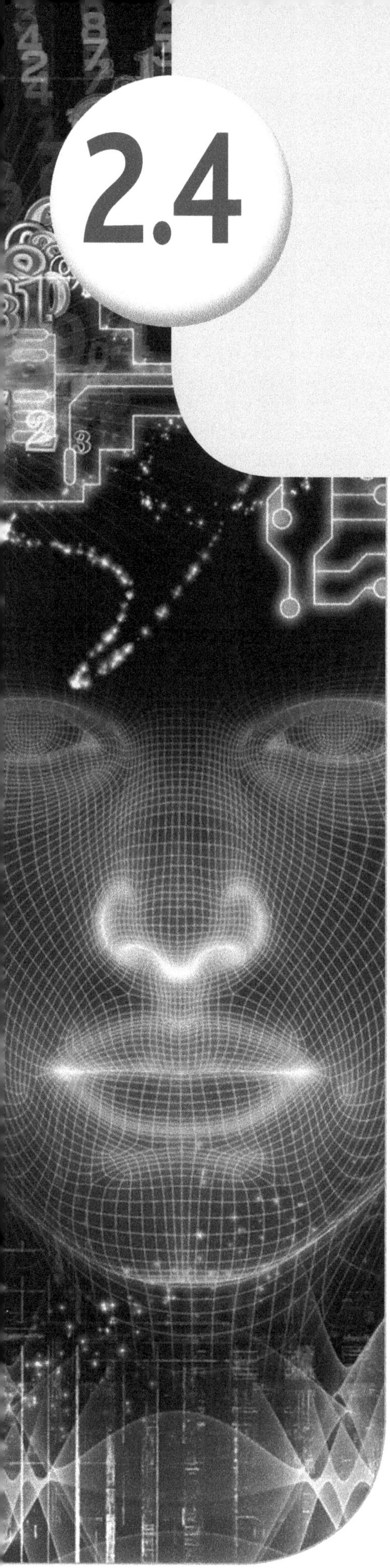

2.4 Design processes

LEARNING OUTCOMES

By the end of this section you should have developed a knowledge and understanding of:

- the stages of a range of design processes in order to apply personal judgement and relevant criteria in the appraisal of products and systems.

If you are studying at A-level you should also have developed a knowledge and understanding of:

- the stages of the design processes used in the NEA
- how to demonstrate the development of a prototype from design proposals
- how different design methodologies are used by designers in the corporate world when designing products, including collaborative working and the cyclic nature of commercial design and manufacture.

Through studying either AS or A-level, you will gain first-hand experience of working through a design process which will involve you designing, making and testing prototypes in response to a design problem that you have identified. You will use a wide range of research and investigation methods, different designing and modelling techniques, and testing and evaluation methods. You learned about the requirements for product design and development in Chapter 1.8 and you learned about design methods and processes in Chapter 2.1. Here we will look in more depth at the stages of design processes, and the different design methodologies that are used by designers in the corporate world when designing products.

The use of a design process

A **design process** is a system that organises how a prototype (in a school or college) or a product (in the commercial world) is developed. There are several different design process **models** that can be used, but they all start with the identification of a design problem (needs or wants of a specific user, or target market).

Following the identification of a design problem, the next step in the design process is to write a design brief.

The design brief can include the following:

- a description of the problem or need (sometimes known as a situation)
- images and details of the context, situation or problem
- an explanation of why existing products are not suitable or do not meet the needs of the user
- details of the client or user group (who the product is intended for and how it may help)

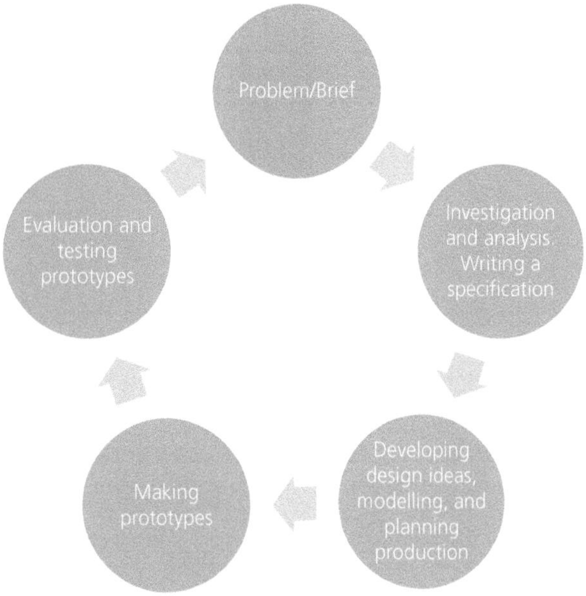

Figure 2.4.1 The cyclic design process.

- an outline description of the type of product that is to be developed
- a description of the key functions of the product
- a list of the key areas of focus for the product to be a success.

Following the writing of a brief, the process continues with investigation and analysis, writing a **specification**, generating ideas (drawing and modelling), developing and planning, the production of prototypes, and evaluation and testing. In producing your NEA project, the process will end with the testing and evaluation of your prototype. In a commercial environment, the process may continue with further development of the prototype using the results of testing and evaluation. This process of continual improvement is known as a cyclic or iterative design (see Chapter 2.1).

Investigations and analysis

Designers carry out investigations (research) using a wide range of techniques. Usually, the most productive methods are **primary research** techniques. Designers often observe people in similar environments using similar products. This can be useful in identifying the weaknesses of existing products and potential areas for improvement. They will also seek the opinion of potential users by using focus groups or questionnaires.

In carrying out your NEA project, you may undertake such primary research to better understand the needs and wants of potential users. In addition to this, you may carry out visits to shops in order to analyse existing products.

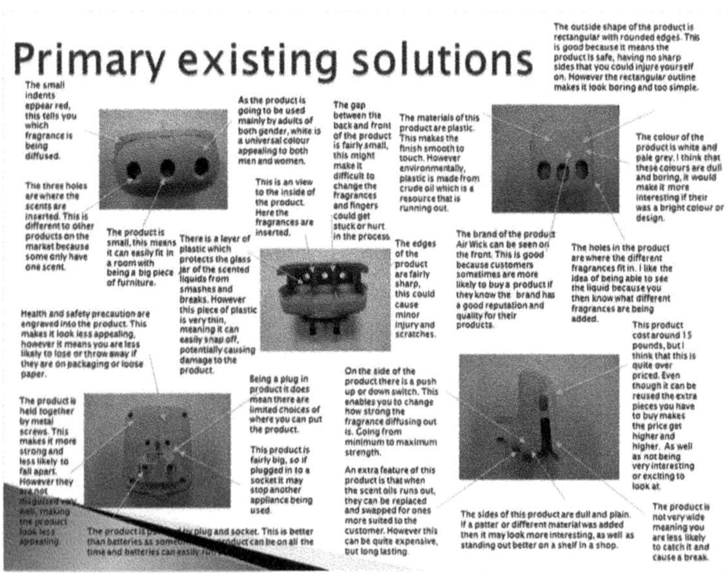

Figure 2.4.2 Analysis of chosen product.

Analysis of existing products

When analysing existing products, where possible, you should use them. Then, if you can, they should be taken apart. Work out how the product functions. What are the vital components that make the product work? What are the materials used and how is it manufactured? Are there any ergonomic features that make the product easier to use?

Try to identify what consumer needs are met by the product and what key criteria were used when the product was developed. You can also critically analyse the product by considering its aesthetics and cost. Where possible, you should compare the product with others and comment on function, suitability of materials and manufacturing method, ergonomics, aesthetics and cost.

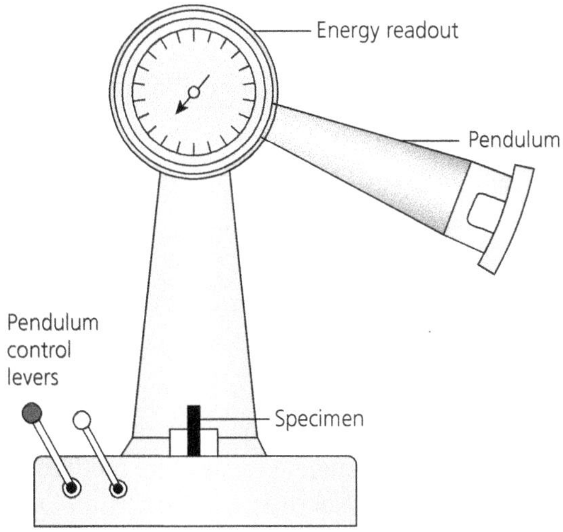

Figure 2.4.3 Izod impact test.

Investigation and analysis of materials

Sometimes designers may need to carry out research into materials, construction methods and finishes so that they can select the most appropriate ones to use in a project. In the commercial world, designers are unlikely to be experts in all materials, components and technologies that go into products, so it is vital to carry out such practical investigations. In the commercial world, this type of testing might be carried out by specialist researchers, who then make the results available for designers to use.

An example of a test that might be carried out in industry is testing the toughness or brittleness of materials. This is done using the Izod impact test, which strikes a heavy pendulum into a notched test piece. The amount of energy absorbed by the notched test piece is recorded and a comparative toughness/brittleness can be calculated.

A similar test can be simulated in a school or college workshop by striking samples of materials that are held in a vice, using a hammer from a set distance away.

The 'wear resistance' of materials can be tested using an abrasion test. Here, the sample material is fixed to a wooden block and a second block is covered in glass paper. The two blocks are rubbed together to determine wear on the sample material.

In a school or college, similar practical investigations and testing can be carried out.

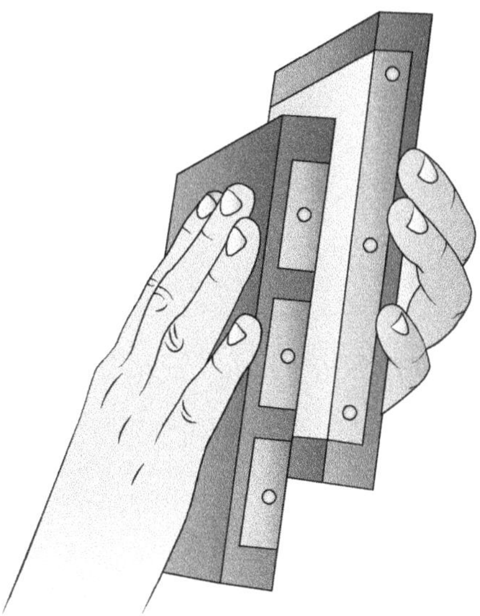

Figure 2.4.4 Abrasion test.

Other forms of investigation

There is a wide range other investigation techniques that commercial designers use, which you could find useful. This includes:

- meeting with a client or user group to discuss their requirements
- using internet forums to gain public opinion about the product being developed
- talking to 'experts' or specialists who work with the type of product being developed. This can identify key design criteria that will be included in a design specification
- investigating the work of other designers (analysing products to identify useful features, materials, methods of manufacture and so on that can be used in a new product)
- investigating historical influences which may inspire product styling
- looking at current trends and styles – fashion influences
- identifying design constraints such as size, the environment the product is to be used in and how it is to be used
- using British Standards that may relate to the product being developed
- investigating the safe use of materials or components that are to be used, ensuring compliance with COSHH
- analysis of anthropometric data to ensure the product will 'fit' the relevant consumers
- consideration of relevant social, moral, cultural and environmental factors.

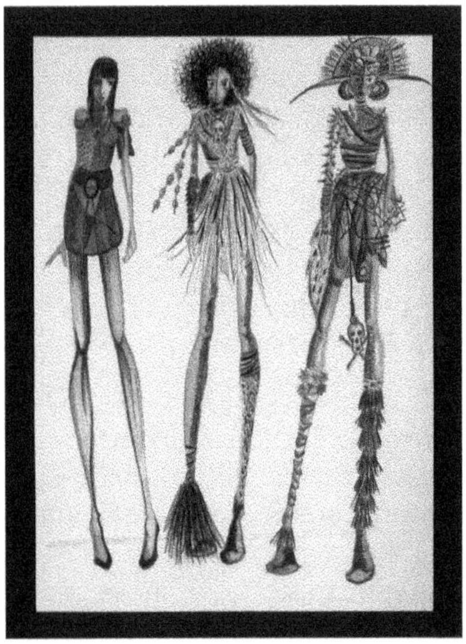

Figure 2.4.5 Fashion clothing – inspiration board with notes.

Using inspiration materials

When conducting investigation and analysis, designers often collect materials to inspire their design ideas. They might use:

- 'mood boards' or inspiration boards – a collage of images relating to the product, which might include colours, styles and other inspirational images
- 'inspiration boxes' which might include swatches, colour samples, sample materials and existing products
- 'job bags' – a collection of cuttings from newspapers, magazines, sketches, material samples, components and products.

Ideas generation

There are many ways in which designers can generate ideas. One method that professional designers and students alike may start with is a mind map. Mind maps are useful in helping you think about the factors that influence the design, and what you might consider while developing the product. These could form a checklist that is worked through when developing designs.

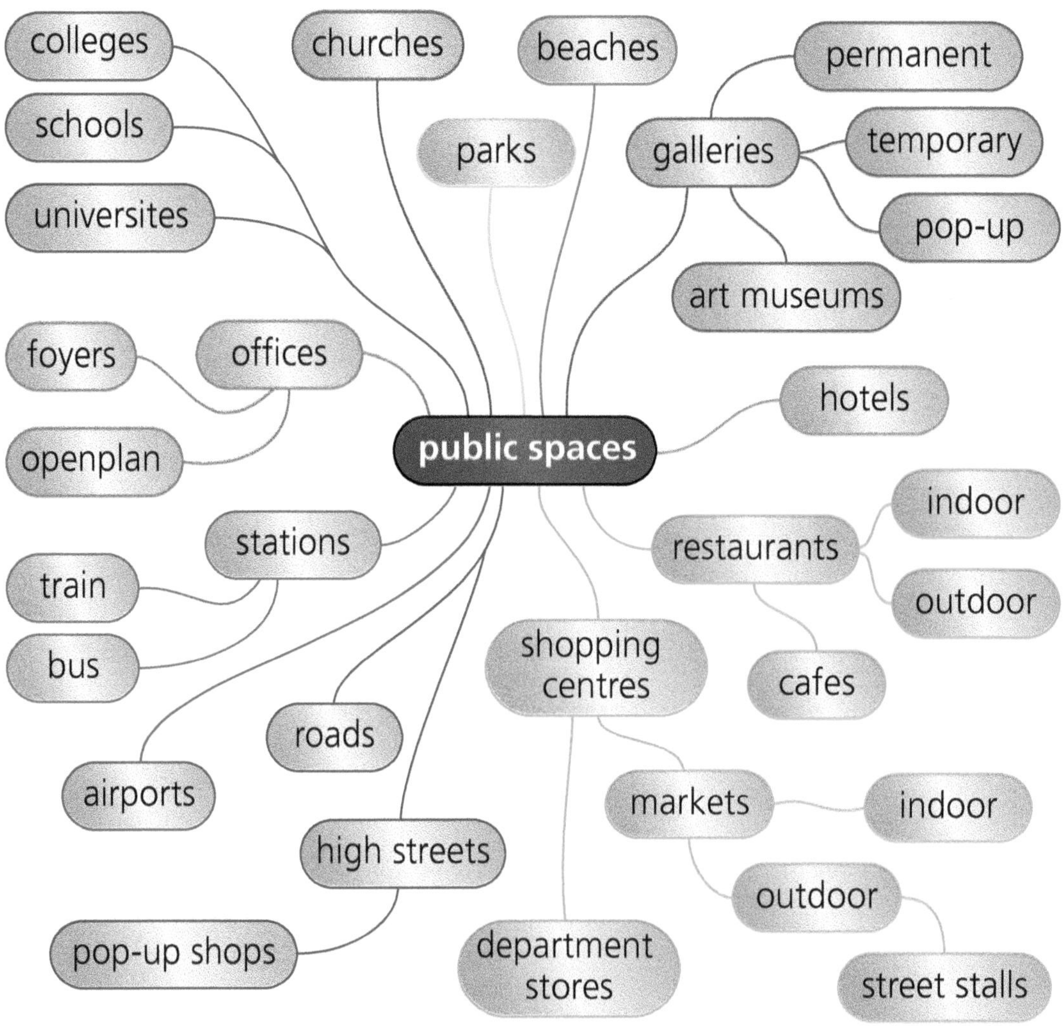

Figure 2.4.6 Concept map exploring concepts of a public space.

Another method of developing ideas is the **SCAMPER** technique:

- Substitute – change materials, components, finishes and so on.
- Combine – mix ideas or parts of ideas together.
- Adapt – alter the design, use part of another idea, change the function, adapt mechanisms or useful parts.
- Modify – change the shape of the design or part of it, increase or decrease the size.
- Put it to another use – is there an alternative use for the design? Could it have different functions?
- Eliminate – remove and reduce parts. Simplify the design.
- Reverse – turn the design inside out or upside down.

Sometimes designers work collaboratively in teams. This is common practice in industry where no single designer will work on a project on their own; rather, the design work for a project will be given to various designers, each one adding to the design and developing it in turn. This is done so that the expertise and creative minds of the whole team can be utilised. It normally results in a much better product than one created by a single designer. Design teams working like this might work together in group meetings, work collaboratively using video conferencing (if they are located in different places) or by passing the design on to each designer in turn. See page 248 for more on collaborative working.

In schools and colleges, the 4 × 4 Design Activity technique, which was first introduced by the Design and Technology Association (DATA), can be used. It starts with one person drawing a design idea in the centre of sheet and four boxes in a grid around the outside. The first person then passes the sheet to a second who will use the original drawing as inspiration to develop it further. The process continues until there are four completed development drawings. This is a good way to work collaboratively with class mates. You will benefit from getting their ideas and viewpoints about your project, and it might make you consider things that you would not have thought about if you worked entirely on your own.

Thinking hats

This is a technique developed by Edward de Bono as a method of improving creative discussion about ideas.

The thinking hats technique uses six different coloured hats. Each colour represents a different way of looking at things. The groups take it in turns to wear each hat so that they analyse the idea from different viewpoints.

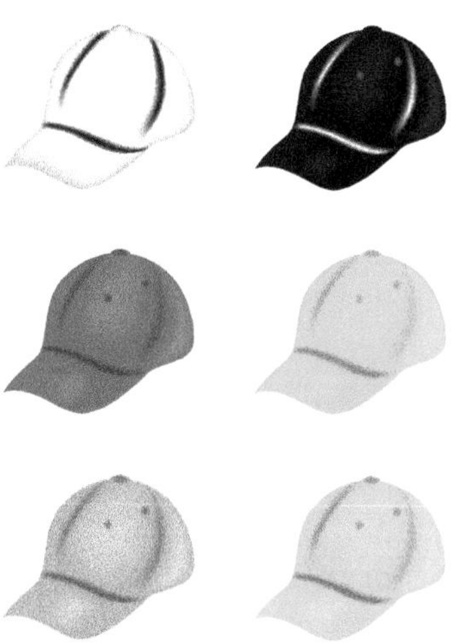
Figure 2.4.7 Thinking hats.

The different types of thinking are represented by the colours as follows:

- The white hat represents the facts and essential information. It also encourages the identification of information that needs to be sought to solve the problem, and thinking about how the information can be found.
- The red hat represents feelings and emotions. Anyone wearing the red hat is encouraged to express their feelings about the idea without any justification. Their thoughts can be subjective and do not need to be rational.
- The yellow hat encourages the wearer to think positively about an idea. They will need to support the idea and justify their statements.
- The black hat is the opposite of the yellow hat. The wearer is to discuss the negative aspects of the idea. They are to put opposing views forward and justify their points.

- The green hat is for creativity. This encourages the wearer to think about unconventional solutions and ideas. Creative thinking may develop the design in unexpected ways.
- The blue hat is for directing the discussion and telling the group when to change hats. The wearer of the blue hat will summarise the key points from the discussion and make decisions.

The thinking hats technique is another way of involving others in sharing ideas about projects. As with the 4 x 4 technique, thinking hats may help you to generate and develop more ideas than if you were working on your own.

Developing ideas through discussion

One of the main methods that can be used to develop ideas is to discuss designs with a client, or with people from the target market. The best way to do this is to sketch or model some initial concept ideas and then obtain feedback on the designs. The feedback, which is usually suggestions about how the design can be improved, should be written down. The feedback can then be used to further develop the most popular design.

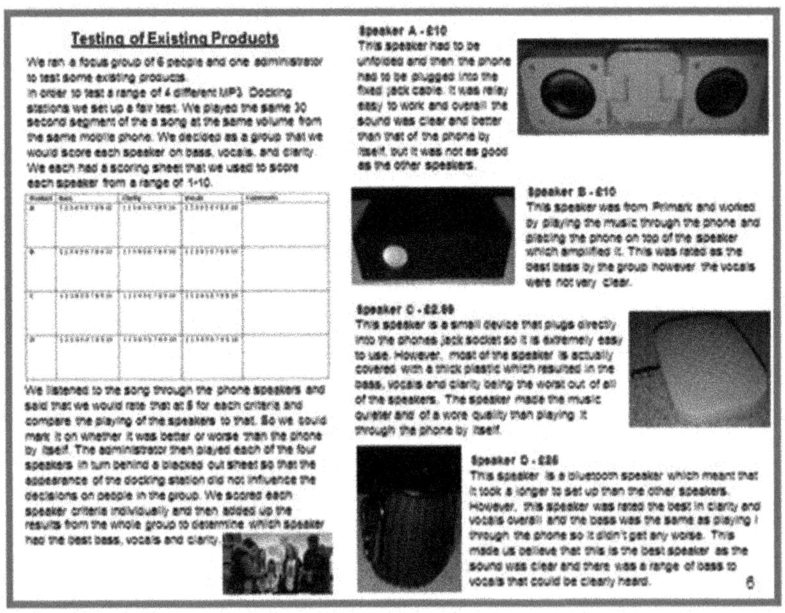

Testing of Existing Products

We ran a focus group of 6 people and one administrator to test some existing products.
In order to test a range of 4 different MP3 Docking stations we set up a fair test. We played the same 30 second segment of the a song at the same volume from the same mobile phone. We decided as a group that we would score each speaker on bass, vocals, and clarity. We each had a scoring sheet that we used to score each speaker from a range of 1-10.

We listened to the song through the phone speakers and said that we would rate that at 5 for each criteria and compare the playing of the speakers to that. So we could mark it on whether it was better or worse than the phone by itself. The administrator then played each of the four speakers in turn behind a blacked out sheet so that the appearance of the docking station did not influence the decisions on people in the group. We scored each speaker criteria individually and then added up the results from the whole group to determine which speaker had the best bass, vocals and clarity.

Speaker A - £10
This speaker had to be unfolded and then the phone had to be plugged into the fixed jack cable. It was relay easy to work and overall the sound was clear and better than that of the phone by itself, but it was not as good as the other speakers.

Speaker B - £10
This speaker was from Primark and worked by playing the music through the phone and placing the phone on top of the speaker which amplified it. This was rated as the best bass by the group however the vocals were not very clear.

Speaker C - £2.99
This speaker is a small device that plugs directly into the phones jack socket so it is extremely easy to use. However, most of the speaker is actually covered with a thick plastic which resulted in the bass, vocals and clarity being the worst out of all of the speakers. The speaker made the music quieter and of a wore quality than playing it through the phone by itself.

Speaker D - £25
This speaker is a bluetooth speaker which meant that it took a longer to set up than the other speakers. However, this speaker was rated the best in clarity and vocals overall and the bass was the same as playing it through the phone so it didn't get any worse. This made us believe that this is the best speaker as the sound was clear and there was a range of bass to vocals that could be clearly heard.

6

Figure 2.4.8 Example of student work.

Illustration

Methods of illustration are covered in detail in Chapter 1.14. The following is a summary of techniques that can be used in the design process:

- isometric drawing: used to sketch design ideas
- one-point perspective: used to show buildings or room interiors
- two-point perspective: used to draw objects from a range of different viewing angles
- exploded views: used to show the relationship of parts of a product and how they assemble
- marker rendering: used to represent colour and materials/finishes on design drawings
- sectional views: used to show the cross section of an object – useful in showing hidden detail such as holes and slots

Figure 2.4.9 Ideas and modelling #1.

- orthographic: used to show the front, plan and end view of an object – usually drawn to scale
- flow chart: used to show the sequence of a manufacturing process – particularly useful in illustrating the QC stages
- 3D CAD: used to give an artist's impression of designs or to make a virtual model
- 2D CAD: used to draw parts for cutting out/machining on computer controlled equipment, or to produce working drawings/orthographic projections.

Development of a design specification

After carrying out an investigation and analysing research, it is possible to develop a design specification.

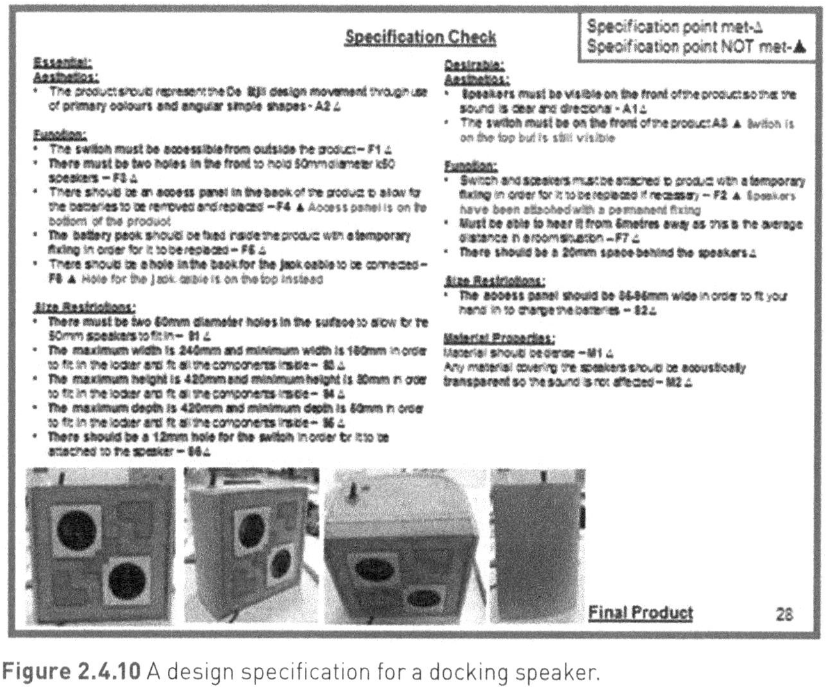

Figure 2.4.10 A design specification for a docking speaker.

A specification is a list of requirements that a design must meet. It can also contain a list of desirable features. In the commercial world, it is highly likely that much of the specification will be determined by the requirements of the client. Typical elements to include in a specification are:

- Function – what will the product do? How well should it perform this task and how often?
- User – are there specific requirements that have been stated by individuals or focus groups?
- Environment – where will the product be used? What external factors might affect it, for example temperature, rain?
- Sustainability – what materials will be used? What manufacturing method will be used? What will the environmental impact of these be? Can the product be easily recycled or reused?
- Maintenance – what is the service life of the product? Will it need to be maintained or will it be a disposable item?
- Size – are there size constraints for the product? Where will the product be used? Does it need to fit into a fixed amount of space?
- Weight – does the item need to be portable? Will it be carried?
- Ergonomics – how will people interact with the product? How will the design take account of different body sizes? Can people easily use the product? Will the product affect posture? Will the product use text, colours, icons to improve interaction?
- Aesthetics – are there specific colours, shapes, forms, textures and finishes to be used?
- Cost – is there a specific target cost for the product? How will choice of material and making method influence this?
- Quality – are there any British or European Standards relevant to the product? How reliable is the product? Are there any specific quality requirements for the product?
- Safety – is there a minimum age at which the product is safe to use? Are warning labels needed? Are safety instructions required? How can potential hazards be minimised?

Modelling

Modelling is an essential tool used in developing design ideas and a final prototype. Modelling is usually done using compliant materials such as paper, card, Styrofoam, clay, wire, polymorph, modelling clay, foamboard, corriflute and balsa. For very quick sketch models, it is a good idea to use cardboard and paper because it can be easily cut, folded and glued.

Models can be made to scale, which can be helpful in communicating the aesthetics of a product. They can also be used to demonstrate the function of a product. Full size models can be used to check the sizes and proportions, for example to test ergonomics, or they can be used to check the function of an idea before committing to using more expensive materials.

When using modelling in a project, it is very important that you photograph the models and explain what you have learned by making them. For example, has it helped you to decide on the final shape and size of the prototype, or helped you to make decisions about the construction method? You may also find it useful to sketch development work over the printed photographs of models so that you can communicate how the design will develop.

Figure 2.4.11 Modelling techniques.

Planning

The success of a design project is dependent on careful planning and preparation.

There are a variety of ways in which designers might produce plans. One of the simplest is to produce a spreadsheet timeline to indicate the main stages of a project and the key dates by which each stage is to be completed. Usually, space is left to add notes and comments on the plan as the project progresses. Other examples of project planning can be seen in the NEA section of this book.

The following points should be considered to ensure good project management:

- Select your project carefully. Your teacher is the best person to advise you on the choice of project.
- Start your project as soon as possible. It will usually take longer to complete than you think.
- Plan and manage your time carefully. Set aside time each week to work on your project, to ensure you make steady progress.
- Keep your drawings and notes organised in a folder.
- If you complete project work on the computer, save each version of your designs so that you can show how the drawings develop. Back up your files.
- Carefully photograph any modelling and making of prototypes, and present these clearly in your folder.
- Keep a log or diary so that you can explain how your prototype was developed using notes, sketches and photos.

Planning in commercial manufacture

In commercial manufacturing, some stages of production can be carried out concurrently. An example of this is in the car industry where different sub-assemblies are prepared before being attached to the car in the correct order.

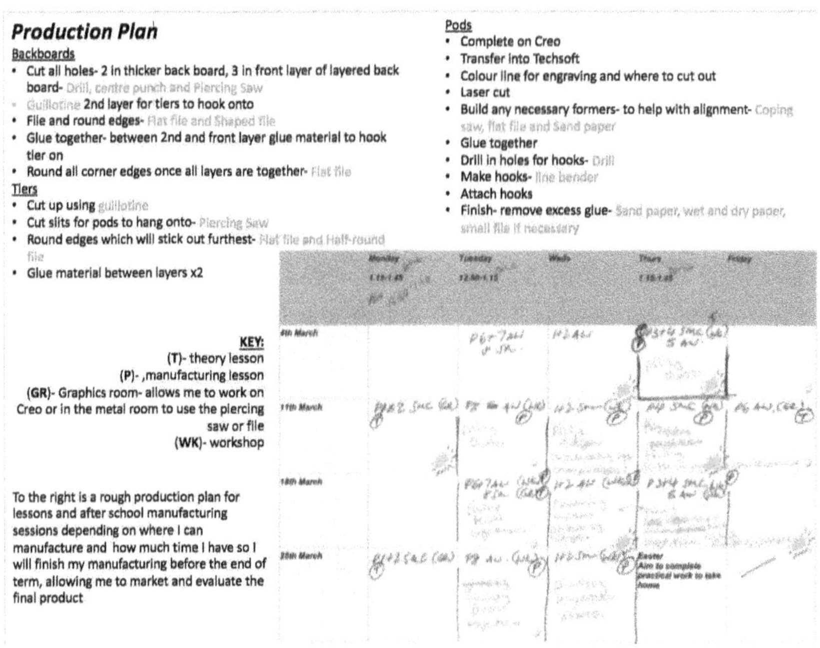

Figure 2.4.12 A production plan.

Production managers for a planning assembly need to consider the timings of operations and the sequence of activities in detail. There are several graphical methods that can be used to plan production. For example, CPA (critical path analysis) diagrams are used to show the key stages and critical points of manufacturing. They can show concurrent manufacturing stages and how they converge into the final assembly. See Chapter 2.5 for more on critical analysis.

Flow process charts are used to plan out production operations, their type and the timings. They are also used to reduce idle time by planning the sequence of operations for maximum efficiency.

Evaluation and testing

Evaluation and testing should be seen as an ongoing activity all the way through the design process. For example:

- After completing some investigation work, such as disassembly and analysis of products, the designer would evaluate their findings and summarise the key points they need to consider in their own designs.
- After completing some initial design drawings, the designer would evaluate them with a client and/or potential users of the product.
- When a design proposal is produced, good quality 3D CAD drawings might be shown to a client and/or potential users in order to gain feedback. Their evaluative comments would be used in a further iteration of the design, or in the development of the prototype.
- When models and prototypes are produced, they can be tested with potential users and the feedback used to further refine the prototype.
- Before making a prototype, it is good practice to experiment with materials and manufacturing methods, testing construction methods and evaluating their suitability.

Finally, when a prototype or product is completed, it could be evaluated as follows:

- A detailed comparison of the product against the specification. The designer would make comments based upon how the product performs against each criterion of the specification.
- Testing the prototype in its intended environment. Here the product or prototype would be tested against key criteria such as function, ergonomics, aesthetics, size and so on.
- Obtaining third party feedback from potential users of the product.
- Obtaining 'expert' opinion from specialist manufacturers or retailers of similar products.
- Using all of the findings from testing and evaluation, suggest how the prototype would be developed further.

ACTIVITY

Take an existing product, such as a school chair, and use the 4 x 4 method to develop it.

Sketch the product in the centre of an A3 sheet. Then draw four boxes – one in each corner of the sheet.

Pass the sheet around four different people and spend about five minutes each to develop the product.

Design processes used in the NEA **A level only**

For more details on design processes used in the NEA, see Section 3 (page 311).

Prototype development **A level only**

For more about the development of prototypes, see Chapter 2.1 and Section 3.

The iterative process in industrial and commercial contexts

A level only

Collaborative working

Designing and making products often requires the expertise, skills and knowledge of a number of people. In industry, design teams consist of people with different areas of expertise and specialist knowledge. For example, one designer may have particular strengths in product styling or aesthetics, while another may have more expertise in engineering components and be able to focus on the functional aspects. Ergonomists may be involved in the development of the product to ensure that it is usable by the target market. Production designers will also be closely involved because they develop the tooling required to manufacture the product. Design teams work alongside product marketing teams who are responsible for promoting the product, and have a clear idea of the needs and wants of potential users of the product.

When producing your own project, you should try to make use of a range of people who might be able to give you specialist advice and practical help. For example, you might be able to contact people who make or sell similar items to the product you are developing. They may offer to give you some guidance or even be involved in evaluating your designs and prototype. You can also use your friends to bounce ideas off and to obtain feedback about your designs.

The cyclic nature of commercial design and manufacture

In commercial design and manufacture, designers constantly evaluate their designs with clients and potential users. Using the feedback from clients and people from the target market, designers create new iterations of their designs and again obtain feedback. Similarly, when prototypes are used, designers can test their designs with potential users and obtain feedback. This might be done with a focus group or selected individuals. In addition to this, prototypes are shown to production engineers, to gain feedback about any design changes that would be required for the prototype to be put into production. Finally, when the product goes to market and consumers start to use the product, designers will already be working on new 'improved' versions of the product, using feedback from customers and data from sales.

Designers working in large businesses and industry are usually involved in a cyclic design process as opposed to a linear model. Working alongside sales and marketing teams, designers will be aware of market trends and will know when their product or product range needs to be refreshed to improve sales.

ACTIVITY

Make a simple model of one of your ideas. Gather a group of friends and ask their opinion about the design. Write down their feedback and two action points for you to further develop the design. Remember to take photos and record the feedback.

KEY TERMS

Design process: a system to organise designing and making a prototype of a product.

Model: either a 3D CAD drawing or a physical mock-up used to communicate designs to others, or to improve particular aspects of a design.

Specification: a list of key points that a designer follows when developing designs and making prototypes.

Primary research: investigation techniques that use first-hand sources such as interviews, observation, disassembly of products.

SCAMPER: a technique of modifying existing designs to create new ideas.

KEY POINTS

- The design process should not be seen as a linear model that starts with a design problem and ends with a product.
- The design process should be seen as iterative or cyclic – once a prototype is developed, it should be tested and evaluated, and the findings should become the starting point of a new design process.
- Planning and preparation are key to the success of a design project.
- Working collaboratively can help you to develop more innovative products, and you can tackle more challenging designs by using the expertise of others.
- When designing, try to avoid being fixated on one idea. Use techniques such as SCAMPER and thinking hats to generate a range of designs.

Check your knowledge and understanding

1 Explain why designers in industry often work collaboratively.

2 Describe the advantages of using techniques such as SCAMPER when developing design proposals.

3 Describe why designers might use models in the development of products.

4 Explain how designers might create a specification and what the purpose of one is.

5 With reference to one of your own projects, explain how a designer might test and evaluate a prototype.

Further reading

This website has a brief summary of the design process, and a downloadable worksheet to fill in to help define the steps needed to tackle a project: www.discoverdesign.org/handbook.

A source of inspiration and ideas, combining cutting edge with low-impact living through the use of responsible, imaginative design: *The Eco-Design Handbook* by Alastair Fuad-Luke.

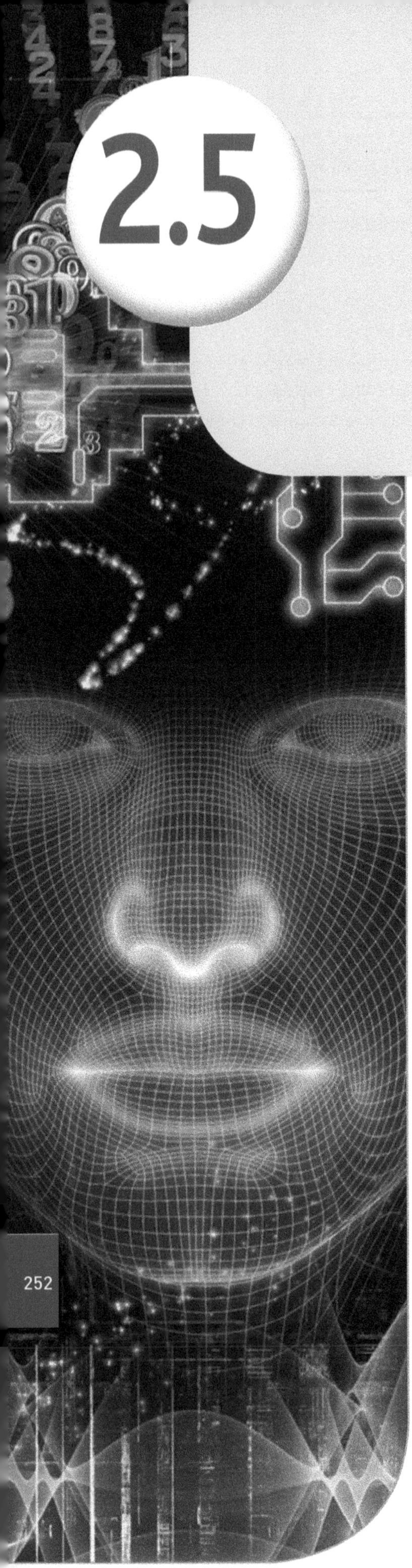

2.5 Critical analysis and evaluation

LEARNING OUTCOMES

By the end of this section you should have developed a knowledge and understanding of:

- how to discuss your own and commercial products, leading to possible improvements/modifications of the original idea
- how products are required to undergo rigorous testing – and the testing methods used – before they become commercially available for sale
- how the use of third-party feedback and testing informs the evaluation process, including:
  - informing future modification and development
  - the importance of ensuring the views of other interested parties are obtained in order to have objective and unbiased feedback.

Success in the design of products is dependent on them fulfilling the **specification** set out in the initial stages of the design process. You can enhance your design skills by identifying the strengths and weaknesses of a range of commercial products, as well as your own designs and outcomes.

It is particularly important for consumer products to be critically analysed and evaluated by their designers and manufacturers, to ensure that they are safe, attractive and comply with the requirements of user-centred design (UCD). Testing, such as the use of wind tunnels and anechoic (echo free) chambers for vehicles, as well as impartial **evaluation**, promotes high-quality design and can help to bring about improvements in future generations of products, as well as enhance their potential for commercial success.

Figure 2.5.1 An engineer testing a vehicle for noise in an anechoic chamber.

Figure 2.5.2 Vehicle aerodynamics being tested in a wind tunnel.

Critical analysis and evaluation

Critical analysis is an important element of the design process; it ensures that an in-depth, objective study is made of elements of a design, often linked to related research. Evaluation of design ideas and finished products is similar to critical analysis and there is some overlap, but evaluation is more concerned with testing – in various ways – how effectively ideas and products satisfy the requirements of the specification.

Your own products

During the design process, an investigation is carried out, using techniques such as those outlined in Chapter 2.1, in which a wide variety of factors relevant to the project are researched and the relevance of this research to the original design brief is analysed. This investigative work plays an important role in generating the product specification, which is a set of key points that must be addressed in the final design.

Examples of product specification points include:

- ease of storage
- spending limit for materials
- appropriate strength to withstand loading, with a margin for safety
- aesthetic considerations to match intended location
- manufacturing methods compatible with workshop facilities
- dimensions appropriate for anthropometric data of intended users.

All of these points can be checked as part of ongoing analysis and evaluation of the designs that are generated. If this is done methodically and thoroughly, it will reduce the possibility of deficiencies being discovered when final evaluation of the product is carried out. It is very challenging to produce a design that is fault free when the final evaluation takes place, particularly when the views of others are sought to provide objective criticism.

During development of your ideas, models and prototype(s) for the NEA, you should make constant reference to the specification and, where possible, you should obtain the views of others. In the final evaluation, well-reasoned critical analysis of the product, which highlights aspects where the specification has not been fully met, provides valuable opportunities to demonstrate advanced design skills. The final evaluation section of a portfolio should, therefore, include a fully justified range of detailed, annotated design ideas for modifications that address key issues that have been highlighted for attention.

Example: modifications to address an issue

Here are some modifications that might be used to address an issue involving 'ease of storage' – a key point in the specification.

The outcome of critical analysis and evaluation is that the client has discovered the child's wooden helicopter toy which has been made does not fit easily in the toy storage box, and has limited play value.

Figure 2.5.3 Toy helicopter too large for storage box.

The possible modifications that could be made include:

- Suggest more appropriate, specific dimensions to ensure it fits in the toy box.
- Incorporate a simple folding mechanism.
- Modify the design so that it is in several parts that need to be assembled by the child. This will facilitate easier storage, while adding educational play value and the possibility of creating additional models.

Figure 2.5.4 Child assembling toy helicopter redesigned as a construction kit.

The third potential modification identified here provides the opportunity to enhance the product design in a way that shows greater understanding of the educational value of the toy. There is also the additional challenge of designing it in kit form, along with the associated understanding of additional manufacturing processes that are required. This example also highlights the importance of ongoing evaluation of designs to try to prevent issues like this arising.

Commercial products

The study of commercial products is a primary investigation and evaluation technique that we briefly looked at in Chapter 2.1, which facilitates learning from the work of professional designers and manufacturers. This type of investigation would typically take the form of:

- considering the target market and purpose of the product, to gain an awareness of the specification likely to have been used when developing the design
- an analysis of the product based on how well it satisfies the points in this specification
- disassembly of the product (where safe and practicable) to study the manufacturing techniques involved. (Particular care and appropriate supervision will be required with some products.)

Example of the potential of studying commercial products

Figure 2.5.5 Disassembly of an electric drill.

The electric drill in Figure 2.5.5 could be studied in the following ways:

- Anthropometric data for users' hands could be gathered and compared to the dimensions of the drill handle.
- Safety features such as double insulation, cable grip and entry sleeve could be studied.
- Maintenance considerations could be observed, such as the replaceable motor brushes.
- The manufacturer's use of integral injection moulded features – reducing the need for additional parts and manufacturing processes – could be identified.
- The type and properties of the materials that have been used could be established and subjected to appropriate testing, if practicable, to establish their suitability.

Annotated sketches of suggestions for modifications or alternatives to elements of the existing design provide a useful way of improving design skills and understanding. Comparison of similar products from economy and luxury ranges provides enhanced opportunities for analysis and evaluation. This makes it possible to study how designers and manufacturers consider budgetary constraints, and the requirements of different consumers.

ACTIVITY

Select a product that you have made in school or college, such as a clock, toy, storage box or lamp. Now choose a commercially available product of a similar type and carry out an evaluative comparison of the two. Establish appropriate specifications and produce analytical, annotated sketches to compare and contrast the features of the two products. The student project pencil case shown in Figure 2.5.6 as well as the commercial version for pencils and other artists' materials shown in Figure 2.5.7 provide an example of the kind of comparison that might be made.

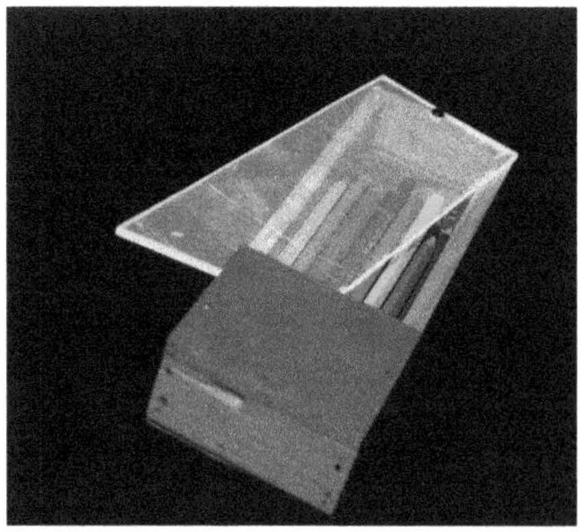

Figure 2.5.6 Student-made pencil case.

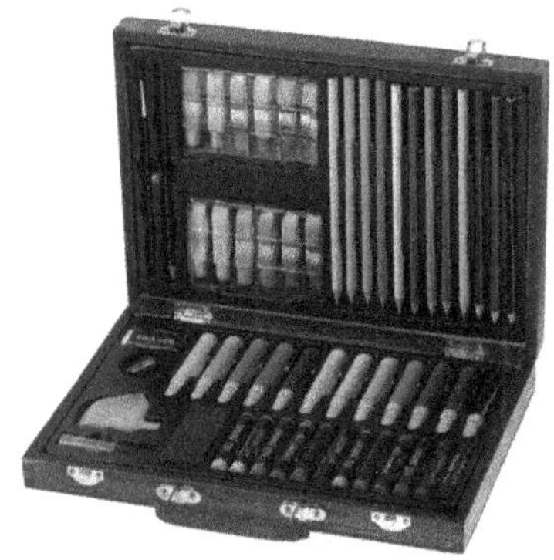

Figure 2.5.7 Commercial case for pencils and other artists' materials.

Testing and evaluating in industrial or commercial contexts

The design process for commercial products is generally an iterative one, as explained in Chapter 2.1. This means that information gained during the evaluation of the performance of products can be incorporated in improvements made to future versions of those products. It is still, however, very important that careful testing and evaluation of products is carried out before they are made commercially available.

Figure 2.5.8 The vice president of Boeing with a model of Boeing's 787 battery design improvements.

This is particularly relevant to flaws in designs that affect product safety, as these can result in harmful situations occurring and the possibility of **product recalls** that are sometimes very expensive. The cost of rectifying faults at the design stage is small compared to implementing changes during manufacture, and pales into insignificance compared to the large sums that can be involved when products have to be recalled for repair or to be scrapped for safety reasons. One of the most expensive recalls was due to the discovery of problems with lithium-ion batteries catching fire in Boeing's 787 Dreamliner aircraft, which led to the entire fleet being grounded. Manufacturing and testing procedure errors were blamed and the cost to Boeing was several hundred million dollars.

Many manufacturers have their own facilities for carrying out sophisticated tests on their products, such as the anechoic chamber and wind tunnel tests for cars shown in Figures 2.5.1 and 2.5.2.

Chapter 2.9 explains a range of quality assurance methods employed by designers and manufacturers to avoid defective products being sold. Some products are also subject to scrutiny by independent, third party organisations in order to further reduce the likelihood of problems, to comply with legislation and receive appropriate certification.

Use of third party feedback in the testing and evaluation process

Third party feedback consists of **objective views** about products from people and organisations that are not involved in their design or manufacture.

A number of international organisations exist to monitor the standards of products and services. In the United Kingdom, the United Kingdom Accreditation Service (UKAS) is responsible for checking and monitoring

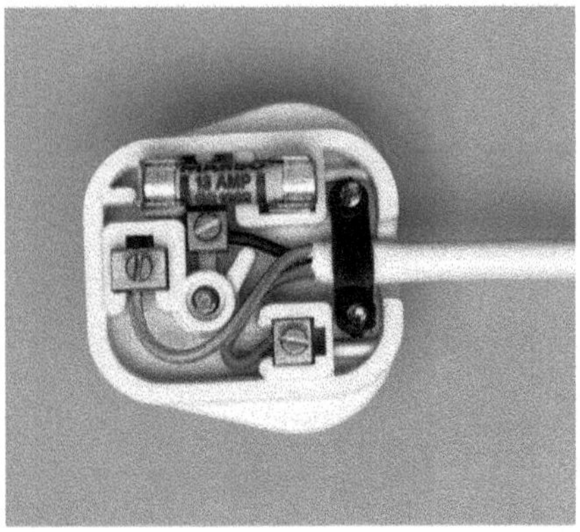

Figure 2.5.9 BS 1363 certified 13-amp plug fitted with a BS 1362 certified fuse.

Figure 2.5.10 Checking the chemical content of materials and finishes used in the manufacture of children's toys.

TM

Figure 2.5.11 BSI kitemark label.

the work of the many agencies that certificate testing and inspection of products, and their manufacture, along with many other sectors. The BSI is the UK's most significant UKAS **accredited** organisation, since it publishes a range of **standards** that are used in the design and manufacture of products. Here are some examples of important standards available from BSI that are used to ensure the quality and safety of products:

Table 2.5.1 Important standards available from BSI that are used to ensure the quality and safety of products.

Standard	Products/Processes
BS 8887	Product manufacture, assembly, disassembly and end-of-life processing
BS EN ISO 6506	Brinell hardness testing of metals
BS 1363	13-amp plugs, sockets and related equipment
BS 1362	Cartridge fuses for domestic plugs
ASTM F3078	Lead content in paint used for toys and other products
BS EN 62031	LED lighting safety specifications

Each of these standards specifies exactly how manufacture and/or testing must be carried out. To comply with ASTM F3078, for example, samples of paint are scraped off products and subjected to high energy x-rays, which facilitate the accurate analysis of all the elements present in the sample. In this case, it ensures that the lead content used in paint for children's toys is not too high, since lead is poisonous and a particular danger to children.

A number of organisations are accredited by UKAS to enable them to carry out the testing necessary to award certification for compliance within standards. The award of appropriate certification enables manufacturers to display the **CE mark** or **BSI kitemark** (the latter is shown in Figure 2.5.11) on their products to let consumers know that they have received certification.

So far we have dealt mainly with aspects of product testing and evaluation that are related to safety, but designers need to be aware of a much wider range of issues if they are to ensure the success of their products.

The UCD (User Centred Design) approach, as explained in Chapter 2.1, is employed to ensure that products are appealing for consumers to buy and use. An important part of this approach is to obtain information from **market research** organisations. Market research organisations are used because they are totally independent of the client companies that use their services. They obtain information relevant to the proposed new product from a number of sources: targeted questionnaires and independent research into a variety of factors such as competitors' products, the current state of the market, retailing constraints and brand identity. One of their key services is to host **focus groups** in which members of the public, from appropriate **demographic groups**, are invited and paid to attend a meeting where they can express their views on existing and proposed products. Members of focus groups may participate in a variety of activities, which will be recorded in some way. They could even be observed through a one-way mirror as shown in Figure 2.5.12.

Typical focus group activities include:

- answering direct questions
- physical interaction with products

Figure 2.5.12 A focus group being observed by clients through a one-way mirror.

- watching videos and presentations
- sketching ideas and logos
- making suggestions for product improvements and desirable features
- role playing
- creating mood boards
- group discussions.

Examples of what might be learned from a focus group include:

- their attitude towards colours for product finishes
- the level of comfort experienced while holding products
- how easily they can change the batteries in electronic products
- how easy they find it to navigate control panels or screens
- their level of enjoyment when playing a computer game
- what they feel is an appropriate pricing level for a particular product
- the strength of brand identity that they recognise.

ACTIVITY

Visit the Chartered Trading Standards Institute's website and find the latest products to be recalled (www.tradingstandards.uk /consumers/product-recalls). Analyse the reasons for the recall of one of the non-food products, and produce a plan of action that might have helped to avoid it.

Ask a member of your Product Design class to invite a few friends along to evaluate one or more of your designs. Formulate the questions you would like to be asked, and seek their permission to make an audio recording. When you have received the feedback from the meeting, generate annotated sketches in response to their views and suggestions.

KEY TERMS

Standards: legislative requirements for a range of processes and products.

Accredited: having the qualifications necessary to carry out a function such as awarding certification for conformity.

BSI kitemark and CE mark: symbols used on products that conform to appropriate standards.

Market research: independently executed inquiries relating to products and their use by consumers.

Focus groups: market research activity utilising the views of a particular demographic group.

Demographic group: a specific category of the population, such as single men between the age of 21 and 40 who live with their parents.

Conformity: reaching appropriate standards to obtain certification.

Specification: a set of key points summarising the main requirements to be incorporated in a design.

Evaluation: report on the effectiveness of a design – or elements of it – in meeting its specification.

Critical analysis: in depth, objective study of a design, or elements of it.

Product recall: taking products off the market and asking consumers to return them, due to faults that must be rectified, usually for safety reasons.

Objective (third party) views: opinions that are independent of any influence from interested parties.

KEY POINTS

- Critical analysis is an in-depth, objective study of a design, often linked to research.
- Evaluation of design ideas and finished products is similar to critical analysis, but is more concerned with testing how effectively the requirements of the specification have been met.
- During the design process, analysis of investigative research will be a major factor in the production of a specification containing key points to be checked in the evaluation of ideas and particularly the final design.
- Identification of flaws in designs when conducting evaluations of your own work can result in valuable opportunities to display advanced design skills through additional development of ideas for design modifications.
- The study of commercial products, particularly through activities such as disassembly, can be very informative in understanding a wide range of design and manufacturing techniques.
- Thorough testing and evaluation of commercial products is particularly important in order to maintain their success in the market place, and to avoid the large-scale difficulties resulting from a product recall.
- A range of organisations and companies exist to implement nationally and internationally recognised schemes, to ensure **conformity** with agreed standards in the design and manufacture of products.
- Compliance with these standards facilitates the use of appropriate symbols such as the BSI kitemark and the CE mark.
- Market research, particularly involving focus groups, is a key element in ensuring that designs are as user-centred as possible.

Check your knowledge and understanding

1. Why is the establishment of a clear specification a crucial factor in carrying out design evaluations?
2. How can deficiencies identified during the project evaluation process contribute towards improving the standard of your work?
3. What is a product disassembly and why is it a useful exercise?
4. Explain the potentially serious consequences of failing to identify and correct flaws in a design at an early stage.
5. Explain how various organisations contribute to formal systems for setting and maintaining appropriate design standards.
6. Explain why focus groups are a particularly important element of market research for companies introducing new products.

Further reading

Find out more about product recalls and what else the Chartered Trading Standards Institute is responsible for: www.tradingstandards.uk/consumers.

For further information on the United Kingdom Accreditation Service: www.ukas.com.

Find out more about what the British Standards Institution does: www.bsigroup.com.

A useful selection of the British Standards, showing how they are applied to a range of products: Design and Technology: Compendium of British Standards for teachers of design and technology in schools and colleges (BSI).

This book provides you with a comprehensive, relevant and visually rich insight into the world of research methods specifically aimed at product designers: *Research Methods for Product Design (Portfolio Skills)* by Alex Milton.

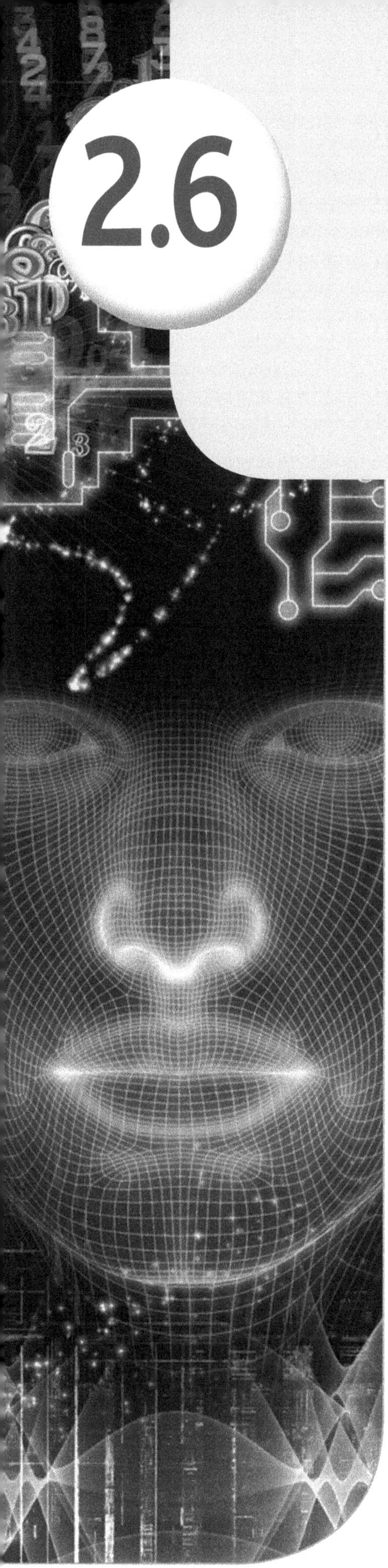

2.6 Selecting appropriate tools, equipment and processes

LEARNING OUTCOMES

By the end of this section you should have developed a knowledge and understanding of:

- how to select the correct tools and equipment for specific tasks
- safe working practices for completing your own project work
- how to select the most appropriate manufacturing processes to realise design proposals.

If you are studying at A-level you should also have developed a knowledge and understanding of:

- how designs are developed from a single prototype into mass produced products
- the effect on the manufacturing process that is brought about by the need for batch and mass manufacture
- the importance of health and safety in a commercial setting, including workforce training and national safety standards.

Successful manufacture is influenced by many factors such as the material and material cost, product function, scale of production and proposed retail cost. Designers need to have a sound knowledge of materials and how they are processed, so that final manufacturing decisions can be made in conjunction with the manufacturing team. The main aim is to select a manufacturing method that is both suitable for the material and product, as well as economic in terms of materials and scale of manufacture.

For example, yoghurt pots can be made in high volume via the thermoforming or vacuum forming process. Having multiple moulds on one sheet helps to keep material wastage to a minimum, while providing the required draft angles for product function and ease of mould removal. Vacuum forming is less suited to a one-off or prototype manufacture due the time taken to produce the mould, as well as the amount of material wasted when the mould is removed from the sheet.

Selecting the correct tools and equipment

Selecting the correct tooling involves knowledge of manufacturing processes as well as the materials used for each manufacturing process. For example, if the manufacturer wants to produce a casing for a games console controller made from a polymer, then polymer processes would be the only processes considered. The manufacturer could consider blow moulding to create a hollow shape for the controller circuit to fit into, but this would not be appropriate because blow moulding doesn't produce the high level of detail that is necessary and would limit the product's ability to be taken apart for fitting and maintaining circuitry. Vacuum forming may also be considered in order to make two halves of the product for the circuitry to fit inside. However, vacuum forming would involve excess waste as well as a uniform wall thickness, which would not protect the controller in the event of a fall from a table. Vacuum forming two halves could be used if another adhesive was used to join the two sections together, but this would not be accurate and repeatable every time due to the thin wall thickness of the polymer.

For the games controller casing, injection moulding would be the suitable process due to the ability to create different wall thicknesses, as well as creating click fittings and circuit board holders within the mould. Injection moulding uses the same mould repeatedly, so each product would be identical in quality and accuracy. If the incorrect process was chosen, for example soft soldering aluminium components, the process would be deemed inappropriate as the oxidation that occurs when aluminium is heated makes it very difficult to achieve a satisfactory soft soldered joint.

Safe working practices in a workshop situation

Safe working practices serve to keep the person doing the work – as well as those in the vicinity – safe and free from harm. These safe working practices may include a set of guidelines for the person(s) in the workshop to follow (see Chapter 1.9 for more details).

Risk assessments are a good way to initially consider what may cause harm to people, and the ways in which people can be protected as much as is practicable from hazard or harm. Remind yourself about risk assessments by looking back at Chapter 1.9.

Maintaining safety in commercial manufacture

A level only

Commercial manufacture typically involves more staff than bespoke manufacturing being undertaken by a designer maker, and employers must maintain safety standards across a wide variety of potential situations with a varying number of people with different experience levels. The employer should comply with the HSWA (Health and Safety at Work Act), COSHH (Control of Substances Hazardous to Health) and the Personal Protective Clothing Regulations 2002. Risk assessments for all processes undertaken should also be carried out, and regularly reviewed and updated to protect those involved both directly and indirectly in the manufacturing process.

The safe and appropriate selection of tooling is critical to ensure safe manufacture and safety for employees as well as suitability for the chosen material. When selecting the right process or tools for the job, the following checklist may be considered for safe manufacture:

- The duration of the job: will workers be carrying out the same repeated task over a short or long timeframe, and could this lead to RSI (repetitive strain injury) or safety implications from boredom?

- What hazards could be controlled before manufacture starts?
- What hazards has the process introduced to the workplace for both the manufacturer and others?
- Machinery and equipment: is there a safe place to load and unload materials, machinery and equipment? Are the safe zones clearly marked?
- Is there a manual handling issue and has training been provided?
- Is machinery and tooling properly maintained in accordance with the manufacturer's specifications?
- Are the appropriate safety guards fitted and in good working order?
- Electrical safety: are power leads tested and tagged? Are RCDs (residual current devices) used with all electrical power tools? Are power leads a trip hazard?
- PPE (personal protective equipment): is all the necessary PPE supplied to protect workers and others in the vicinity? Are there systems in place for maintenance and replacement of PPE as required? Is the necessary signage displayed within the workplace for advisory notices and to warn of hazards?
- Emergency equipment: are fire extinguishers, first aid kits and eye baths available? Has training been provided for emergency procedures such as building evacuation in the event of fire?

A checklist such as this, used in conjunction with HSWA, COSHH and the Personal Protective Clothing Regulations 2002 will help protect employees and aid good safety practices. Further to this, employers should provide training on the safe use of machinery, the correct procedures in the event of chemical spillages and for waste disposal; ensure machines are regularly maintained in accordance with manufacturers guidelines; and provide appropriate extraction systems, PPE and clear guidelines and procedures for accident reporting and emergency procedures. These measures combined should ensure employees are safe in the workplace, are clear how to carry out tasks safely and can safely conduct themselves in the event of an emergency without putting themselves or others at risk.

Development of designs from single prototypes to mass produced products

A level only

In commercial manufacture, prototype products are often the starting point for future productions of batch or mass-produced products. A prototype is the first generation of a product, which is used to fault find and evaluate prior to future production runs. The car industry makes use of prototypes to test performance and aesthetics, and to gather market feedback prior to manufacturing a car for retail purposes.

Prototypes usually take three main forms prior to production: visual, proof of concept and production.

- Visual prototype: this is a sample or model that shows the overall shape and size of the product, but does not usually have any working parts. The materials will not be the materials that would be used if the product were to be mass produced. Often the visual prototype is made from traditional modelling materials such as Styrofoam, or it could be 3D printed. Visual prototypes provide an opportunity to test the visual impact that a product form may have, and enable designers and clients to suggest improvements.

- Proof of concept prototype: this shows the key functionality and main technical aspects of the design. It is not intended to look like the final product, and will be a functional model, which may make use of existing 'off the shelf' components. A proof of concept prototype would not usually be made from production grade materials, but allows the design team to test the functions to see if they work as intended. There may be many proof of concept prototypes made and tested prior to a viable solution being found. Proof of concept prototypes can also be given out to trial or test groups to gain feedback of real world use rather than lab-based experimentation.
- Production prototype: this gives a representation of how the mass-produced product would look and shows how the product would function. 'Off the shelf' parts may still be used at this stage, but it is more likely that it will be built from bespoke parts. The materials used will be as close to production grade as possible, unless it is not economically viable, in which case a substitute material may be used. Manufacturers will make use of a presentation prototype to assess any alterations that may be required prior to mass production. These could be materials, positioning of components for a more economic manufacture layout as well as the most suitable manufacturing processes. Production prototypes will also generally be marketed in small user trials prior to a full production run. In some instances, a company may produce an initial batch of products for testing, trialling and further refining prior to the final product being put into full production.

Figure 2.6.1 James Dyson makes use of many prototypes before perfecting his final products.

James Dyson, who invented the first bagless vacuum cleaner, made use of a huge number of prototypes prior to the first Dyson vacuum cleaners being launched in the retail sector. Over a five-year period, he made 5,127 prototypes before he was satisfied he had a successful and optimally functioning product.

Batch or mass manufacture and the effect on the manufacturing process

A level only

The number of components or the complexity of the product and the volume of production will determine the type of machinery used to create a product. Before considering the machinery types, the designer must be familiar with the number of products expected from a manufacture process. If a one-off dining table were being made in a workshop, standard tooling such as bandsaws, pillar drills and mortice machines may be utilised. If the dining table were required to be manufactured in a batch of 50, the manufacturer would start to utilise jigs to ensure ease of repeatability and speed of production. Jigs could be used for many processes such as making a mortice and tenon joint for the legs, and the use of a gluing jig could ensure that all the frame joints were assembled quickly and 'square'.

There are three key terms for volume of production: one-off, batch and mass. You will have come across these in Chapter 1.6.

One-off production

A '**one-off**' is a product that has been designed and manufactured for a single, specific situation. Most one-off products are hand manufactured and are described as made to measure or bespoke.

Examples of one-off products include a shelving unit made to fit a specific gap, a handmade fitted kitchen, a racing bicycle for an elite athlete or a commissioned sculpture. For one-off manufacture, machinery is usually set up for the task required as the machine is required, and then adjusted for the next task. A circular saw may be used for wooden table legs; a strip is cut from the plank to the correct width, the circular saw would then be adjusted to cut the strip of wood to the correct length required for the legs. One-off production is labour intensive and is primarily the domain of skilled craftsmen and manufacturers.

Batch production

Batch production is where a set number of products are made at one time. A useful analogy is in food production, where a baker will make a batch of bakery products at one time and then later in the day make another batch of the same product. In the furniture industry, batches of tables, chairs and stools would be made, for example a manufacturer may make a batch of 20 dining table and chair sets. Once the manufacture is complete, another batch of products is made.

To ensure consistency throughout the batch, manufacturers use control devices such as jigs, **templates** and CNC machinery. In batch production, machines are often set up to do specific tasks. For example, a pillar drill could be set up to one depth stop, resulting in all holes drilled for a specific batch of parts being the same depth. You may also find in batch production that different workers carry out different tasks on a job rotation between machinery.

Mass production

Mass production is where large quantities of the same standard product or component are made. Mass production is used for products that we use every day, and for which there is therefore a high demand, such as light bulbs, polymer sandwich containers and bottles.

Mass production utilises the automation of manufacturing processes wherever possible, and may make extensive use of dedicated machinery as well as CNC machinery such as laser cutters to manufacture the same product many times over. Products such as blow moulded polymer milk cartons, screws, nuts and bolts are all made in mass production. Due to the reliance on automation, mass production processes do not require highly skilled workers for product manufacture.

Table 2.6.1 shows how a manufacturer may alter a manufacturing process based on the number of products required – in this instance moving from prototype to batch production.

Table 2.6.1 How a manufacturer may alter a manufacturing process based on the number of products required.

Process	Prototype	Batch of 50 products
Drilling holes in a wooden product for a dowel joint	Mark out each individual hole. Set up a pillar drill to the correct depth and drill all holes.	Use a drilling jig set up on a dedicated pillar drill within the workshop. The component is placed in the jig and the drill is set to a predetermined depth. Each component is fed into the jig, until all required components for the batch have been drilled.

ACTIVITY

Make a table or chart to show how the manufacturing method for your coursework product would change depending on the volume of production.

KEY TERMS

Safe working practices: methods and guidelines that an employer will use to ensure the safety of employees, visitors and others within the workplace.

Risk assessment: a document taking consideration of what might cause harm to people and if reasonable steps are being taken to prevent that harm.

One-off: a product that has been designed and manufactured for a single, specific situation.

Batch production: a set number of products are made at one time. To ensure consistency throughout the batch, manufacturers use control devices such as jigs and templates, as well as dedicated and CNC machinery.

Mass production: large quantities of the same standard product or component are made.

KEY POINTS

- Successful manufacture is influenced by factors such as the material, material cost, product function, scale of production and proposed retail cost. The main aim is to select a manufacturing method that is both suitable for the material and product, as well as economic in terms of materials and scale of manufacture.
- Ensuring the safety of employees, visitors and others within the workplace is key in terms of good practice and legislation. Many employers make use of safety and process checklists to consider aspects such as the duration of the job, machinery and equipment, electrical safety, provision of PPE and emergency equipment and safe usage.
- The number of components, product complexity and the volume of production will determine the type of machinery used for manufacture.

Check your knowledge and understanding

1. With reference to a specific, named product of your choice, describe the safety requirements that should be considered during manufacture to ensure the safety of workers.
2. Create a checklist that a manufacturer may use when assessing the potential risks within the manufacturing facility.
3. List three factors that a designer would have to consider when selecting tooling for the manufacture of a new product.
4. Name the three types of prototype used and explain the purpose of each prototype.

Further reading

The James Dyson Foundation provides frequently updated information regarding product development and the use of prototypes at Dyson: www.jamesdysonfoundation.co.uk.

In 1913, Henry Ford designed a moving assembly line to manufacture his Ford Model T. This website gives some history and further reading about Ford and the mass production concepts: www.ford.co.uk/experience-ford/Heritage/EvolutionOfMassProduction.

Process Selection: From Design to Manufacture by K. G. Swift is a practical guide to selecting manufacturing processes, and is suitable for students.

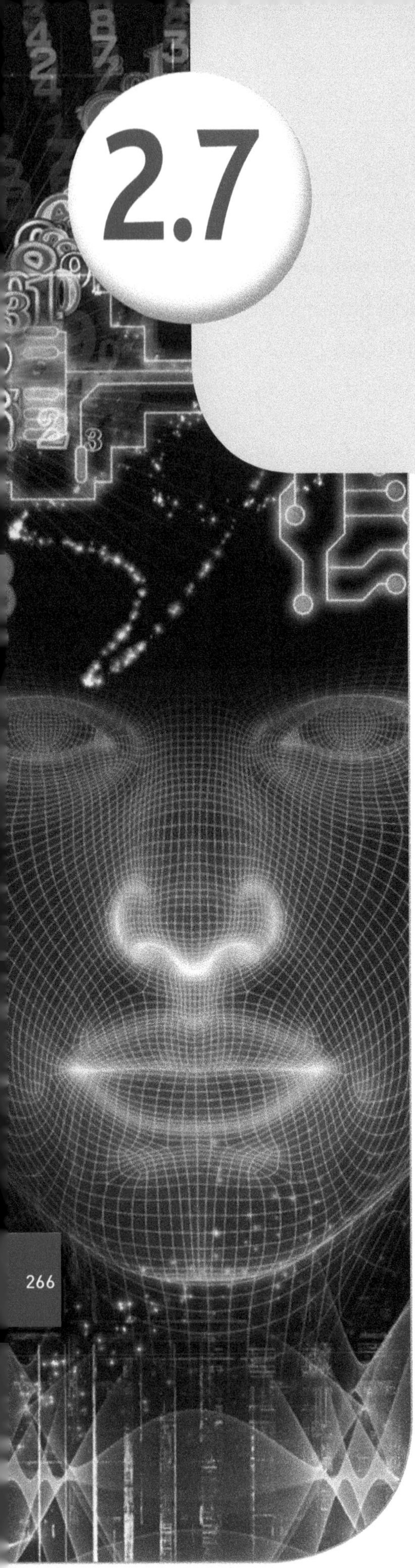

2.7 Accuracy in design and manufacture

LEARNING OUTCOMES

By the end of this section you should have developed a knowledge and understanding of:

- how to use a range of measuring and marking out equipment
- the importance of accuracy in manufacturing
- how testing can eliminate errors
- the value of using measuring aids to ensure consistency of accuracy and the reduction of human error.

Accuracy is essential in order to design, manufacture and test products within the **tolerance** specified so that they will function effectively. This can only be achieved by the correct use of appropriate measuring and marking out tools. Additional equipment is also used to reduce the possibility of measuring and marking out errors, particularly in large-scale production.

Using measuring and marking out equipment

Measurement is the process of using tools such as rules, tapes and gauges to check existing **dimensions**. **Marking out** is the transfer of designs onto pieces of material or parts to indicate where features such as edges, holes, slots, recesses and bends have to be made during manufacturing processes.

Tools commonly used for measuring and marking out

Marking metal usually requires the use of a high carbon steel scriber, which is harder than the metal it is marking, so it is able to scratch the surface. If the scratched lines are difficult to see, it is common for special ink such as engineer's blue to be applied to the surface being marked out, so that the lines are more visible. If circles or arcs are required, dividers are used, which have two points and function in the same way as a pair of compasses. Beam compasses, sometimes called trammels, can be used for larger radii. Various additional tools, such as external and internal calipers, are used for checking dimensions.

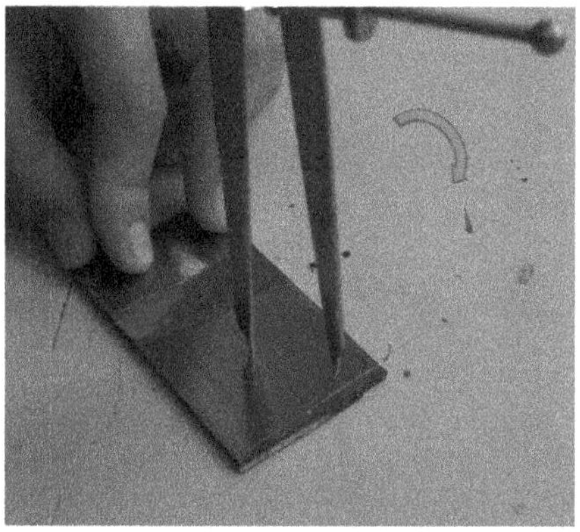
Figure 2.7.1 Marked out steel plate using engineer's blue to make scribed lines clear.

Figure 2.7.2 Beam compasses (trammels) can be used for larger radii.

Figure 2.7.3 Scriber being used on metal.

Timber is usually marked out with a marking knife or pencil, and there are carpenter's pencils specially designed for the purpose, with a thicker lead to avoid breakages. A marker pen is usually used on polymer sheet, although a scriber is sometimes used if the material is to be cut rather than bent.

Measurement for marking out is normally carried out with a steel rule, however distances over a metre usually require the use of a measuring tape. The metric international system of units (SI) is used in every major country in the world, with the USA being one of the last to adopt it for measurement. The metre (m) is the base SI unit, but millimetres (mm) are usually used in product design.

Laser devices such as cross line levels are becoming more commonly used for measuring and marking out when large and complicated fabrications are being undertaken, as this method projects a perfectly horizontal and vertical reference line onto objects of any shape.

Figure 2.7.4 Steel rule.

Figure 2.7.5 Cross line laser level in use.

The marking device that is being used usually needs to be guided in some way, since marking out with precision cannot be achieved freehand. A rule is often used, but a number of additional tools are available to provide improved precision.

Surface plates and gauges fitted with digital or dial readouts, angle plates and v-blocks are often used when accurate measurement and marking out are required in engineering applications. For this type of work, it is necessary to establish **datum** surfaces, so that measurements can be accurately referenced from them, in the same way that a face side and edge are marked on timber to ensure squareness.

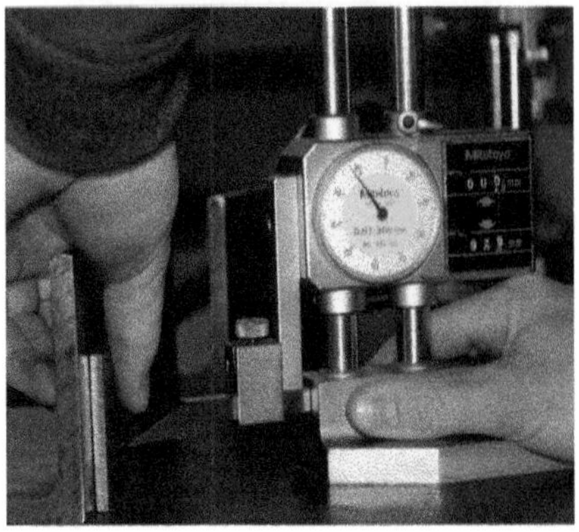

Figure 2.7.6 Surface plate, angle plate and gauge with dial readout being used to mark out a piece of steel.

Figure 2.7.7 Marking gauge being used on a piece of timber.

A marking gauge can be used to mark a line at a constant distance from an edge of a piece of timber.

It is commonly required to mark lines perpendicular (at a right angle) to an edge, for which there are many types of try square available, such as the one being used in Figure 2.7.8. The combination square shown in Figure 2.7.9 has the additional advantage of being able to mark out a mitre (45°), and it also has a spirit level to help with checking the horizontal and vertical accuracy.

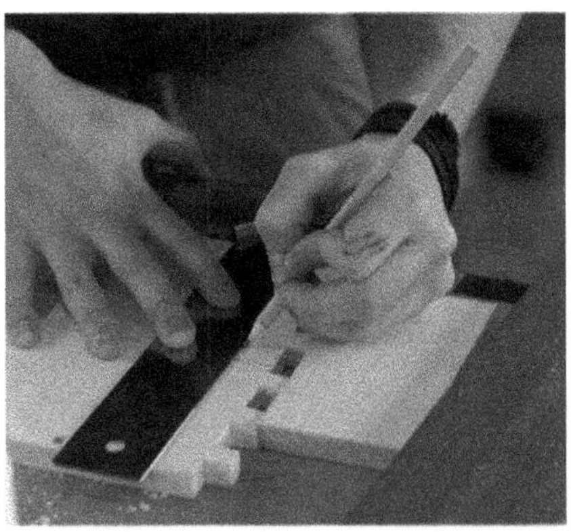

Figure 2.7.8 Try square being used to mark out joints for timber.

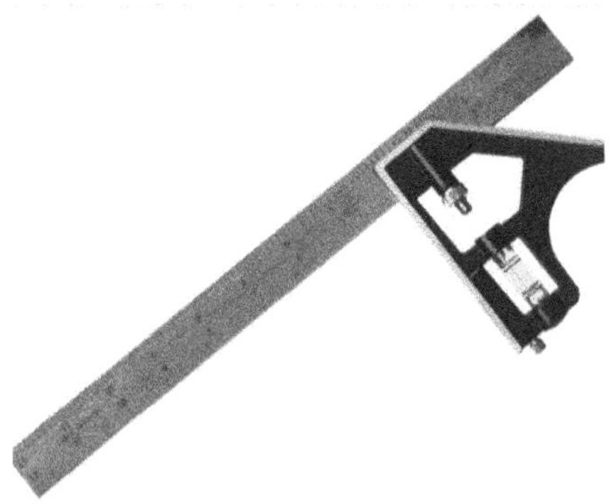

Figure 2.7.9 Combination square.

For angles other than a right angle or mitre, a protractor of some kind is normally used. These are available in many forms, from a simple protractor to the highly accurate vernier protractor shown in Figure 2.7.10.

When marking out holes for drilling, it is normal practice to use a centre punch mark to ensure that the drill centres itself correctly and does not 'wander'. Similar marks are often used to make lines stand out for processes involving heat, such as forging, since scribed lines can be difficult to see when metal has been heated.

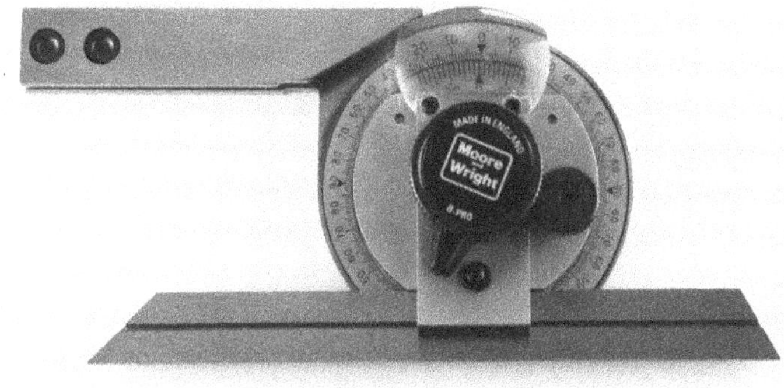

Figure 2.7.10 Vernier protractor.

Figure 2.7.11 Digital test gauge being used to measure a bearing.

Figure 2.7.12 Lathe slide handle and dial.

High precision work

Micrometers, vernier calipers and co-ordinate measuring machines (CMMs) are examples of tools used for precise measurement involving fine tolerances, and are covered in Chapter 2.9. Other equipment is also available for highly precise measurement, such as the profile projector (see page 272), and the digital test gauge. The digital test gauge is a single axis measuring device that is used extensively for precise measurement. It has largely taken over the role of the mechanical dial indicator because it has the advantage of an easy to read display, and the potential for providing direct input to computer systems.

At the manufacturing stage, the dials fitted to the handles of machines facilitate direct input of measurements for machining operations such as milling and turning. These graduated dials indicate the precise linear movement of the tool or workpiece for each revolution or fraction of a revolution of the handle, and can typically achieve an accuracy of 0.01 mm, which is the linear distance travelled by the slide for one graduation rotation of the dial. The dial can be set to zero at the start of a cut so that the distance moved can easily be gauged, although care needs to be taken when using a lathe cross slide to check whether the dial is calibrated for diameter or radius.

Impact of computer numerically controlled machining

Computer numerically controlled (CNC) machining processes do not require marking out since the G-codes generated by the CAD/CAM software control the co-ordinate of the path of the cutting tool in relation to the workpiece. It is, however, necessary to ensure that the optimum layout for all of the parts has been established to avoid wasting material. This applies to all situations where decisions must be made regarding the optimum size of components in relation to stock forms of material that are available. An example of this is given in Chapter 1.11.

The importance of accuracy

Accuracy is defined as the level of conformity of a measurement to the required value. It is an increasingly important feature of modern product design to ensure that parts fit and products function correctly, particularly in the light of more prevalent miniaturisation of many devices. For each part and product, the designer must consider what will be an acceptable range of accuracy. The measure of the level of accuracy appropriate for particular situations is known as the tolerance and can vary considerably depending on factors such as material, size, function and need for interchangeability. Here are some examples of situations where there is a need for accuracy to ensure that the manufactured outcome is fully functional:

- cutting the gears for a watch movement
- ensuring that threads on nuts and bolts fit correctly
- fitting a new glazing unit in a window frame
- positioning holes for KD fittings in self-assembly furniture
- selecting the correct washer to repair a dripping tap
- positioning holes in a wall for securing shelf brackets.

Figure 2.7.13 Gears in a watch movement.

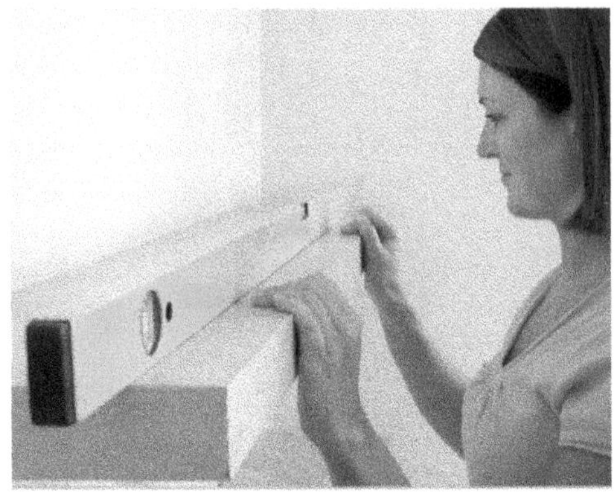

Figure 2.7.14 Using a spirit level to check that shelf brackets have been positioned accurately.

The tolerance, which is the specification for the allowable upper and lower accuracy limits of the dimension involved, is usually expressed as X ± Y, where X is the exact size required and Y is how much error can be tolerated. Usually components that are outside the tolerance have to be rejected as part of the QA and QC measures.

The system normally used by designers and manufacturers to specify the required dimensions of a part or product is called geometric dimensioning and tolerancing (GD&T). This is a standardised system of symbols and figures used on CAD model or engineering drawings (sometimes referred to as working or technical drawings) that express the ideal distance and angular dimensions required, along with the amount of variation that is allowable. BS ISO 8015 and BS 8888 are the standards used for presenting drawings, to ensure the adoption of this common method of representation.

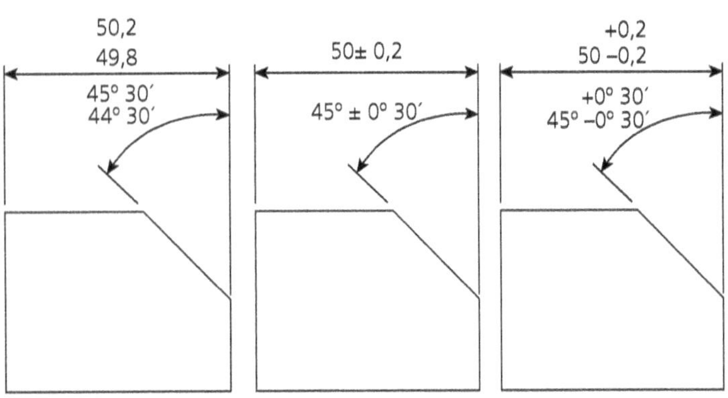

Figure 2.7.15 BS ISO 8015 methods of indicating tolerances on an engineering drawing.

Sometimes tolerances are added to a drawing as a general indication of what is acceptable for all dimensions, such as 'Tolerance unless otherwise stated is: linear ±0,2 angular ±0° 30'. Where specific dimensions require tolerances, they are indicated by including two figures, with the largest acceptable dimension above the smallest, or by using the + and – symbols to indicate the extremes. Three acceptable methods of indicating tolerances for linear and angular dimensions are shown in

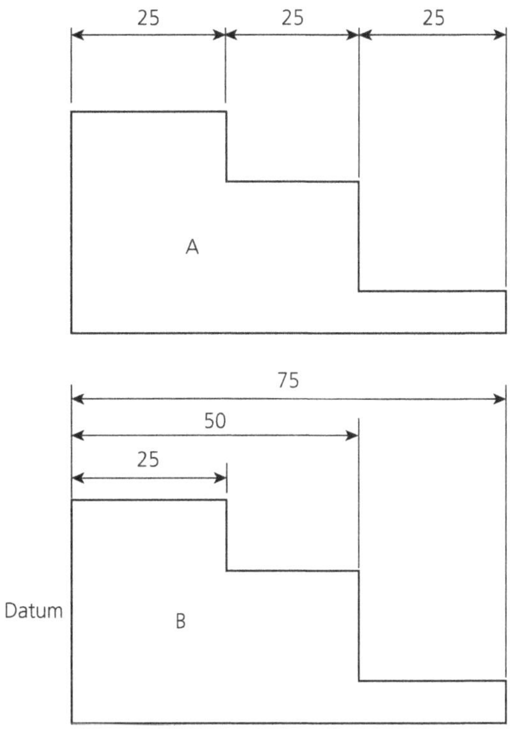

Figure 2.7.16 Using a datum for dimensioning.

Figure 2.7.15. The acceptable convention for indicating a decimal marker on an engineering drawing is a comma.

An important factor that can impact considerably on the accuracy of dimensioning and measurement is the use of datum surfaces or edges from which other dimensions are taken. Datum surfaces and edges are usually machined or ground for accuracy. Drawing B in Figure 2.7.16 shows the preferable, accurate method of dimensioning from a datum. This reduces the chances of cumulative errors that can occur when the method shown in drawing A is used, where a small error occurring when marking out each of the dimensions can add up to a large error overall.

Testing and the elimination of errors

Before undertaking any measurement or marking out in connection with manufacturing a product or component, it is important to consider the tolerance that has been specified in that particular situation. This will have been dictated by a number of factors, as you will see in the following examples of manufacturing situations:

1. hand-forged iron or steel decorative scrolls for a bespoke gate
2. a rectangular piece of MDF for mounting a picture; to be cut from a board
3. mass produced glass bottles for carbonated drinks
4. a mould for blow-moulding a glass bottle.

These situations vary in the amount of tolerance that is acceptable and the method that is used to establish the level of accuracy. We will look at each situation in turn.

Figure 2.7.17 Blacksmith hammering metal on an anvil.

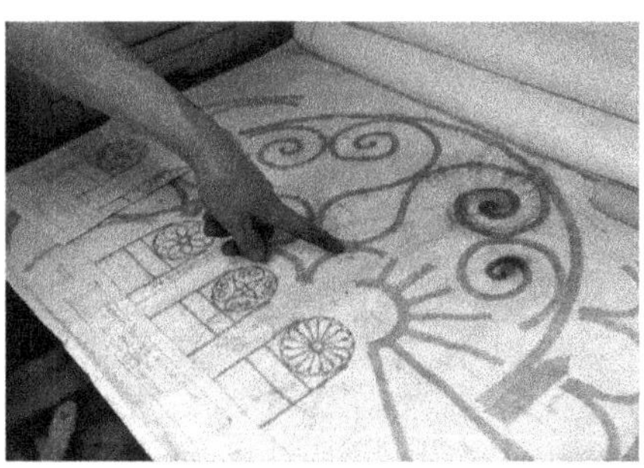

Figure 2.7.18 Blacksmith's drawing showing scorch marks.

1 Hand-forged iron or steel decorative scrolls for a bespoke gate

In this example, there is no need for great accuracy and it can be argued that a lack of accuracy might be a good thing, within reason, in that it will result in each of the decorative pieces having a slightly different shape and thus provide a handmade, bespoke look. This is often much sought after by clients who wish to avoid identical machine-made products.

In Figure 2.7.19, a drawing is being used for guidance in the manufacture of a gate. Scorch marks can be seen where the blacksmith has held the hot metal scrolls next to the drawing to test that the shape is in line with what is required. The tolerance in this case can be many millimetres since there is no need for a precision fit. The drawing is providing dimensional guidance to the blacksmith so little marking out is required.

2 A rectangular piece of MDF for mounting a picture; to be cut from a board

It would be necessary to measure the photograph and decide how close the board needs to be to the edges. Typically in this situation, the tolerance would be around ± 0,5 mm, but it would not be critical if it went slightly beyond this, as there are no working parts involved and a small amount of protrusion would be acceptable. A rigid ruler would be the best tool for measuring, and this would be accompanied by a large try square and sharp pencil for the marking out. In practice, however, there is often a fixed scale on a circular saw that would facilitate setting up the cuts accurately without the need for marking out.

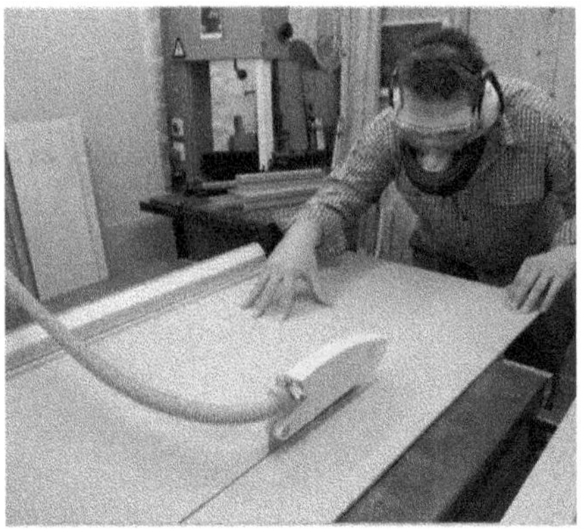

Figure 2.7.19 Cutting a piece of MDF on a circular saw.

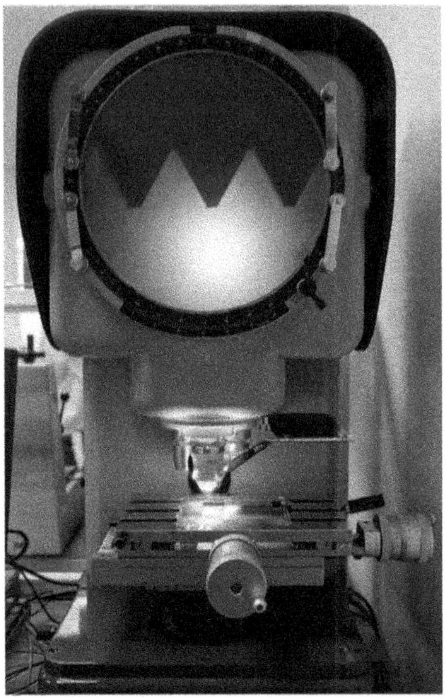

Figure 2.7.20 Profile projector being used to check the accuracy of a thread.

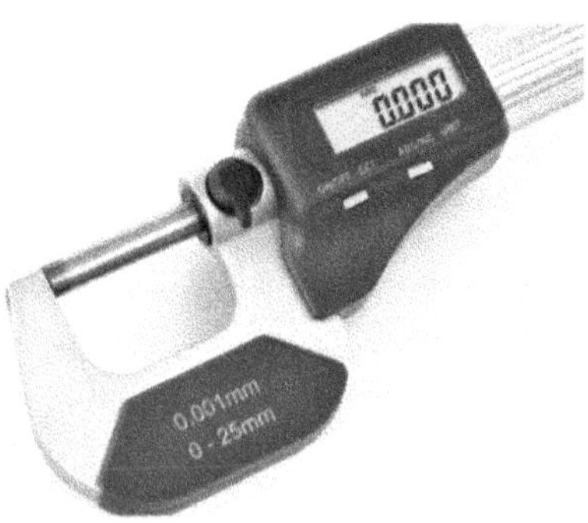

Figure 2.7.21 Digital micrometer.

3 Mass produced glass bottles for carbonated drinks

The glass bottle provides some significant challenges as far as measurement is concerned, since the shape of the bottle will be dictated in the main by the mould that is used. Checking dimensional accuracy is crucial for several reasons:

- A good fit with the screw top is essential to ensure no fluid or gas leakage.
- Sagging may take place during moulding, resulting in a distorted shape.
- Correct volume is paramount to comply with weights and measures legislation.

This means that a series of checks need to be carried out on the dimensions of the bottles and, because of the large number of products involved, methods have to be used which are either very easy to use or automated in some way.

Various camera/computer scanning systems are used, which analyse digital images of the bottles to check the level of compliance with the acceptable dimensional tolerances.

The thread profile for the screw top fitting is checked using a profile projector, which has the capability of achieving a tolerance of ± 0.015 mm. This testing would need to be carried out in a suitably equipped QC laboratory, and would therefore be applied to samples taken from the production line at appropriate intervals.

Tolerance checks can also be carried out using a go/no go gauge, which can be used quickly and easily. It is used by trying both ends of the gauge; if the 'go' end fits in and the 'no go' end does not, then the dimension being checked (for example, the diameter of the bottle neck opening) is within the acceptable range. This type of testing requires very little skill, and the gauge itself is very robust so it can easily be used in the manufacturing plant itself.

The weight of the bottle is also checked on a regular basis; being overweight results in thick glass and too small a capacity, and the reverse for low weight. A tolerance of ± 5 g is typical.

4 A mould for blow-moulding a glass bottle

The mould will dictate the shape and form of the bottles that are produced, and these are often made by using electrical discharge machining (EDM) that can achieve a tolerance of ±0,002 mm. Highly trained engineers would be involved in checking the moulds and particularly ensuring that they have not become too worn by the moulding process. Accurate equipment, such as a digital micrometer or Vernier, would be required. These can work with an accuracy of 0.001 mm and provide flexibility in measuring situations because they can be used for a wide range of dimensions, unlike go/no go gauges, which are only able to check one specific tolerance range.

More information on the use of go/no go gauges, vernier calipers, micrometers and other measuring equipment used in QC for the manufacture of glass bottles can be found in Chapter 2.9.

Using measuring aids

Some of the most important measuring tools were dealt with at the start of this section, but in some cases, accuracy and speed can be improved and potential human errors eliminated by using **jigs**, **fixtures** and **templates**. This is particularly appropriate in large-scale manufacturing situations where accuracy, consistency and time are crucial.

The use of non-contact measuring devices employing lasers and other technologies, such as material thickness sensors, laser micrometers and alignment systems, is also becoming more common.

Jigs

These are devices used to control the motion of a tool relative to a workpiece. The most common type is a drilling jig, used to ensure that holes are drilled in the correct location every time. They are normally locked onto the workpiece using clamps; and replaceable, hardened bushes guide the drill. Using a jig means that the need for marking out is removed, and the work does not require high skill levels.

Fixtures

These are robust frames with holding points and clamps that are used to hold workpieces firmly in place while machining, welding and other manufacturing operations take place. It is essential that every component or assembly is held in exactly the same position each time the manufacturing process is carried out, to ensure consistency.

Templates

A template usually consists of a rigid shape or pattern, often made of thin manufactured board or sheet metal, which is used to transfer a design onto a workpiece. It is normal practice for a template to be held securely in position while the outline is traced onto the material below with an appropriate marking tool. By using a template to mark out all parts, manufacturers can ensure that the same shape will be transferred each time.

Figure 2.7.22 Using a simple template.

ACTIVITIES

List the tools required and the stages that are necessary for marking out a piece of sheet steel to make the part shown in Figure 2.7.23.

Use a piece of manufactured board to check the accuracy of a try square by marking out superimposed lines with the stock of the try square held in both directions.

Find out about the need for a face side and face edge on a piece of wood, and use a simple annotated diagram to explain how these can serve as datum surfaces when marking out other features and executing manufacturing processes.

Draw a design for a simple jig to guide a drill used for making dowel joints.

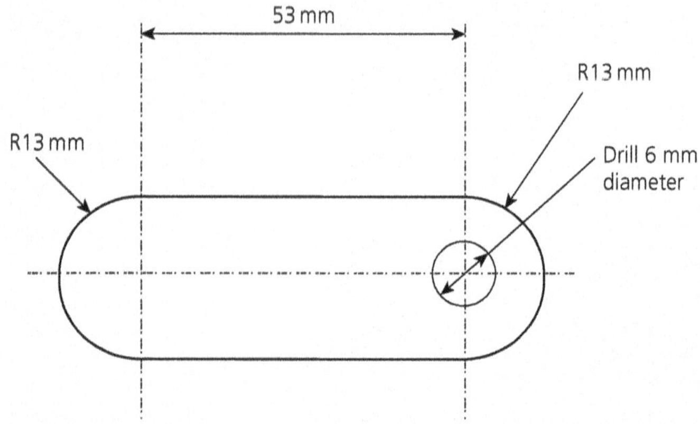

Figure 2.7.23 Sheet steel part.

Draw a fixture that could be used to maintain accuracy in the welding of a bike frame.

Make a template that a jeweller could use to mark out identical sheet copper shapes that would link together to make a neck decoration; then use the template to help you make a card mock-up of the full neck decoration design.

KEY TERMS

Accuracy: the amount of conformity of a measurement to the required value.

Tolerance: the acceptable upper and lower limits of accuracy of a measurement.

Measurement: the process of using tools to check existing dimensions.

Dimension: the measured distance in a straight line between features of a part or product, such as height or length.

Marking out: the transfer of designs onto pieces of material or parts.

Datum: a surface or edge used as a reference from which measurements are taken to improve accuracy.

Jig: a device used to control the motion of a tool such as a drill relative to a workpiece.

Fixture: a robust frame with holding points and clamps to hold product components firmly in place for manufacturing processes such as welding.

Template: a rigid shape or pattern, used to transfer a design onto a workpiece to ensure consistency.

KEY POINTS

- A range of measuring and marking out equipment is available. A tool needs to be chosen carefully to suit the material and level of accuracy required.
- It is important that parts and products are made accurately to facilitate correct function and fit.
- An appropriate tolerance needs to be established for each dimension of a part or product, depending on factors such as size, function and need for interchangeability.
- Relatively unskilled workers using devices such as go/no go gauges can quickly execute tolerance checking of mass produced products.
- Skilled workers use vernier calipers, micrometers and other equipment designed for greater flexibility and accuracy when measuring.
- Accuracy can be improved by using datum surfaces and edges as a reference for measurements.
- Jigs, fixtures and templates can be used to improve accuracy and consistency of production.

Check your knowledge and understanding

1 Name three measuring or marking out tools used for general applications where high levels of precision are not required.

2 Name three measuring or marking out tools used by skilled workers for very precise measurement.

3 What is 'accuracy' and why is it important in the manufacture of products?

4 Define the term 'tolerance' and explain how it might apply in a situation where very detailed parts of a product have to fit accurately.

5 Why is marking out unnecessary when using CNC machines?

6 How do graduated dials help a manual machine operator control a lathe or milling machine accurately?

7 How can a datum surface help to make measurement and marking out more accurate?

8 What is the difference between a jig and a fixture?

9 What is a template used for?

Further reading

The Essential Guide to Technical Product Specification: Engineering Drawing by Colin Simmons (BSI)

Measuring and Marking Metals by Ivan R. Law (Argus)

Woodwork Step by Step by DK books

Taunton's Complete Illustrated Guide to Jigs & Fixtures by Sandor Nagyszalanczy (Taunton Press)

This website provides links to accuracy information guides from the National Physical Laboratory: www.npl.co.uk/publications/guides/

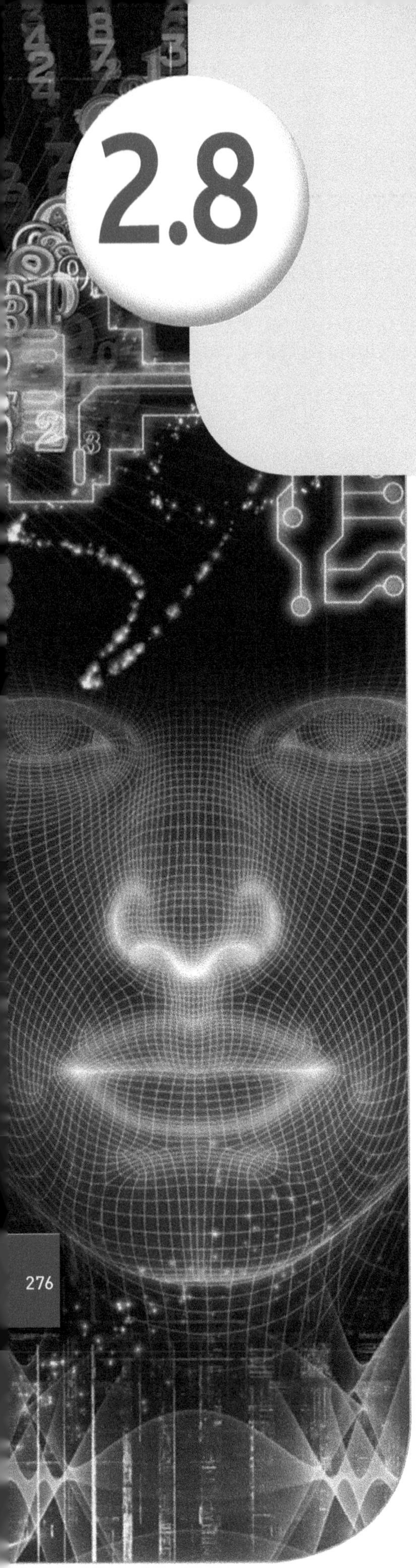

2.8 Responsible design

LEARNING OUTCOMES

By the end of this section you should have developed a knowledge and understanding of:

- the responsibilities of designers and manufacturers to ensure products are made using sustainable materials and components
- the environmental impact of packaging
- how products are designed to conserve energy, materials and components
- the concept of a circular economy
- how to design products for minimum impact on the environment
- sustainable manufacturing, including the use of alternative energy and methods to minimise waste.

Designers have an ever-increasing responsibility to design products that have minimal environmental impact. Designers not only have to consider how to conserve materials but also how to conserve energy during product manufacture, with the aim of making products that are as sustainable and environmentally friendly as possible.

In 1987, the World Commission on Environment and Development (the Brundtland Commission) introduced a concept called sustainable development, which helped to shape the international community's attitude towards economic, social and environmental development.

The Brundtland Commission defined sustainable development as 'development which meets the needs of current generations without compromising the ability of future generations to meet their own needs'.

Sustainable development has three strands:

- economic
- social
- environmental.

Designers of the future must consider all three strands when designing and manufacturing products. The three strands are interlinked and involve decisions about material choice (such as the cost, use of finite resources and end of product life recycling), processing implications (such as energy use, pollution and the impact on people's health) and manufacturing (such as energy use and working conditions for those involved in the manufacturing process).

Industrial designer Phillipe Starck said, 'Designers should define their role broadly as agents of good in the world, and limit their work to "legitimate" products. This means products that are needed, and those that can be made without damage to nature or – through the unethical actions of manufacturers and investors – damage to people'.

Environmental issues

The starting point for many environmental decisions regarding product design is the **six Rs of sustainability** (you will remember this from Chapter 1.11). The six Rs are:

Reduce

Cut down the amount of material and energy used to make and package the product.

Figure 2.8.1 Excessive product packaging which could be reduced.

For example, companies producing bottled water constantly strive to use as little polymer as possible, while retaining the integral strength properties required for the product. Material savings can be minute, but when added up over a multiple production run of thousands of bottles per day, the savings can be huge. Reductions can be made in terms of raw materials, energy use in processing (if the polymer is thinner, it will take less time and energy to reach blow moulding temperature) and energy use when transporting the product from factory to distribution centre.

Reuse

At the end of a product's life, reuse the product for the same or another purpose.

An example would be buying a glass jar of instant coffee and then reusing the glass jar and buying a refill pack. Filter coffee makers use a paper filter; some manufacturers offer metal or polymer filters, which can be reused time and again. Reuse can also cover using a product for a different purpose, such as using an empty glass jam jar to store pens and pencils.

Designers need to consider the reuse potential of their product or product components; this may influence decisions such as ease of disassembly when designing.

Recycle

Convert of waste products into new materials for new products.

End of life disposal is a key area for designers to consider, not only to conserve finite resources such as crude oil for making polymers, but also for environmental impact such as scarring from metal extraction, loss of animal habitat and plant life when cutting trees for wooden product manufacture, as well as the environmental and aesthetic issues associated with mountains of rubbish in landfill sites.

Many manufacturers actively advertise the use of recycled waste in their products, for example many packaging products will have a label informing consumers that the product is made of 100 per cent recycled paper, or a wooden furniture product will display an FSC stamp.

Legislation also exists for manufacturers to meet recycling targets. Car manufacturers must identify all recyclable parts in new cars. This relates not only to polymers but also textiles, fluids and electronic components. Any manufacturer that makes a product that is recyclable must consider ease of disassembly for recycling purposes. For example, car manufacturers will use click fittings to attach dashboards and interior door panels, thus removing the need for additional joining components such as screws.

Recycling happens in three ways:

- Primary recycling: the use of functioning second-hand products, which the first user no longer has a need or want for. Charity shops, freecycle initiatives, local council recycling centres and websites such as eBay and Gumtree are the main sources for primary recycling. Primary recycling can also fall under the term 'reuse'.
- Secondary recycling: at the end of a product's life, the materials are recycled to make different products. Boat sails can be recycled to make seating products such as beanbags, seatbelts can be recycled to make courier bags, bike tyres can be made into belts and polymer drinks bottles can be made into plant pots.
- Tertiary recycling: this is completely breaking down a product and reformulating it via a chemical process. For example, polymer drinks bottles can be shredded and spun into fibres to make fleece textile clothing, polymer vending machine cups are made into pencils, car tyres can be recycled into mouse mats or the components of soft paving squares in children's playgrounds.

Products that are able to be recycled should have one of the following symbols printed onto or moulded into the product. For polymers, the number inside the triangle indicates how easy the polymer is to recycle, for example how much heat it needs to reform. The low numbers are the easiest to reform, such as PET used for drinks bottles; the high numbers are more difficult and these include mixed polymers as well as hard, tough polymers such as ABS.

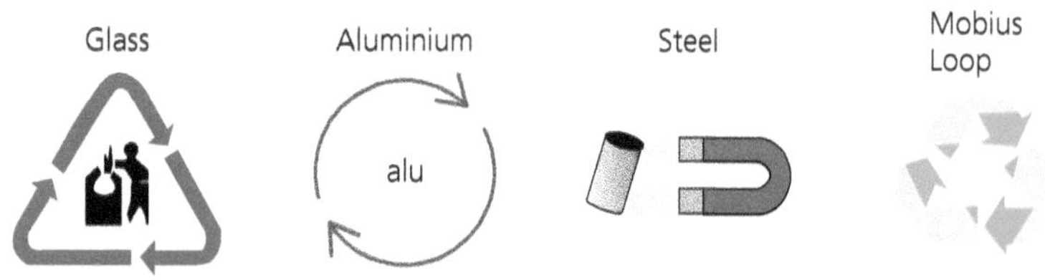

Figure 2.8.2 Recycling symbols commonly seen on consumer products.

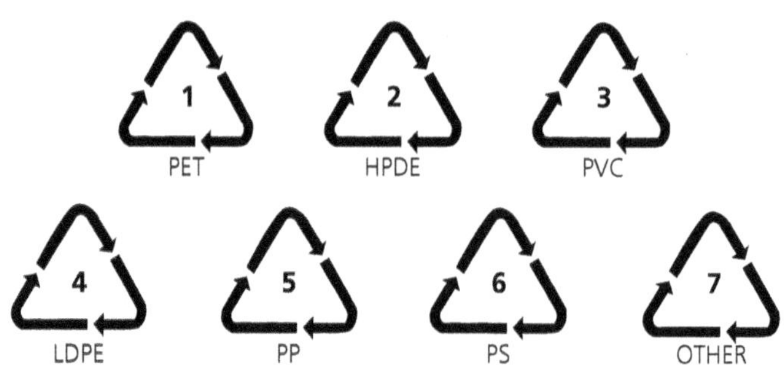

Figure 2.8.3 Polymer recycling codes.

Repair

When a product or component fails, fix it rather than throwing it away.

Designers have a responsibility to consider the use of resources as well as to address the 'throw away' culture that many consumers adopt. For example, if a bicycle tyre has a puncture, it can be repaired rather than throwing the whole tyre or bike away and buying a new one.

Refuse

Exercise consumer choice as to whether to buy a product or not.

The consumer may choose not to buy a product if they consider the product to be bad for the environment or if it is a product that they may want but do not need.

Considerations for the consumer include:

- Do I really need the product?
- Has the product been ethically made?
- Is the product sustainable?

A consumer may choose not to buy bottled water in a single use bottle, but instead may buy a reusable water bottle and fill it with tap water at home. The consumer would consider this a more sustainable choice because they are not contributing to the use of finite resources.

When considering white goods, such as washing machines or refrigerators, the consumer can assess the size of product required to suit their needs and then refuse to buy anything beyond this, or refuse a model that is less energy efficient than a competitor's model.

Rethink

Rethink the way products are designed and manufactured so that they carry out the same function, but more efficiently.

This could include making products more energy efficient in terms of both manufacture and running costs. It also includes design aspects such as the use of click fittings and SMAs (Shape Memory Alloys) to aid product disassembly prior to recycling.

For the consumer, rethinking may be as simple as taking a refillable cup to a coffee outlet rather than using a throw away single use cup. For the manufacturer, rethinking may involve how a product can best perform the function or what may help the consumer, for example making packaging out of a single material that is easy to recycle, rather than using mixed materials.

How a designer could make use of the six Rs when designing everyday products such as a TV remote control

Reduce:

- Make the walls of the remote control as thin as possible to use less polymer.
- Design the product to be space-efficient and smaller.
- Design for only essential buttons such as power, channel and volume. Integrate dual function buttons to reduce the space used.
- Make use of solar powered or human powered chargers rather than batteries.
- Use renewable energy sources when manufacturing the product.

Reuse:

- At the end of the product's life, reuse the casing for another remote control.
- Reuse the electrical components for another product.

Recycle:

- Use recycled polymers for the body and buttons.
- Ensure the product is made from materials that can be recycled at the end of the product's life.
- Use recycled and recyclable copper and solder for the electric circuitry.
- Mould recycling codes into the casing and provide online instructions for how to dispose of the product at end of its life.

Repair:

- Ensure the components can be fixed rather than discarded.

Refuse:

- Ultimately this is the consumer's choice, but advertising the positive environmental aspects could help.
- Use recycled polymer for the casing and recycled copper for circuit boards to attract the consumer.
- Use materials that are not banned under the Restriction of Hazardous Substances directive.

Rethink:

- Make the product as simple to operate as possible using rechargeable, solar or human powered low energy sources.
- Rethink the number of essential buttons that a user requires from the product, for example focusing on the most used function buttons.
- Make the controller multi-purpose so that one product can operate all TV peripherals rather than having separate controllers for each device.

Using sustainable materials and components

The use of sustainable materials and components is an essential objective for any designer. The use of finite resources such as oil and metal ores cannot continue because there is not enough resource to meet future demand. If trees are used and not replanted, this impacts not only the supply of materials but also the supply of oxygen, the prevention of flash flooding by trees absorbing water run-off, the fight against soil erosion and the volume of CO_2 (a contributor to global warming) entering the environment. (Trees act as carbon sinks which prevent CO_2 entering the atmosphere.)

Many companies and individuals will consider their **carbon footprint** and the subsequent impact on the environment. The carbon footprint is the total amount of CO_2 released into the atmosphere as a result of the activities of an individual, a community or an organisation.

Primary carbon footprint measures direct emissions of CO_2 from the burning of fossil fuels, including transport and domestic energy consumption. For example, about 2.3 kg of CO_2 is emitted for every one litre of petrol used.

Secondary carbon footprint measures indirect CO_2 from the products we use. The production of five polymer carrier bags produces about 1 kg of CO_2.

The environmental impact of packaging

Packaging is an important aspect of any product design. Without packaging, some goods and associated components would not get to the consumer. For example, a mobile phone is sold with charging cables, plugs and accessories, which may get lost without packaging. Packaging also prolongs the life of food and helps to prevent waste.

Designers and manufacturers need to consider using the optimum amount of packaging to protect and preserve products and prevent waste. Factors such as making the packaging lightweight, using recycled content and making the packaging recyclable or reusable should be considered. In addition, consideration should be given to how the packaging can be redesigned to minimise environmental impact, such as the use of refills and concentrates.

Figure 2.8.4 Use of concentrate to reduce polymer use in drinks packaging.

Figure 2.8.4 shows two different drinks containers. The bottle on the left contains the same number of servings as the smaller container on the right. The use of juice concentrate leads to a reduction in the amount of polymer required for the same number of drinks.

One of the most commonly used packaging items in a supermarket is the single use polymer carrier bag. Simple measures such as the 5 pence carrier bag charge can have a major impact on reducing carrier bag use, waste and landfill. In the United Kingdom, Wales introduced the charge in 2011 and saw a 71 per cent reduction in polymer bag usage in the first year; Northern Ireland introduced the bag charge in 2013 and saw a reduction of 72 per cent in the first year; Scotland saw a decrease of 80 per cent in the first year of charging. If fewer single use carrier bags are used by consumers, there is a reduced need for manufacture, which in turn leads to a drop in the use of finite resources.

Many supermarkets have also adopted a bag for life scheme, where an individual purchases a carrier bag for 10 pence and at the end of the bag's life, the supermarket swaps it for a new bag. Schemes like this ensure that bags are sent to recycling rather than landfill. Some countries such as Bangladesh have banned the use of thin polymer bags, because they were found to contribute to blocked drainage systems during floods. In 1988, floods submerged two-thirds of the country, and the littered polymer bags added to the existing problem of the country being very flat and suffering regularly from heavy rainfall. In 2011, to try to reduce landfill issues, Italy banned the distribution of polymer bags that are not from biodegradable sources. As of July 2014, there were 20 states and 132 cities in the USA with polymer bag bans in place. The USA uses about 12 million barrels of oil every year to meet the demand of manufacturing polymer bags.

Food packaging in supermarkets has traditionally included examples of overpackaging – the use of excessive amounts of material to contain the product. Designers should consider aspects such as:

- Do bananas, which have a natural protective layer, need to be contained in a polymer bag?
- Do vegetables such as cauliflowers need to be shrink wrapped for sale?
- Do fruits such as oranges need to be contained within mesh bags?
- Do apples need to be grouped in fours, placed on a polymer tray and shrink wrapped in polymer?

Packaging issues such as those above are primarily as a result of consumer convenience rather than a functional requirement. With the rise in consumer awareness of food sources, food miles and sustainability, designers and manufacturers have the opportunity to reduce the environmental issues of packaging in perhaps very simple ways.

In the UK, Marks and Spencer was one of the front runners in addressing the environmental need to reduce packaging waste. One initiative involved redesigning pizza boxes. Pizzas used to sit on a polystyrene (PS) tray, shrink wrapped in LDPE inside a carton board box. The redesign involved changing the carton board box to a narrow, recycled card sleeve. This redesign reduced the overall packaging by 62 per cent, and made a saving of over 480 tonnes of packaging a year.

Figure 2.8.5 Old-style pizza box packaging.

Figure 2.8.6 Redesigned pizza packaging with narrow sleeve.

Figure 2.8.7 Detergent packaging.

In 2012, Marks and Spencer used 25 per cent less Easter egg packaging than in 2009, with over 70 per cent of Easter egg packaging from sustainable sources. About 68 per cent of all polymer used in the range contains 50 per cent post-consumer recycled materials, in place of new, first use polymer.

Another example of packaging redesign to save resources is detergent packaging. The use of concentrated detergent, with reduced water content, allows the package size and material use to be reduced, sometimes by up to as much as 45 per cent on the original bottle. The use of refill bags requires about 80 per cent less packaging in transit to the retail outlet than bottles, thus saving energy use for transportation.

Figure 2.8.8 KeepCup – a reusable coffee cup designed for use within barista coffee machines.

KeepCup is a reusable coffee cup that was designed to reduce the environmental impact caused by disposable cups ending up in landfill. Twenty standard disposable cups and lids contain enough polymer to create one KeepCup. At present, about 500 billion disposable cups are manufactured globally every year, equating to about 75 cups per person across the globe. In the USA, 58 billion disposable cups are thrown away each year, with the majority ending up in landfill.

Taiwan discards about 1.5 billion disposable cups annually, and landfill issues are a huge problem. To reduce the amount of waste going to landfill, a new rule was introduced whereby customers were given a discount if they used a reusable cup. This action is predicted to result in a 30 per cent reduction of disposable cup use annually (450 million cups), saving 110,000 trees from being processed to produce paper cups, and reducing carbon emissions by 15,000 metric tonnes.

ACTIVITY

Find out how many tonnes of waste the average UK household produces each year.

Research your local authority waste management plan and collect data on what specific items they recycle and/or reuse, and the impact this has on the local environment.

Environmental advantages of KeepCup:

- Low energy manufacture: injection moulding polymer is a low energy process.
- Recyclable at end of life: the KeepCup is manufactured from one polymer, PP (polypropylene), which means the parts do not have to be disassembled and sorted at a recycling plant. PP also has a relatively low melt point for recycling purposes and lower energy use. The KeepCup can be recycled into other products with no wastage of material. Currently about half of the polymer used in the world is for single use items.
- Weight and freight: PP is a lightweight polymer, and the KeepCup is designed to be stackable. Being lightweight and stackable reduces the costs of shipping – stacking makes better use of the transport space available and the reduced weight results in less fuel being used.

The KeepCup compared to a disposable paper cup results in annual reductions of 36–47 per cent in global warming carbon emissions, 64–85 per cent in water use for manufacture and a 92 per cent in landfill waste.

Conservation of energy and resources

Energy use is a major factor in both the manufacture and day-to-day running of products. A number of energy resources are non-renewable such as oil, coal and natural gas, and as such, alternative energy sources (renewables) must be found to provide enough energy to meet the demands of the modern world.

When designing products, consideration must be given to the minimal use of materials to save resources and energy in production. A simple example could involve nesting or tessellating components to be cut from a sheet material when CNC laser cutting or routing. The nesting of components minimises waste products and enables maximum material use from each sheet. Where the material initially comes from is also an aspect to consider as some materials are from a finite source. Designers and manufacturers have a responsibility to use materials that can be or have already been recycled, such as using recycled polymers for products such as mobile phones or calculator casings.

Alternative energy sources (renewables)

Advantages of renewable energy:

- The major advantage of renewable energy is that it is sustainable, and as such, it will never run out.
- Renewable energy facilities generally require less maintenance than traditional generators.
- Fuel derived from natural resources reduces operational costs.
- Little or no waste, such as CO_2 or other chemical pollutants, is produced, meaning minimal environmental impact. Renewables are considered clean energy.
- Social and economic benefits – a renewable energy project can bring benefits through employment and use of local services to an area.

Disadvantages of renewable energy:

- It can be difficult to generate the large quantities of electricity that are produced by traditional fossil fuel generators. As such, more renewable energy facilities may need to be built.
- Renewable energy often relies on the weather for its source of power, and if the supply is unreliable or inconsistent, the energy production will be unreliable too. For example, wind turbines need wind to turn the blades, hydro generators need rain to supply water and solar collectors need sunshine to collect heat and make electricity.
- Renewable energy cannot be stored in large quantities for later use due to the impracticalities of having battery power plant backups.
- Renewable energy is currently more expensive than traditional fossil fuel or nuclear energy, primarily due to the large capital cost associated with new technologies.

Renewable energy systems

Table 2.8.1 Renewable energy systems.

Energy type	Method
Wind	Turbines driven by blades that catch the wind.
Hydro	Dams are built to trap water. Stored water is released to turn turbines and generate power.
Solar photovoltaic (solar PV)	Photovoltaic cells convert the sun's energy directly into electricity.
Wave	Energy produced by the constant movement (kinetic energy) of the waves. The wave movement forces air up a cylinder to turn the turbine and generate power.
Tidal barrage	Barrages are built across estuaries; as the tide comes in and goes out, the water movement turns turbines which generates power.
Geothermal	Natural heat from the Earth heats up water to produce steam. The steam spins a turbine and generates electricity.
Biomass	Organic matter such as wood, dried vegetation and crop residues are burned to heat water and produce steam. The steam turns the generator to produce power.

Conserving energy, materials and components

Designers and manufacturers must not only consider conserving energy and materials at source but also within processing and manufacture. Products should be designed so that there is minimal waste material and ideally the waste can be reused or recycled. Injection moulded products are often moulded in groups with sprues to attach each component in the mould (similar to an Airfix model kit). As injection moulding is a thermoplastic process, the sprues can simply be pelletised and put back into the hopper to be part of the next moulding cycle. For sheet materials or complex shapes, tessellation should be used so that components fit snugly onto the sheet with limited areas of waste product.

Apple, the mobile phone and electronic devices manufacturer, knows that due to the volume of products it sells, even minor adjustments to the manufacturing materials can have a big environmental impact. It uses aluminium that is smelted using renewable hydro-electricity energy rather than fossil fuel. It has also re-thought the manufacturing process so that waste aluminium off-cuts can be reused within the process.

Waste material is only one factor that the manufacturer must consider regarding environmental impact. The machinery used, such as injection moulding machines, require mechanical, electrical and hydraulic systems to operate. They may also incorporate water cooling systems which utilise collected water from rainfall source. The components require lubricants such as oils, which need to be used and disposed of in an environmentally positive way. Oils and any chemicals used are treated as hazardous waste and require specialist uplift and disposal, following guidelines such as those set out by the Government and bodies such as the Oil Recycling Association. Specialist disposal can be an expensive additional cost for the manufacturer which cannot be avoided. Many local authorities also charge for water use, and these costs can be substantial if the manufacturer is in a country that has water shortages.

Companies such as Smart, the car manufacturers, are constantly looking at ways to help the environment as well as reduce waste disposal and water costs. Smartville, the factory complex where the Smart car is produced, has introduced a number of environmental measures which contributed to their winning the best factory category in the 'Trophée de l'Excellence Industrielle' in 2012. Smartville recovers its waste water and has its own waste water treatment plant. It also has a gas-fuelled heat and power generation plant to provide heat and about 25 per cent of the factory's power requirements.

Product use may also have energy and environmental implications, which all add to the manufacturer's overall carbon footprint. Consumers must take responsibility for how they use products as well. Companies such as Apple are striving to manufacture products that use less energy, incorporating specialist power saving features such as efficient mobile phones, screen energy saving on a computer when the screen is static between key strokes, or power saving when the hard drive is not being used. These features all help, but without the consumer being aware of how they use energy, the benefits may be reduced.

Product miles

Product miles are the total lifetime distance that a product is transported from its place of production to the place of use by the consumer. The concept is widely used in the food industry (food miles) where there is a market push by consumers for locally sourced and produced foods. Many food producers now include labelling to indicate the number of food miles for a food product.

A typical product might travel as follows:

- raw material source to processing plant
- processed material to the manufacturing facility
- manufacturing facility to a distribution hub
- distribution hub to the retail outlet
- retail outlet to the user's home
- from home to a recycling centre.

As manufacturers are becoming more conscious of consumer awareness of product miles and the impact of CO_2 emissions, they are looking at ways of reducing product miles. One way is to cut out the middle man and deliver products straight from the manufacturing facility to the retail outlet. Not only does this have environmental benefits, such as fewer journeys and reduced emissions, it also has cost savings because there is no requirement for an additional storage facility which would also use heat and light (energy use). Another method is to use the electric train network rather than the road network for transportation of goods.

Using manufacturing facilities in the country of use is also important for reducing product miles. The carbon footprint of companies that manufacture goods abroad, ship them back and then distribute them is far greater than that of companies that manufacture and distribute locally.

In the UK, two-thirds of the wood that is used for product manufacture is imported. Some wood that is harvested in the UK is sent abroad for processing and then shipped back for manufacture and sale. There is an increasing market awareness and drive for locally grown, harvested and processed wood, whereby consumers can trace the journey of a wooden product from seed to store.

Circular economy

The world has finite resources of oil, natural gas and metals, but the current 'throw away' culture and cycle of raw material extraction, processing, use and unconsidered disposal is not a sustainable approach.

The traditional approach is for a linear economy (take, make, dispose) where raw materials are taken from the source, made into products which may last from a few minutes to a few years and then disposed of in landfill. A **circular economy** aims to use materials in a way that ensures a continual cycle of reuse and remanufacture, without utilising wasteful resources or having products end their life in landfill. Resources should be available to keep materials, products and components at the highest level of effectiveness at all points in the life cycle.

The circular economy approach anticipates and designs for biological and technical 'nutrients' to be continuously reused at the same quality, dramatically reducing the dependency on sourcing new materials. It is a restorative and regenerative design and manufacture process, which makes use of systems such as product maintenance, refurbishment, remanufacture and recycling. Manufacturers can adopt several methods to support a circular economy such as product leasing (a popular approach for IT hardware and tool hire), take-back (such as the Dyson method of taking back a Dyson product for recycling at end of its life at no cost to the consumer) and optimising their approach to the entire life cycle of the product and manufacturing systems, including the impact on water resources, energy and ecology. A circular economy is a cradle-to-cradle approach in a product life cycle, rather than the traditional cradle-to-grave approach – designing products to be made again with processing powered by renewable energy.

A circular economy considers two 'nutrient' types for continual product disassembly and reuse:

- Biological nutrients: organic non-toxic materials that can be simply composted, and safely re-enter ecosystems without harming the natural environment, such as plant based and biodegradable materials. Examples include shampoos and cleaning detergents that can safely be washed down a household drain without harm to the local ecosystems, or products made from wood that can naturally degrade to provide compost or nutrients to the ground upon decomposition.
- Technical nutrients: man-made materials including polymers and alloys, designed to be used repeatedly at the same initial high quality with minimal energy and no adverse environmental effects. Products such as cars and washing machines are made from technical nutrients. These products are designed to be used and then, at the end of their life, recycled.

Circular economy principles for designers and manufacturers

- Preserve and enhance natural capital: the control of finite resources, and the use of processes and technologies that use renewable or more efficient resources including energy to power production. Bio-systems are also included in this, such as using compostable elements to feed back into soil systems.
- Optimise resource yields: designing for refurbishment, effective disassembly, remanufacture, reuse and recycling. The aim is to keep the highest quality and value in all resources and components so that there is a continuous cycle, which actively contributes to society and the economy. Resources include the two 'nutrient' types, biological and technical.
- Foster system effectiveness: considering, assessing and rethinking design to reduce negative impacts on factors that could directly affect population, such as how land is used. This could impact food production, building and shelter, how products are sourced and processed, and the impact on air quality, water quality and availability, as well as the release of toxic substances that may contribute to greenhouse gases and climate change. This includes factors such as the use of fossil fuels to make fertilisers for crop growth and process through the food chain to end-product. Resources and systems for managing forests and felled wood should be considered, as should the value that wood decomposition adds to ecosystems and forest floors.

A circular economy is an industrial economy that promotes greater resource productivity, aiming to reduce waste and avoid pollution by design or intention. In a circular economy, material flows are of two types: biological nutrients (designed to re-enter the biosphere safely) and technical nutrients (designed to circulate at high quality in the production system without entering the biosphere, as well as being restorative and regenerative by design). This is in contrast with a linear economy, which is a 'take, make, dispose' model of production.

The importance of a circular economy

As well as creating new opportunities for growth, a more circular economy will:

- work against the unsustainable 'take, make, dispose' culture
- promote and drive greater resource productivity
- reduce the use of finite resources
- reduce waste
- avoid pollution
- deliver a more competitive UK economy
- help reduce the environmental impact of our product manufacture and consumption in both the UK and global markets.

ACTIVITY

For a household product or appliance of your choice, research the environmental measures that the company takes to ensure sustainability.

ACTIVITY

Investigate the costs of standard household refuse collection and hazardous waste collection services in your local area. Create a table to show the cost per substance and weight or volume of refuse.

KEY TERMS

Six Rs of sustainability: reduce, reuse, recycle, repair, refuse, rethink.

Reduce: cut down the amount of material and energy used to make and package the product.

Reuse: at the end of product life, reuse the product for the same or another purpose.

Recycle: convert waste products into new materials for new products.

Repair: when a product or component fails, fix it rather than throwing it away.

Refuse: exercise consumer choice as to whether to buy a product or not.

Rethink: rethink the way that products are designed and manufactured so they carry out the same function, but more efficiently.

Carbon footprint: the total amount of CO_2 released into the atmosphere as a result of the activities of an individual, a community or an organisation.

Primary carbon footprint: measures direct emissions of CO_2 from the burning of fossil fuels, including transport and domestic energy consumption.

Secondary carbon footprint: measures indirect CO_2 from the products we use.

Circular economy: an approach that anticipates and designs for biological and technical 'nutrients' to be continuously reused at the same quality, dramatically reducing the dependency on sourcing new materials.

KEY POINTS

- Global warming: climate change that causes an increase in the average temperature of the lower atmosphere of the Earth. Global warming is commonly associated with the release of excessive amounts of greenhouse gases.
- Greenhouse gases: gases such as CO_2, CH_4, Chlorofluorocarbons (CFCs) and water vapour let the heat from the Sun into the atmosphere, but do not allow the heat to escape back into space. Excessive amounts of greenhouse gases have been linked to global warming.
- Sustainable development has three strands: economic, social and environmental.
- Designers of the future must consider material choice (the use of finite resources and end of product life recycling), processing implications (energy use, pollution and the impact on people's health) and manufacturing (energy use and working conditions for those involved in the manufacturing process).
- Circular economy: use materials in a way that ensures a continual cycle of reuse and remanufacture, without utilising wasteful resources or having products end their life in landfill. Resources should be available to always keep materials, products and components at the highest level of effectiveness.

Check your knowledge and understanding

1 List and explain the meaning of each of the six Rs of sustainability.

2 What factors may a consumer consider when deciding to buy or refuse a product?

3 What are the three strands associated with sustainable development?

4 List the ways in which a design team can reduce product miles.

5 Discuss the methods that a design team may consider to ensure a mobile phone meets the principles of a circular economy.

6 For a specific packaging product of your choice, explain how sustainability has been considered in the design and manufacture.

Further reading

Further information and resources about circular economy: www.ellenmacarthurfoundation.org

The Waste and Resources and Action Programme website contains lots of information regarding the way in which resources are used sustainably: www.wrap.org.uk

Information about how Apple considers the environment in the design and manufacture of its products: www.apple.com/environment

www.philips.com/a-w/about/sustainability/sustainable-planet/circular-economy

www.zerowastescotland.org.uk/our-work/circular-economy

www.theguardian.com/sustainable-business/2015/apr/15/circular-economy-jobs-climate-carbon-emissions-eu-taxation

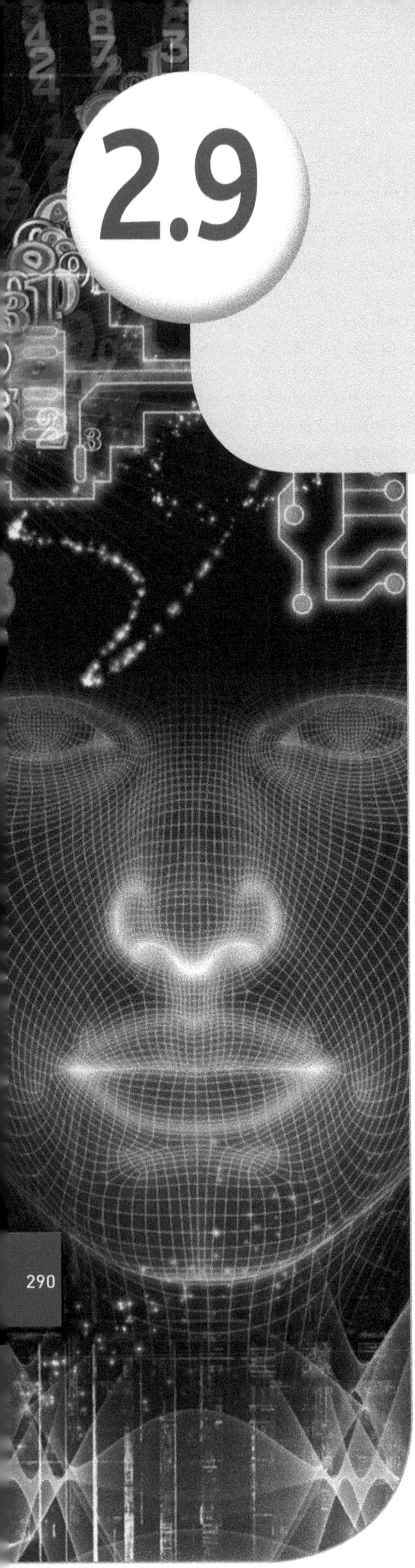

2.9 Design for manufacture and project management

LEARNING OUTCOMES

By the end of this section you should have developed a knowledge and understanding of:

- the importance of planning for accuracy when making prototypes
- quality control, including:
  - the monitoring, checking and testing of materials, components, equipment and products throughout production, to ensure they conform to acceptable tolerances
  - the use of digital measuring devices such as vernier calipers and micrometers.

If you are studying at A-level you should also have developed a knowledge and understanding of:

- the importance of planning for accuracy when making recommendations for small-, medium- and large-scale production
- quality assurance: the procedures and policies put in place to reduce waste and ensure manufactured products are produced accurately and within acceptable tolerances, including systems such as total quality management (TQM), scrum, Six Sigma and their applications to specific industrial examples such as critical path analysis (CPA)
- quality control, including:
  - the use of go/no go gauges, laser or probe scanning and measuring
  - non-destructine testing (NDT) such as x-rays and ultrasound.

The design and manufacture of a successful product relies heavily on appropriate planning and monitoring throughout the process. Effective planning and preparation reduce the risk of an unsuccessful final outcome that may fail to meet key specification criteria.

Designers must ensure that the products they design:

- meet customer and legal requirements
- can be produced with the equipment/technology available
- can be produced to budget
- can be produced within the required timescale.

Often, a feasibility study is carried out to assess the practicality of a proposed project or system (for more on feasibility studies, see Chapter 1.12).

Planning for accuracy and efficiency

When producing a prototype product, it is essential that you plan for accurate production. This requires effective research and development prior to investing time in the final manufacture. A detailed product design specification (PDS) is vital to give a clear list of the client's essential and desirable requirements.

The development of the design must allow the concept to be tested against key specification criteria prior to final production (even of a prototype). This is done using a variety of modelling strategies, possibly including virtual CAD simulations and full-size block models.

The proposed manufacturing procedures must be evaluated for suitability and the achievable level of accuracy. For example, it is difficult to reliably produce a component to an accuracy of ± 0.0001 mm using a manually operated lathe, so the designer must either reduce their expectations or outsource the production to a company that can machine to that accuracy.

The successful manufacture of a product on any scale of production requires a clear schedule of production, with deadlines and QC inspections embedded into the schedule; the type of inspection carried out will vary depending on the scale of production. One-off bespoke products may be manually checked at every stage of production; this can be altered as the scale of production changes.

Ensuring accuracy of prototype designs

Pre-production: accuracy in prototype development relies heavily on a UCD approach (see page 190), with client feedback playing a major part in the success of the prototype. The list below shows a range of pre-production QA (quality assurance) procedures used to ensure accuracy:

- CAD simulations
- working drawings with tolerances
- mock-up models and mechanical systems
- client feedback
- peer review.

Production: during production of a prototype, product accuracy is ensured using a range of QC checks. These are used in conjunction with client feedback to guarantee the prototype meets the PDS. Control checks include:

- visual aesthetic checks
- appropriately accurate dimension checks, using flexible measuring equipment (for example, a vernier caliper), of:
  - individual parts
  - the overall product
- machine tooling and alignment checks
- assembly checks of multiple components
- quality checks of the manufactured finish.

The definition of small-, medium- and large-scale production is very subjective, and therefore the testing procedures for accuracy are mainly governed by a cost analysis to ensure profitability. This could include the rate of production, the time allowed for testing, the regularity of testing/sampling, the testing equipment used and the skill set of the workforce.

Ensuring accuracy in small-, medium- and large-scale production

Pre-production: during product development, a wide range of techniques are used to ensure that the product going into production is accurate compared to the PDS. The cost of rectifying errors identified during pre-production is considerably less than that of rectifying errors identified during production. The techniques used include:

- CAD including process simulations and costings
- working drawings with tolerances
- sample prototypes
- templates, jigs and fixtures produced
- focus groups and surveys
- NDT (Non Destructive Testing) and destructive testing.

Production: when manufacture of the final product begins, accuracy is assured using a range of QC checks to compare the products against the PDS. The specific checks used will depend upon the product, but these are some of the types of checks used:

- visual aesthetic checks as regular sampling, during and at end of production
- sample dimension checks, using flexible measuring equipment (for example, a vernier caliper), of:
  - individual parts
  - the overall product
- tolerance dimension checks using fixed specific measuring systems for the product (for example, a go/no go gauge)
- machine tooling and alignment calibrations
- machine maintenance and tooling replacement
- specific sampling regularity and procedures set by the company, customer or legal requirements
- checks of the quality of the manufactured finish.

ACTIVITY

Use diagrams to explain at least one modification you have made to a project due to the limits of the machinery available to you.

Quality assurance

A level only

Quality assurance (QA) refers to the procedures and policies put in place to reduce waste, and to ensure manufactured products are produced accurately within set acceptable tolerances. By using effective QA procedures, a manufacturer is aiming to produce products 'right first time and every time, which is ambitious. To check that the products being produced conform to the tolerances set, QA checks must be included within the production process.

Some examples of QA policies that companies may use include:

- only sourcing materials and components from suppliers that hold the ISO 9001 Quality Management Standard
- setting specific temperature ranges for product moulding to ensure effective filling of cavities and speeds of cooling
- setting rigid maintenance schedules for machinery to ensure cutters are machining within tolerances.

As the effectiveness of QA procedures increases, the number of QC issues and waste products will decrease.

Figure 2.9.1 The embossed base of an octagonal glass bottle showing mould number 16.

Figure 2.9.2 The bottom plate used to form the embossing on a cylindrical glass bottle showing mould number 14.

Prior to production, designers and engineers can model production procedures using CAD and CAE (computer-aided engineering) technologies to ensure that products are produced accurately first time.

- Using CAD software, components can be modelled and assembled to check that they fit, and to review any possible issues with the assembly process; this can be done with sectional and wireframe views combined with exploded diagrams.
- FEA (Finite Element Analysis) can be used to analyse the stresses that a component will face during its working life, although this may still be tested on the real product because some environmental conditions may not be effectively modelled in the software.
- Specific forms of CFD (Computational Fluid Dynamics) such as MFA (Mould Flow Analysis) can be used to check the flow of materials within moulding machinery. This is essential to ensure that the correct cooling and widths of channel are used to allow the molten material to fill the cavities in the time required.

Along with these pre-production checks, the design of components is often modified to help QC checks. This can be seen, for example, on the base of glass drinks bottles where a number is embossed to record the mould used to form it. This allows manufacturers to trace issues back to specific moulds or cavities.

In Figure 2.9.2, you can see that multiple part moulds mean manufacturers can replace components within the mould independently if errors are found, rather than replacing the whole mould.

Project management systems

Effective project management is essential in all design and manufacture activities, to ensure that they are completed within budget and to agreed time scales. To ensure that both of these elements are addressed, management systems aim to improve efficiency and reduce waste within all the activities. Within industry, there are several different systems used that have developed since the 1950s. These include TQM, **Scrum** and Six Sigma.

Total quality management

Total quality management (TQM) is an approach to project management that has been used since the 1950s, gaining popularity in the 1980s and beyond. The main aim of TQM stems from QA, with the ambition to remove waste and produce products right first time.

Companies that use TQM strive for continual improvement. They value the views of their workforce, and encourage the workforce to participate in teams where individuals can problem solve and contribute to the overall effectiveness of the production process without fear of expressing their views. All procedures are clearly communicated to the workforce, and the understanding of these procedures is essential.

Scrum

Scrum (agile manufacture) is a method of project management first used within software development. The main focus of Scrum is to work in a team to reach goals in short timescale 'sprints'. The team works on the specified team goal, and attends daily scrum updates where individuals feed back their progress towards the team goal, as well as any issues that are stopping them progressing. The distribution of tasks can be updated based on these issues, and the team can quickly respond to changing customer demand due to the regularity of feedback meetings.

Six Sigma

In the 1980s, Motorola introduced the Six Sigma system – a set of techniques and tools for process improvement which is designed to minimise defects. The aim of the system is to reduce the number of defective products produced to less than 3.4 in every million. The system requires the implementation of a DMAIC procedure (define, measure, analyse, improve and control) to assess and improve each stage of the design and manufacture activity.

The five key stages of Six Sigma:

- Define: what is the issue within the process?
- Measure: take steps to measure the extent of the issue.
- Analyse: determine where the issues measured occur.
- Improve: introduce procedures to rectify the issues identified.
- Control: ensure the modified procedures are implemented and maintained through effective QA.

Each Six Sigma project undertaken in an organisation follows these defined stages, and has a specific target, for example 'reduce costs' or 'increase profits'.

Lean manufacture

Six Sigma is primarily aimed at reducing the number of defective products through monitoring. Lean manufacture, however, is a systematic approach to production which aims to eliminate all waste from product production. Waste is identified as anything that does not benefit the client and is given the name 'muda'. There are seven forms of muda, which are given the acronym TIMWOOD:

Transport: when a product is being transported, it is at risk of damage or loss. By reducing the distance a product is transported, you reduce these risks and the chance of production being delayed due to traffic issues and so on.

Inventory: the main aim of JiT manufacture is to reduce the inventory (raw materials and finished goods) on site at any time. Any stock held on site is again at risk of damage and loss in value, and may delay final sales.

Movement: unlike transport, movement refers to employees and their equipment. When assembling products, any unnecessary movement increases the production time, and should be minimised.

Waiting: if processes within the manufacture are not split evenly in regards to time then products or workers may be waiting as others catch up.

Over production: again, a major part of JiT manufacture is the lack of storage requirements. If a manufacturer was to over produce products 'just in case', extra storage would be required, and the excess products may never be required for sale, wasting investment in materials and production.

Over processing: it is essential that the correct equipment is used for each manufacturing process. If a company invests heavily in machinery, this investment must be justified. This can be likened to the purchase of a new smartphone – if the user is not going to use all of the functions available on that particular model then a cheaper version would be sufficient.

Defects: any defective products must be removed. This is key within Six Sigma and relies on effective QC and quality QA procedures.

ACTIVITY

Choose one of the case studies below and summarise the key benefits that applying a lean approach have made for the company:

- www.epa.gov/lean/lockheed-martin
- www.epa.gov/lean/general-motors-corporation
- www.epa.gov/lean/apollo-hardwoods-company.

Lean Six Sigma

The DMAIC approach of Six Sigma is a structured framework that can be applied to all areas of the workplace to reduce variation in performance. The TIMWOOD waste reduction strategy of lean manufacture gives clear guidance on the forms of inefficiency that need to be addressed. By combining both strategies, companies aim to improve operational and manufacturing excellence that has maximum benefit for the customer.

Six Sigma is used within a wide variety of industries, including banking, logistics, manufacture and hospitality.

Critical path analysis

Critical path analysis (CPA) is a project management method used to analyse all individual stages within a project, and to plan the effective and time efficient completion of each element within the desired schedule.

Using a task analysis, the project/process can be split down and each individual element can be arranged in time order. The basic order is sequential, with each element being completed one after the other.

This method often identifies wasted time within the process, when the individual completing the process is waiting unnecessarily to complete the next task.

By analysing dependencies, the process time can be reduced. Table 2.9.1 shows a simple example of how this can be done for making a cup of tea.

Table 2.9.1 Making a cup of tea.

Task		Time (seconds)	Dependent on	
A	Fill kettle with water.	30		
B	Boil water.	180	A	Fill kettle with water.
C	Place teabag in teapot.	10		
D	Retrieve milk from fridge.	30		
E	Choose cup.	10		
F	Add boiled water to teapot.	20	B	Boil water.
G	Let tea brew in pot.	180	F	Add boiled water to teapot.

Task		Time (seconds)	Dependent on	
H	Pour tea into cup.	10	G	Let tea brew in pot.
I	Add milk to cup.	10	E	Choose cup.
J	Return milk to fridge.	30	I	Add milk to cup.
	Total time	510		

It is clear from the list of tasks that if all the tasks are completed one after the other (sequentially), there would be a lot of wasted time.

By reorganising the tasks using the dependencies to structure the process, then the overall time required to make the tea can be shortened considerably.

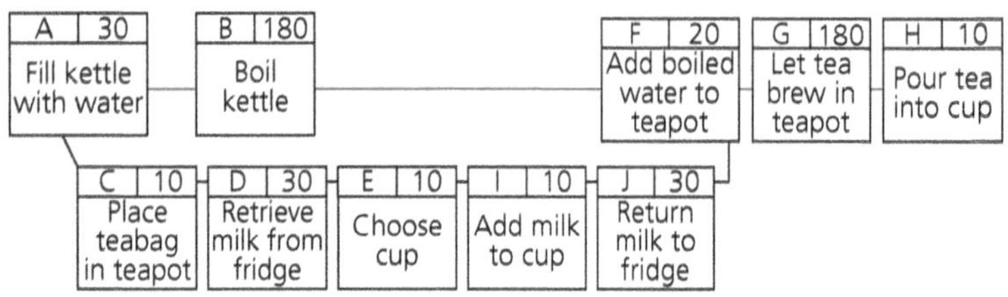

Figure 2.9.3 A critical path for the tea making process.

Figure 2.9.3 shows the tea making process with the critical path (longest time path) shown in red. This shows that the minimum time required to make the cup of tea is 420 seconds, a saving of 90 seconds in comparison to the sequential process shown in Table 2.9.1. This is due to parallel processing, where two tasks are undertaken at the same time.

As making a cup of tea will only ever need one person, parallel processing can only happen while waiting for the kettle to boil or the tea to brew.

In industry, where there are many people working on a single project, tasks can be processed simultaneously (in parallel) more easily. In the manufacture of products, this can be aided by the creation of sub-assemblies, where sections of a product can be assembled and then brought together on the final product. If we consider a simplified example of the mobile phone, this can be split into several sub-assemblies:

- the electronics – PCB and central processing unit (CPU)
- the screen
- the external case
- the battery.

Each of these sub-assemblies can be created separately at the same time, and then the final assembly of all four elements can take place.

Further example

To complete a CPA, several elements are required:

- a list of all activities within the project
- an estimate of the time each stage will take
- an understanding of how each stage relates/depends on the completion of other stages in the plan (dependencies)
- specific deadlines for individual stages/items.

The overall aim of CPA is to enable the completion of a project in the least possible time. To do this, you must produce a critical path network (CPN), which shows each individual stage and its relationship to all other stages within the project.

Below is an example CPA for the production of a table lamp base:

Table 2.9.2 An example CPA for the production of a table lamp base.

Stage	Description	Dependent on	Duration (hours)
A	Produce dimensioned drawings for each component.		5
B	Order materials for individual components.	Completion of stage A	1
C	Prepare mould for concrete base.	Completion of stage B	8
D	Cut and weld low carbon steel cage.	Completion of stage B	3
E	Drill and prepare low carbon steel cage.	Completion of stage D	2
F	Assemble finished low carbon steel cage within mould for base.	Completion of stages C & E	2
G	Cast concrete base.	Completion of stage F	3

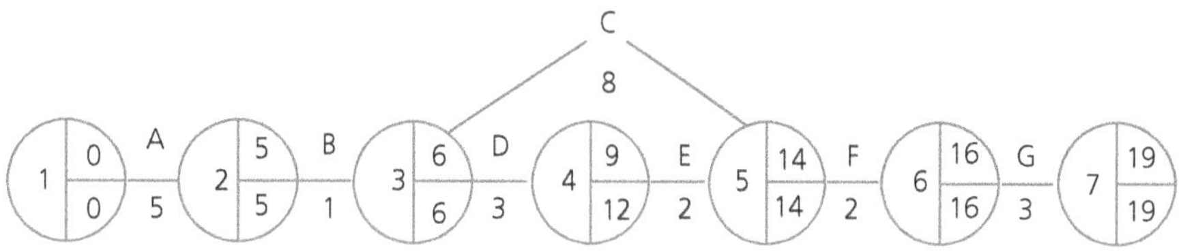

Figure 2.9.4 The CPA for a lamp base.

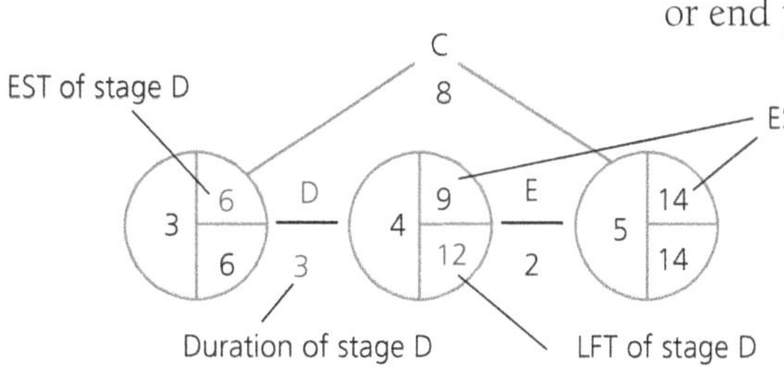

Figure 2.9.5 CPA activities C–E.

Each activity is given a letter. A circular node, which represents the start or end point of a stage, is included, with the stage letters shown above the joining lines, and the time of each stage shown below the line.

Figure 2.9.4 shows the stages within the lamp base construction. The left-hand side of each circle shows the node number. The upper right quadrant of each circle shows the earliest start time (EST) in hours from the beginning of the project. The lower right quadrant shows the latest finish time (LFT) for the previous stage so as not to impact on the overall finish time.

As you can see in Figure 2.9.5, the only time where the EST of one stage and LFT of the previous stage differs is in node number 4. This is due to stage C taking longer than the combination of stages D and E. As stage E must be completed by the same time as stage C, then it must be started no later than two hours before the EST of activity F.

Table 2.9.3 An example CPA showing LFT and EST times.

Stage	LFT	Duration	EST	Slack
A	5	5	0	0
B	6	1	5	0
C	14	8	6	0
D	12	3	6	3
E	14	2	9	3
F	16	2	14	0
G	19	3	16	0

ACTIVITY

Consider the manufacture of the most difficult component of a project you have worked on. Using the five key stages of Six Sigmas, explain how you would modify the production process to improve QA.

The two stages showing slack in Table 2.9.3 can be delayed without affecting final project completion, and therefore do not lie on the critical path.

By using CPA within project management, a designer or manufacturer can ensure that they are focusing on the key elements of the process (those that lie on the critical path). This is essential to prevent delays in the final production. This works well with the Scrum approach to project management because the regular meetings allow the team to prioritise tasks on the critical path and redistribute resources to complete these on time.

Quality control

Quality control (QC) refers to the monitoring, checking and testing of materials, components, equipment and products throughout production to ensure they conform to acceptable tolerances specified within the QA policies within the company.

QC inspection checks take place throughout the production process, and are performed in conjunction with strict guidance documentation produced by the company and the client, to ensure that the products fit the specified requirements of the client.

Material checks

At the beginning of any production process, the materials must be checked for their compliance with the manufacturer's specification. The check carried out at this point can consist of:

- simple visual checks on the materials and data sheets supplied
- chemical analysis on small volumes of the material, checking for levels of moisture or specific elements that may affect the product
- colour matching may also be required for polymer pigments or paint finishes.

Dimensional accuracy checks

Throughout the production process, relevant QC checks will be implemented to ensure the equipment is working accurately. This can include dimensional accuracy checks on formed components, where they are measured to check they fit within set limits.

Digital measuring devices

Dimensional accuracy checks may be carried out with flexible measuring equipment, such as a digital vernier caliper or micrometer, where a range of measurements can be checked and exact readings can be recorded. This tends to be done with interval sample testing, where a small sample

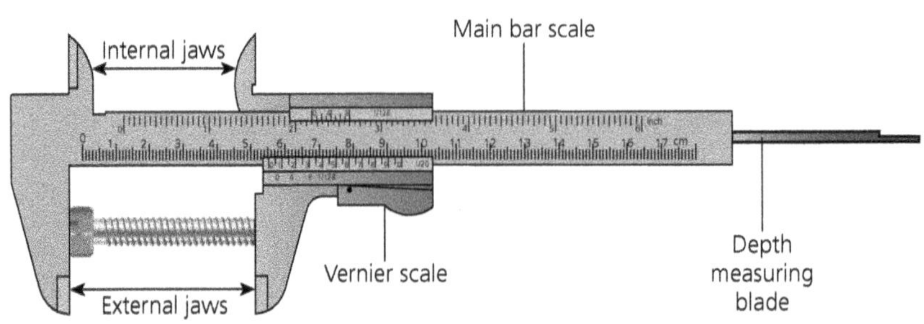

Figure 2.9.6 A manual vernier caliper.

Figure 2.9.7 Car engine crankshaft being checked and measured with a micrometer.

of products are removed from the production line and checked in the QC department. This is because the check cannot be performed quickly enough to keep up with the production line.

Vernier calipers are a precision measuring instrument for making a wide range of distance measurements which will give readings accurate to ± 0.02 mm. They can be used to measure external, internal and depth measurements as seen in Figure 2.9.6.

Micrometers are used to check dimensional accuracy in much the same way as a vernier caliper, although they tend to be used on small components due to the form of the jaws restricting the maximum measurement possible. The level of accuracy possible with a micrometer is higher than with a vernier caliper due to the screw thread movement of the jaws, which gives them an accuracy of ± 0.01 mm. However, due to their shape they are limited to measuring external dimensions, usually diameters only.

Go/no go gauge

A level only

When checking dimensional accuracy on a production line, a specific measuring instrument is often used, such as a go/no go gauge, which checks whether a single measurement fits within a tolerance range, giving a simple pass/fail reading. This form of measuring is quicker than a vernier caliper or micrometer because the operator does not have to check for an accurate reading, and the checking procedure can be easily given to another operative with little training. The gauge does not need adjustment or recalibrating.

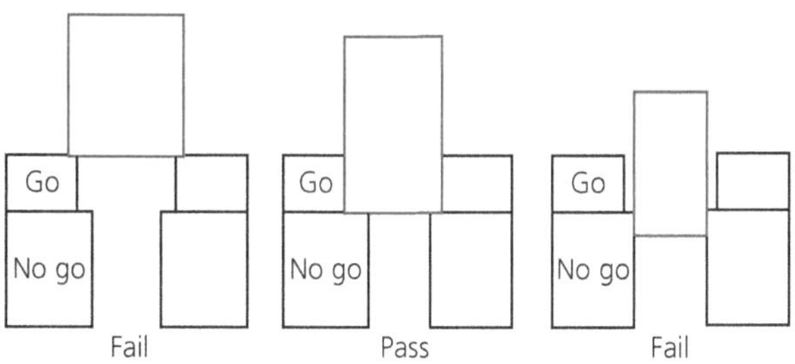

Figure 2.9.8 How a go/no go gauge works.

These gauges have two dimensions – the maximum allowable size given by the dimensional tolerance and the minimum allowable size given by the dimensional tolerance for the part. Figure 2.9.8 shows three parts being tested with a go/no go gauge. For the part to be accepted, it must be within the dimensional tolerance required. This means it must be smaller than the go opening and bigger than the no go opening.

Co-ordinate measuring machinery

A level only

The use of co-ordinate measuring machinery (CMM)s such as a probe scanner, allows a manufacturer to check a range of predefined measurements on finished components. The accuracy of this technique is extremely high, and the data is automatically compared against a 3D CAD model and logged on a computer system.

More recently, the introduction of non-contact laser scanning allows manufacturers to log a far greater number of dimensions, with scanners recording many thousands of readings per second. This allows a true comparison with a CAD model, and 3D CAD models can be produced from the scan itself. This relies on the reflection of a laser off the surface of a product to record accurate dimensions.

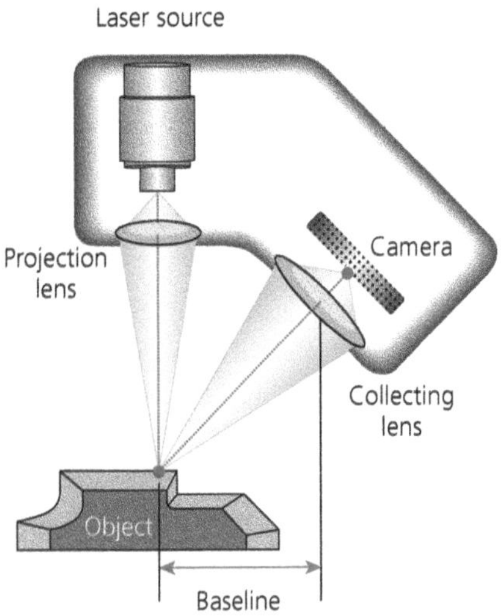

Figure 2.9.10 CMM using a non-contact laser scanner.

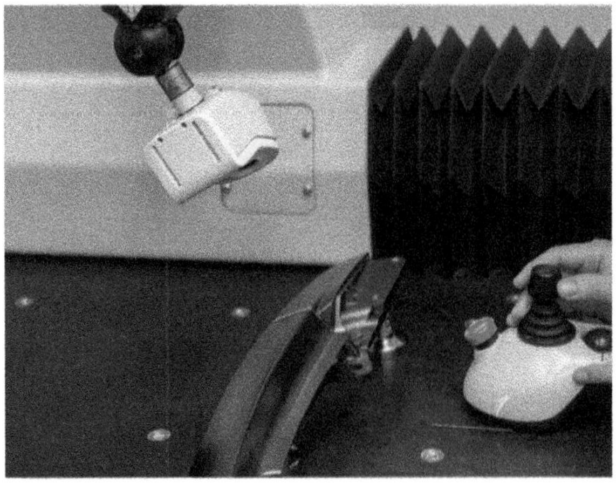
Figure 2.9.9 3D laser scanning a metal casting.

CMM can be used to check tooling for dimensional accuracy during maintenance, and the results may be used to update QA procedures, setting the regularity of tool changes.

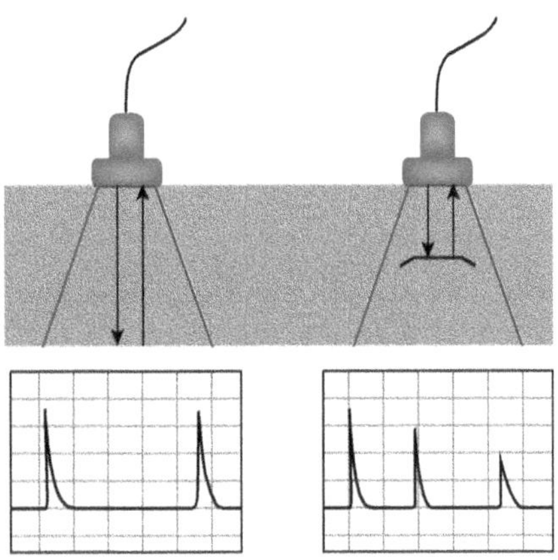
Figure 2.9.11 Ultrasonic testing diagram.

ACTIVITY

With reference to a specific industry you are familiar with, suggest a suitable use for each of the four measuring devices listed below:

- vernier calipers
- micrometers
- go/no go gauges
- CMM.

Non-destructive testing

A level only

Non-destructive testing (NDT) refers to methods used to check the internal structure of materials, often after joining through processes such as welding. Two of the main methods used for this are x-ray and ultrasound analysis. The material is subjected to radiation or ultrasound waves to check for refraction of the signals, which can indicate any faults not visible to the naked eye. This form of testing is essential on products when a joint failure could be a major safety issue. Ultrasonic testing is safer than x-ray testing because it uses sound waves rather than radiation. The test records defects where the reflecting signal indicates something other than a solid material.

ACTIVITY

Consider a product you have chosen to manufacture for a project. If you were to produce it on a production line, where may you use a go/no go gauge and what would the gauge look like?

Monitoring, checking and testing throughout production

Case study

Allied Glass is a world producer of high-quality glass containers. Its success over the last 150 years is largely down to its ability to facilitate a wide range of varying designs. Currently it produces in excess of 600 million containers per year for the premium brands drinks market. To ensure that these containers meet the high level of standards expected by its customers, it employs a wide range of QA procedures and QC controls.

Stage 1: Customer specifications

The initial design of a glass container is developed closely with the prospective customer, who will give clear and specific criteria that Allied Glass must follow. These details will include aesthetic elements such as the type and clarity of glass to be used, maximum sizes, details of the market where the product will be sold and the volume of liquid to be held within the container.

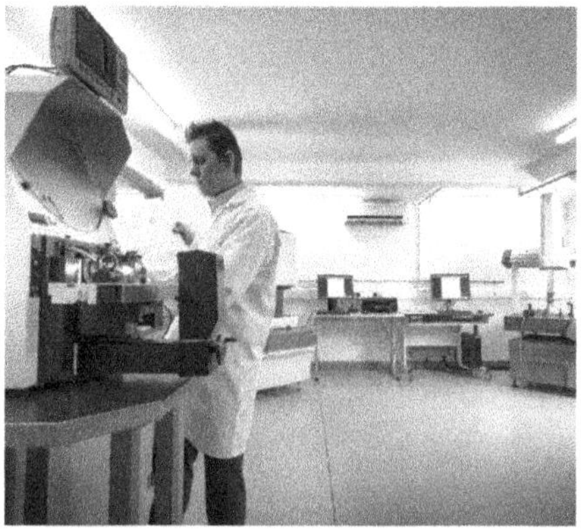

Figure 2.9.12 A QC lab used for testing sample products taken from a production line.

Stage 2: Concept development

Design development is done in-house. This means that the lead time (time from initial concept to final production) is reduced because the new product development (NPD) team employs a concurrent approach, guiding the customer through the production limitations during concept generation. Concepts are modelled in solid form using 3D CAD software which is used to drive two rapid prototyping machines. Prototype models are produced as full-size visual representations using either powder or PLA polymer, allowing the customer to hold and critique the form of the container.

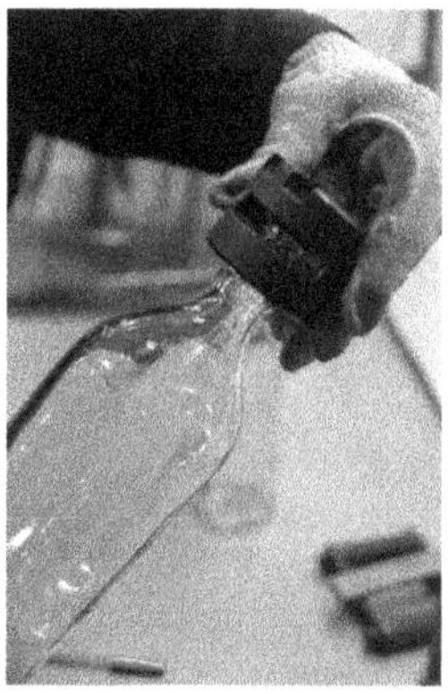

Figure 2.9.13 Checking the dimensions of a bottle neck using a go/no go gauge.

Stage 3: Mould design and sampling

Using 3D CAD software, the mould for the container is developed to fit with the existing production equipment. This means that the design must conform to the specific size constraints of the machinery. The bottle mould is constructed from many components, several of which are standardised across a range of designs. This improves QA due to the experience of using tried and tested designs. Once the mould has been designed, a sample set of mouldings is produced for testing. The testing regime is thorough; some of the main stages are listed below:

- Working drawings with tolerances: a set of specification drawings and criteria are issued for each container design, and the tolerances are used to adjust the production equipment for the specific job.
- Sample prototypes: a sample set of moulds is used to produce a batch of the containers for QC testing.
- Sample dimension checks using flexible measuring equipment (for example, vernier calipers): due to the low volume of sample containers, this form of checking is preferred because all aspects of the container are checked for dimensional accuracy, allowing the company to specify the limits of acceptability for the full-scale production.
- Visual aesthetic checks: as the sample bottles are formed, they are checked visually for faults in the forming process, such as 'shorts' where the glass has not filled the mould. These faults can sometimes

be corrected by adjusting the temperature of the glass entering the mould or the cooling rate of the mould itself. The faults are checked to see if they are common across all moulds of the same product. This may indicate a possible issue with the design, whereas an issue on an individual mould may indicate a machining issue. All of the data gained at this stage allows the programming of temperature ranges and mould cooling times for full scale production.

- Machine tooling and alignment calibrations: the production line tooling is calibrated for the production of the sample, and adjustments to these settings are made during the sample production, with visual inspection of the product guiding these adjustments. When the settings are finalised, these are recorded within the specification report. When full scale production commences, the settings are modified specifically for the container being produced.
- Machine maintenance and tooling replacement: Allied Glass has a pre-determined programme for machine maintenance, and the regularity of tooling changes to prevent unnecessary wear from the molten glass. This may be adjusted based on the performance of a container mould during sampling. Any modifications would be recorded on the specification report ready for full scale production.

Figure 2.9.14 Glass bottles leaving the forming moulds before entering the annealing oven.

- NDT and destructive testing: the sample containers are tested in the QC department by checking internal volumes, stress levels in the glass to prevent shattering, stability of the container and pressure resistance (specifically for carbonated drinks).
- Template, jigs and fixtures produced: every different container design has a specific set of gauges machined to test the product against set tolerances. These are produced based on data from the specification drawings and sample dimension checks.
- Specific sampling regularity and procedures set by the company, customer or legal requirements: based on all the sampling data and reports, the testing procedures and regularity are planned and communicated to relevant personnel. This sets specific checks to be performed on every container, on random containers hourly and on QC machines to check reliability.

Stage 4: Production

Combined with the QA procedures employed for each individual container, Allied Glass must monitor the overall quality of the glass being used during the production of all containers.

The raw materials used for clear glass production are sand, soda ash, limestone and cullet (recycled glass). One of the main issues with these materials is the colour of glass produced: the higher the iron content, the greener the glass. This means that all the materials must be sample tested before they are accepted on site.

Suppliers must provide data sheets and comply with the tolerances of acceptability set by Allied Glass. Once the materials have been accepted, they are stored in silos ready for production.

As the materials are melted in the furnace, the temperature is tightly controlled within pre-determined tolerances for the specific containers being made (larger containers need slightly lower temperature glass because they retain the heat longer during forming). The furnace feeds several production lines, each producing different containers. The temperature of the glass at this point is monitored and controlled remotely specific to the individual container being produced.

Once the glass enters the production line, it is formed through a glass blow moulding process and then fed through an annealing oven to reduce the internal stresses and prevent brittleness.

During production, the QC checks are too numerous to record. However, below is an explanation of a range of common checks and how they are used to ensure quality and efficiency.

When full scale production begins, the new container moulds are installed on a production line and the specification report from sampling is used to calibrate all machinery and tooling.

All QC checks and gauges are issued to the relevant stations on the production line, and all relevant personnel are briefed in advance of production. The company operates a process of continual improvement, and all employees are encouraged to contribute their expertise in specific areas to improve efficiency.

Figure 2.9.15 Glass bottles passing a light box on a production line for manual visual inspection before camera inspection.

QC testing

Every container is checked, using camera analysis, for glass imperfections; this is checked against the bottle labelling on the base, which records the individual mould where that container was produced. Any issues with specific moulds will be fed back to the forming end of the production line and investigated.

Every hour, each production line provides a sample for the QC department to check volumes, stresses within the glass, dimensional accuracy and several other specific QC checks. Random sampling also takes place on the production line using specific gauges for the container being produced. These are checked and any faults are recorded, with the results feeding back to relevant points on the production line.

KEY TERMS

Quality assurance (QA): the procedures and policies put in place to reduce waste and ensure manufactured products are produced accurately within set tolerances.

Scrum: a project management method, which focuses on the organisation of workforce teams to reduce the lead time of projects.

Total quality management (TQM): a project management system focusing on QA through the full involvement of the workforce in continual improvement.

Critical path analysis (CPA): a project management method used to analyse all individual stages within a project and plan the effective and time efficient completion of each element within the desired schedule.

Quality control (QC): the monitoring, checking and testing of materials, components, equipment and products throughout production to ensure they conform to acceptable tolerances.

KEY POINTS

- When producing a prototype product, it is essential that you plan for accurate production; this requires effective research and development prior to investing time in the final manufacture.
- The development of the design must allow the concept to be tested against key specification criteria.
- Modelling strategies may include virtual CAD simulations and full-size block models.
- Manufacturing procedures must be evaluated for suitability and the achievable level of accuracy.
- By using effective QA procedures, a manufacturer is aiming to produce products 'right first time and every time'.
- Prior to production, designers and engineers can model production procedures using CAD and CAE technologies, to ensure that products are produced accurately first time.
- Effective project management is essential in all design and manufacture activities, to ensure that they are completed to budget and within the agreed time scales.
- QC inspection checks take place throughout the production process, and are performed in conjunction with strict guidance documentation produced by the company and the client.
- Dimensional accuracy checks may be carried out with flexible measuring equipment, such as a digital vernier caliper or micrometer. A range of measurements can be checked and exact readings can be recorded.
- When checking dimensional accuracy on a production line, a specific measuring instrument is often used, such as a go/no go gauge, which checks whether a single measurement fits within a tolerance range, giving a simple pass/fail reading.
- NDT refers to methods used to check the internal structure of materials, often after joining through processes such as welding.

Check your knowledge and understanding

1 How many defects per million products produced would be acceptable for a Six Sigma compliant process?

2 Explain three QA procedures that may be used in the manufacture of injection moulded components.

3 What is meant by the term 'critical path'?

4 Explain three QC checks that may be implemented in the production of a turned aluminium component.

5 Why is ultrasonic testing used in preference to x-ray testing?

Further reading

For more information on TQM: asq.org/learn-about-quality/total-quality-management/overview/overview.html

For an introduction to Six Sigma: www.isixsigma.com/methodology/total-quality-management-tqm/introduction-and-implementation-total-quality-management-tqm/

www.isixsigma.com/new-to-six-sigma/getting-started/what-six-sigma/

For more information on Scrum: www.scrumalliance.org/why-scrum

For greater detail on NDT: www.asnt.org/MinorSiteSections/About/ASNT/Intro-to-NDT

For more information on measuring for accuracy: www.qualitymag.com/articles/85023-quality-101-caliper-basics

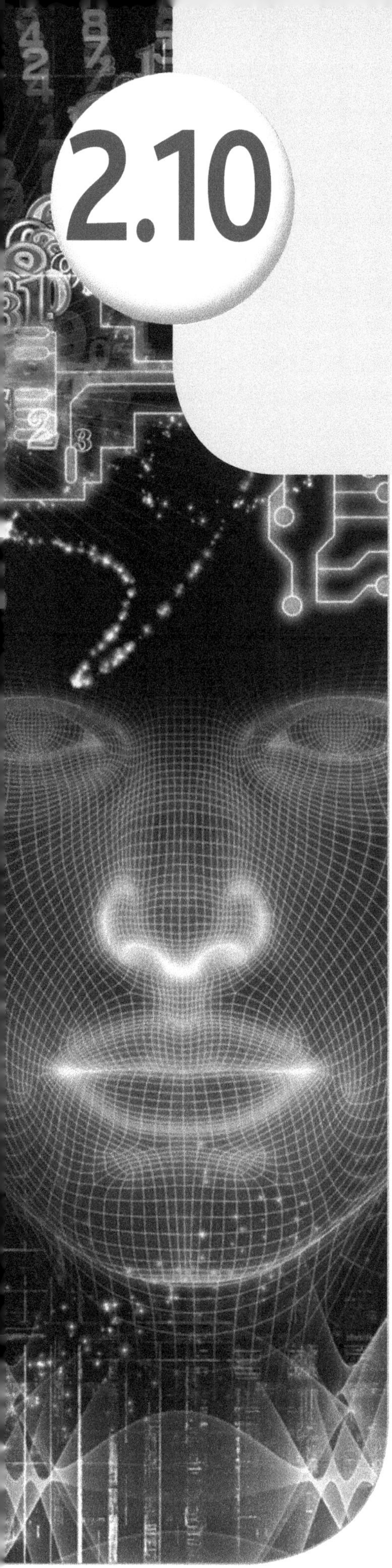

2.10 National and international standards in product design

A level only

LEARNING OUTCOMES

This topic is for A-level only. If you are studying at AS level, you do not need to cover the content in this section.

By the end of this section you should have developed a knowledge and understanding of:

- the importance of national and international standards in product design.

For products to be commercially successful, they must conform to strict national, European and international standards. These standards cover a wide range of areas. Safety standards must be complied with if the product is to be sold within a specific region. Certification with standards such as ISO 9001 (quality management) is not compulsory, but can be essential when working with other companies that may, as part of their QA procedures, only deal with compaines that are ISO 9001 certificated.

Please note: from time to time, European standards may become obsolete and are superseded by new standards. You should be aware that European standards for certain products do exist. You should aim to know about the following standards and directives.

British Standards Institution

The **British Standards Institution (BSI)** is a national organisation formed to devise agreed standard procedures for performing a wide range of tasks. The range of current standards in the BSI portfolio exceeds 30,000 (January 2017). The recognised kitemark shown in Figure 2.10.1 for British Standard Verification or conformity can only be applied where the relevant standard has been met.

Figure 2.10.1 BSI kitemark.

If a British standard has been accepted by a European standardisation organisation, it will carry the prefix 'BS EN'.

Examples of current British standards:

- BS EN 71-1:2014 Safety of toys. Mechanical and physical properties.
- BS EN 62115: Electric toys. Safety.
- BS EN 50361: 2001 Basic standard for the measurement of specific absorption rate related to human exposure to electromagnetic fields from mobile phones (300 MHz – 3 GHz).

The presence of a BSI kitemark can be vital, because consumers often use this when deciding on a product to purchase.

International Standards Organisation

BSI is one of over 150 national standards bodies that are part of the **International Standards Organisation (ISO)**, where internationally recognised standards are agreed and put in place. With many companies increasingly looking to extend markets and trade in other countries, it is essential that they conform not only to national but also international standards.

Standards for management services such as ISO 9001, which deals with quality management, is applied worldwide, with many companies only dealing with others who conform to the standard.

Many standards refer to management systems, one of the most well-known being ISO 9001.

The presence of a CE mark means a product conforms to all relevant European safety standards. Its presence is mandatory for a product to be sold within the EU.

Key international standards include:

- BS EN ISO 9001 quality management
- BS EN ISO 14000 environmental management
- BS EN ISO 50001 energy management
- BS EN ISO 31000 risk management.

Companies can opt out of conforming with these standards. However, the aim of the standards is to improve safety, productivity and reliability.

Restriction of Hazardous Substances directives

The **Restriction of Hazardous Substances (RoHS)** directive (2002/95/EC) is a European directive that restricts the use of specific hazardous materials found in electronic and electrical products. As of July 2006, all relevant products sold in the EU were required to be RoHS compliant. This directive aims to prevent hazardous substances from entering the production process in order to prevent damage to human health and the environment.

Out of the ten specific substances that RoHS restricts, four are metals: lead, mercury, cadmium and chromium.

Battery directive

The Battery directive (2013/56/EU) is an amendment of the directive (2006/66/EC) which, in combination with the RoHS and Waste from Electrical and Electronic Equipment (WEEE) directives, deals specifically with the restriction of hazardous substances and safe disposal of batteries

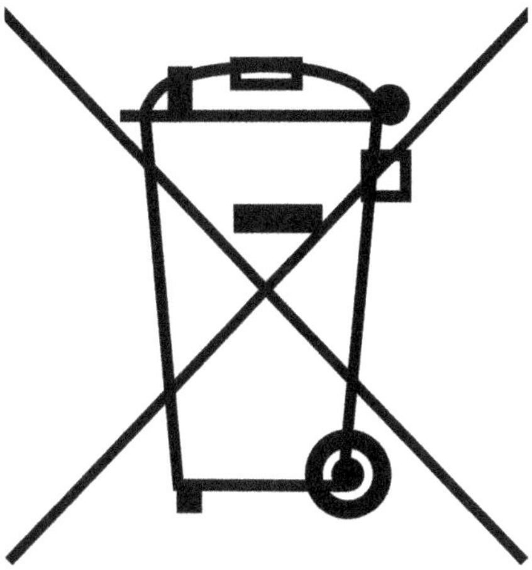

Figure 2.10.2 Battery directive symbol.

and accumulators (used for storing energy or power such as a capacitor or electromechanical cell used in rechargeable batteries). The directive states that a limit of 0.0005 per cent mercury is allowed in batteries and accumulators, including button cells such as watch batteries. It also restricts the volume of cadmium in portable batteries and accumulators to 0.002 per cent. As of 1 January 2017, this restriction also includes the previously excluded cordless power tools.

The battery directive requires all batteries and accumulators to show a crossed-out wheelie bin to explain that batteries should not be disposed of in normal waste. Clear instructions for safe removal and disposal must be provided with the product for it to conform with the directive.

Polymer codes for identification and recycling

The Mobius Loop is an internationally recognised recycling symbol of three arrows, which shows that a product may be recycled. The loop may include a percentage or, on a polymer product, an SPI (Society of Plastics Industry) code is used to state the polymer resin used in its production, so that during recycling materials can be effectively separated.

Figure 2.10.3 Polymer recycling Mobius Loop.

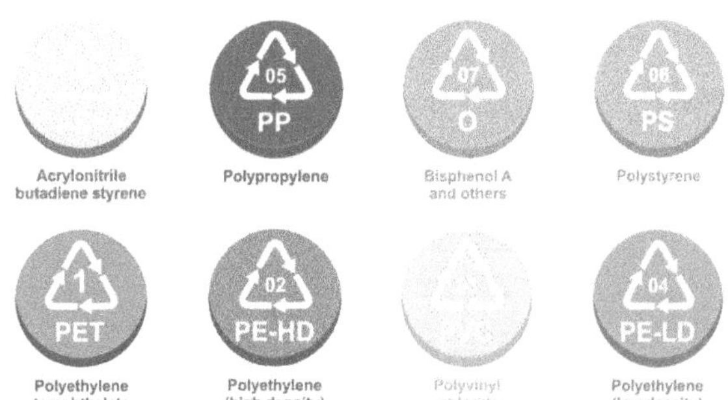

Figure 2.10.4 Polymer SPI codes and descriptions.

Packaging directives

The EU packaging and packaging waste directive (94/62/EC) aims to limit the production of, and prompt the recycling and reuse of, packaging materials. The directive covers all areas of packaging from commercial to household.

Figure 2.10.5 EC energy label.

The most recent amendment states that:

- at least 60 per cent (by weight) of packaging waste must be recovered or incinerated at waste incinerators with energy recovery
- all packaging must be marked with the specific materials used to assist in identification and classification. The marking must be clearly visible on the product or its label.

Each member state has to ensure that all new packaging complies with the following three requirements, and is required to monitor packaging and waste against the targets set in order to:

- keep the weight and volume of packing materials to a minimum
- reduce hazardous substances and materials within packing materials
- design reusable or recoverable packaging.

WEEE directive

The **Waste from Electrical and Electronic Equipment (WEEE)** directive (2002/96/EC) is a mandatory European directive that covers the end of life of electric and electronic equipment, whereas RoHS details the materials that can be used in a product. The directive came into force in August 2006, and all relevant products must also carry the crossed through 'wheelie bin' sticker to show compliance with the directive.

The EC energy label

The EC energy label is a compulsory system required on a wide range of household appliances. The label must be displayed by manufacturers and retailers to assist customers in making purchasing decisions. The use of a coloured scale from A (most efficient) to G (least efficient) gives consumers an easy method of comparison. The compulsory nature of the label, combined with the high visibility in retail outlets means that manufacturers have been forced to develop the efficiency of these products.

Since 2010, a new EC energy label has been introduced, as shown in Figure 2.10.6, which uses pictograms to show key information. For household goods, such as washing machines, the scale has changed to A+++ to D, due to improvements in energy efficiency.

Eco-labelling

Eco-labels are a wide range of voluntary environmental certifications given to companies/products that are seen to meet specific environmental targets set for a product category. By displaying these eco-labels on their products, manufacturers are able to provide potential customers with a greater level of information.

Recognised across Europe, the EU Ecolabel is a label of environmental excellence that is awarded to products and services meeting high environmental standards throughout their life-cycle, from raw material extraction to production, distribution and disposal. Table 2.10.1 shows some of the criteria for washing machines showing the European energy label.

Figure 2.10.6 EU Ecolabel.

Table 2.10.1 Some of the criteria for washing machines showing the European energy label.

Life cycle step	Criterion	Expectations
Manufacturing	Design for environmental use	Clear indication of the appropriate settings according to fabric type and laundry.
		Clear indication of the energy and water saving programmes and options.
Use	Energy saving	Electrical energy consumption ≤ 0,17 kWh/kg of washload (for the same standard 60 °C cotton cycle as chosen for Directive 95/12/EC).
	Water saving	Water consumption ≤ 12 L per kg of washload (for the same standard 60 °C cotton cycle as chosen for Directive 95/12/EC).
	Reduction of noise	Information about the noise level of the machine provided in a clearly visible way to the consumer.
End of life	Eco-design to facilitate recycling	Easy disassembly of the machine taken into account in the design.
		A disassembly report shall be provided.

The energy efficiency label and logo

The energy efficiency label is a certification mark issued by the energy saving trust as a recommendation for only the most energy efficient products within specific categories. The certification is available for appliances such as washing machines and fridges, light bulbs, central heating boilers and also insulation products.

Forest Stewardship Council®

The **Forest Stewardship Council (FSC)** logo can be found on products made from timber, paper or other forest products which are sourced from well-managed forests and/or recycled materials. FSC certified forests are managed in an environmentally appropriate, socially beneficial, and economically viable manner, protecting the wildlife and people who live there and rely on the forest for their way of life. The workers within the forest must be local, and FSC guarantee that they are trained and work in a safe environment for a fair wage. Most importantly, when trees are harvested they are replaced or allowed to regenerate, preventing deforestation. FSC is the only forest certification scheme endorsed by WWF, Greenpeace and The Woodland Trust.

Figure 2.10.7 FSC logo.

EU ENERGY STAR®

The EU ENERGY STAR® program was developed from an agreement between the EU and US to standardise how IT equipment was labelled to show the energy used. Products are assessed on their power usage when idle and in sleep mode, and these figures are collated into a database to allow customers to make informed decisions on product choices.

The scheme covers the following types of equipment:

- computers: including desktops, laptops, and tablets
- displays: including monitors and signage displays
- imaging equipment: including printers, scanners and copiers.

You can have a look at the qualifying desktop computer database here: www.eu-energystar.org/database/.

Figure 2.10.8 EU energy star logo.

ACTIVITY

Find examples of consumer products that display a range of the labels described in this chapter. Explain why each of them applies.

Design packaging that would be suitable for transporting a recent project, and display the relevant labelling required for it to comply with the EU packaging directive.

KEY TERMS

British Standards Institute (BSI): a national organisation formed to devise agreed standard procedures for performing a wide range of tasks.

International Standards Organisation (ISO): a federation of national standards institutions that devise international standards to improve safety, productivity and reliability.

Restriction of Hazardous Substances (RoHS): a directive aimed at preventing hazardous substances from entering the production process in order to prevent damage to human health and the environment.

Waste from Electrical and Electronic Equipment (WEEE): a mandatory European directive that covers the end of life of electric and electronic equipment.

Forest Stewardship Council (FSC): a council that manages a certification system to identify and label timber and timber-based products from sustainable sources.

KEY POINTS

- British Standard Verification or conformity can only be applied where the relevant standard has been met.
- If a British Standard has been accepted by a European standardisation organisation it will carry the prefix BS EN.
- The presence of a CE mark means a product conforms to all relevant European safety standards and its presence is mandatory for a product to be sold within the EU.
- As of July 2006, all relevant products sold in the EU were required to be RoHS compliant. For products to comply with the battery directive, clear instructions for safe removal and disposal must be provided with the product.
- Eco-labels are a wide range of voluntary environmental certifications given to companies/products that are seen to meet specific environmental targets set for a product category.

Check your knowledge and understanding

1 Explain the benefits for both the manufacturer and consumer of the EC energy label.

2 Why do manufacturers apply for voluntary eco-labels?

3 State a specific product where you may find the FSC logo and what its presence on the product means.

4 How should you dispose of batteries safely?

5 What does ISO 9001 refer to and why is compliance with this particular standard so important?

Further reading

Details of the EU Battery Directive: data.energizer.com/pdfs/eubattdirectivesummary.pdf

FSC: www.fsc-uk.org/en-uk

BSI: www.bsigroup.com/en-GB/

ISO: www.iso.org/home.html

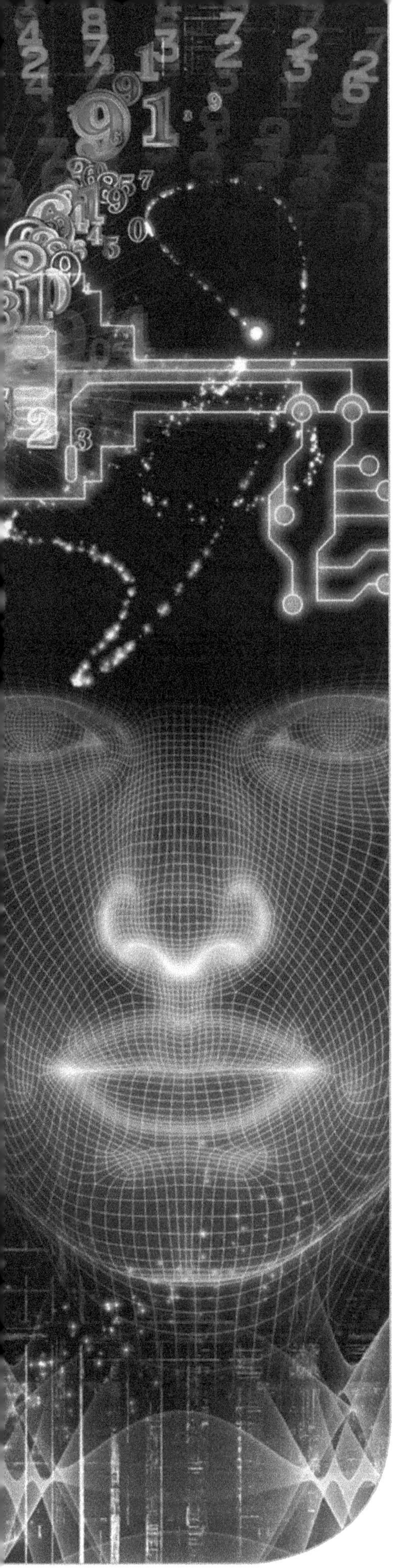

Section 3 Non-exam assessment (NEA)

In this section you will learn about the non-exam assessment (NEA) part of the product design qualification:

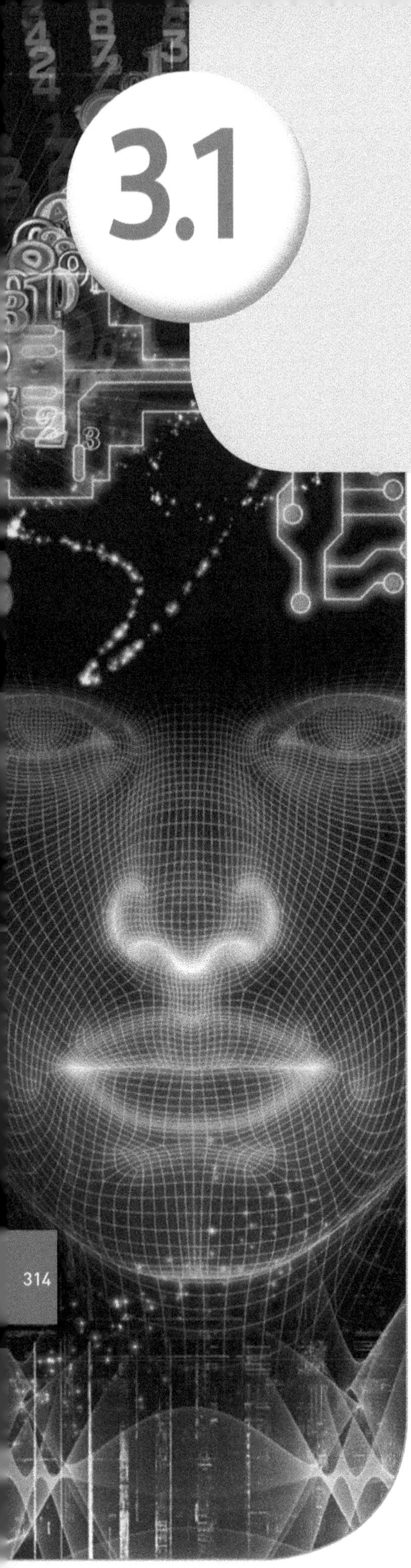

3.1 AS non-exam assessment: design and make task

Introduction to the design, make and evaluate project

In the AS non-exam assessment (NEA) you will be required to carry out a single design and make activity, where you develop a design brief in response to a realistic contextual challenge. The context of this task will be set by AQA.

You will be expected to identify a design problem that will be used to produce a design brief which meets the needs and wants of a user, client or market. It is important that you investigate all of the design contexts that are published by AQA, and select one that you will find interesting, as this will give you the best chance of success.

How long should I spend on the NEA?

You are expected to spend around 30–35 hours working on this NEA project.

You will be able to work on your folder independently (outside of school or college), but you should complete the manufacture of your prototype under the direct supervision of your teacher. You will be allowed to make use of specialist facilities outside of school/college, for example, to weld or powder coat your prototype. However, any work that is done by an outside specialist will have to be detailed on a candidate record form (CRF). The CRF is a document that you and your teacher will sign to authenticate that the NEA is your work, and your teacher will record their marks on it.

What format will the NEA take?

You will create a design portfolio, which includes photographic evidence of your final prototype. Your portfolio can be written or electronic. It is important that you do not produce too much material because you will need to allow time to make a prototype product.

When you produce your design brief, you need to resist the temptation to have just a single, specific design in mind. To be successful in the NEA, you will need to produce a range of design ideas before choosing a final design to develop.

How will I be assessed?

The NEA is worth 50 per cent of your total AS level mark.

You will be assessed against the three assessment objectives (AOs) shown in the table below.

Table 3.1.1 Assessment objectives.

Assessment objective (AO)	Marks
AO1: Identify, investigate and outline design possibilities to address needs and wants.	24
AO2: Design and make prototypes that are fit for purpose.	40
AO3: Analyse and evaluate: • design decisions and outcomes, including for prototypes made by themselves and others • wider issues in design and technology.	16
Overall weighting	80

There are 80 marks available for the project. You will be marked in the five sections shown in the table below.

Table 3.1.2 Assessment criteria.

Assessment criteria	Maximum marks available
A: Identifying and investigating design possibilities	16
B: Producing a design brief and specification	8
C: Development of design proposal(s)	20
D: Development of design prototype(s)	20
E: Analysing and evaluating	16
TOTAL	80

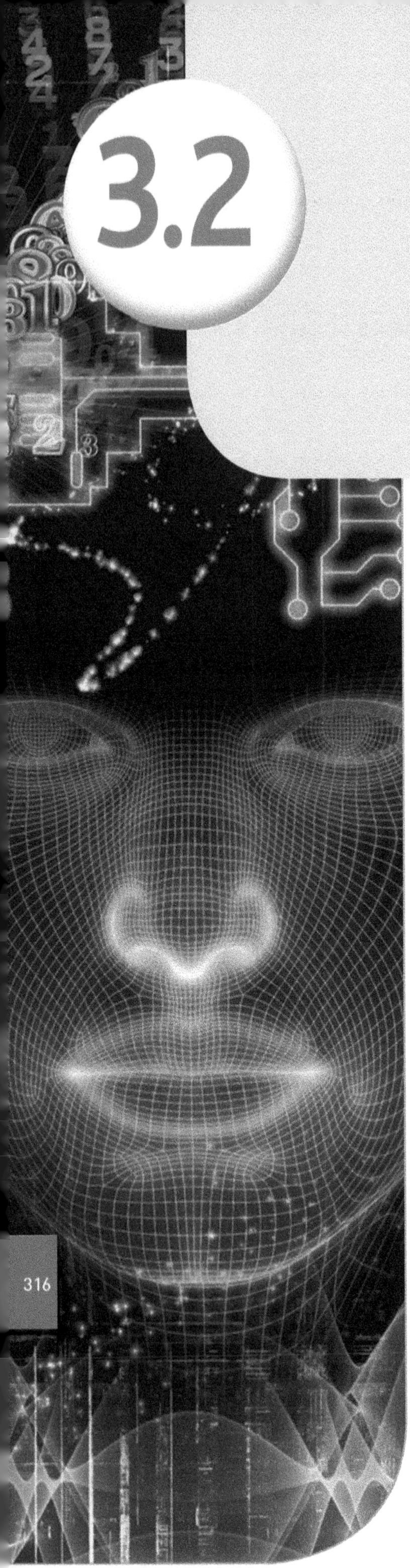

3.2 A-level non-exam assessment: substantial design and make task

Introduction to the substantial design and make task

In the A-level NEA, you will undertake a substantial design and make task, and produce a final prototype based on a context and design brief you have developed yourself.

A-level students are free to choose any design context and are not restricted to choosing one from those offered by AQA. However, the choice of context must be given careful consideration to ensure that you will be able to meet the assessment criteria for the NEA.

How long should I spend on the NEA?

You are expected to spend approximately 45 hours working on this NEA project.

You will be able to work on your folder independently (outside of school or college) but you should complete the manufacture of your prototype under the direct supervision of your teacher. You will be allowed to make use of specialist facilities outside of school/college, for example, to weld or powder coat your prototype. However, any work that is done by an outside specialist will have to be detailed on a candidate record form (CRF). The CRF is a document that you and your teacher will sign to authenticate that the NEA is your work, and your teacher will record their marks on it.

What format will the NEA take?

You will create a design portfolio, which includes photographic evidence of your final outcome. Your portfolio can be written or electronic. It is recommended to be around 45 pages.

When you produce your design brief, you need to resist the temptation to have just a single, specific design in mind. To be successful in the NEA, you will need to produce a range of design ideas before choosing a final design to develop.

How will I be assessed?

The NEA is worth 50 per cent of your total A-level mark.

You will be assessed against the three assessment objectives (AOs) shown in the table below.

Table 3.2.1 Assessment objectives.

Assessment objective (AO)	Marks
AO1: Identify, investigate and outline design possibilities to address needs and wants.	30
AO2: Design and make prototypes that are fit for purpose.	50
AO3: Analyse and evaluate: • design decisions and outcomes, including for prototypes made by themselves and others • wider issues in design and technology.	20
Overall weighting	100

There are 100 marks available for the project. You will be marked in the five sections shown in the table below.

Table 3.2.2 Assessment criteria.

Assessment criteria	Maximum marks available
A: Identifying and investigating design possibilities	20
B: Producing a design brief and specification	10
C: Development of design proposal(s)	25
D: Development of design prototype(s)	25
E: Analysing and evaluating	20
TOTAL	100

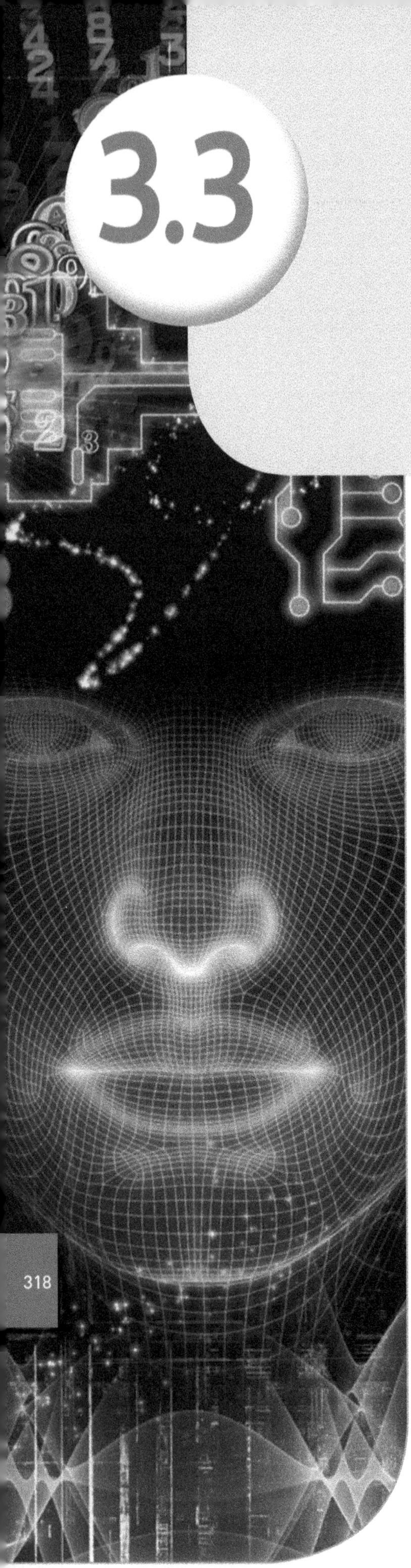

3.3 The design and make task: a closer look

Section A: Identifying and investigating design possibilities

At AS level, you will need to select a design context from the list provided by AQA. At A-level, you will have the freedom to choose your own context that will form the starting point of your NEA project. You need to take guidance from your teacher and be careful when choosing your design context, making sure that it will give you enough scope to generate evidence to meet all of the assessment criteria.

Once you have chosen a context, you will need to identify a potential user(s) for your product. When this is done, you will need to carry out a thorough investigation of all the aspects of the context, and the requirements of the user.

There are a range of investigation methods that you could use; these are divided into primary and secondary techniques. Chapter 2.1 contains useful information on these techniques.

Primary techniques

Primary techniques include interviews, questionnaires, observation of people using products, visits to shops, museums and exhibitions and practical testing and disassembly of existing products.

Through carrying out interviews with a client or potential users of the product, it may be possible to identify their preferences or specific requirements of the product. By observing people using a product, it is possible to identify problems or good features in existing products, and use these to inform how you develop a new product. By disassembling products, you can identify components or useful features which can be incorporated into a new design. Through practical testing of materials, components, construction methods and finishes, you can make choices about which ones will be the best to use in developing your prototype.

In Figure 3.3.1, the student is investigating the context of a primary school and the problem of moving teaching resources between classrooms. The sheet shows how a site visit can be used to identify key points such as the space available.

Figure 3.3.2 shows an example of a client interview presentation. Here you can see that the student has written discussion points that were raised during the interview, as well as some initial sketching as early ideas were explored.

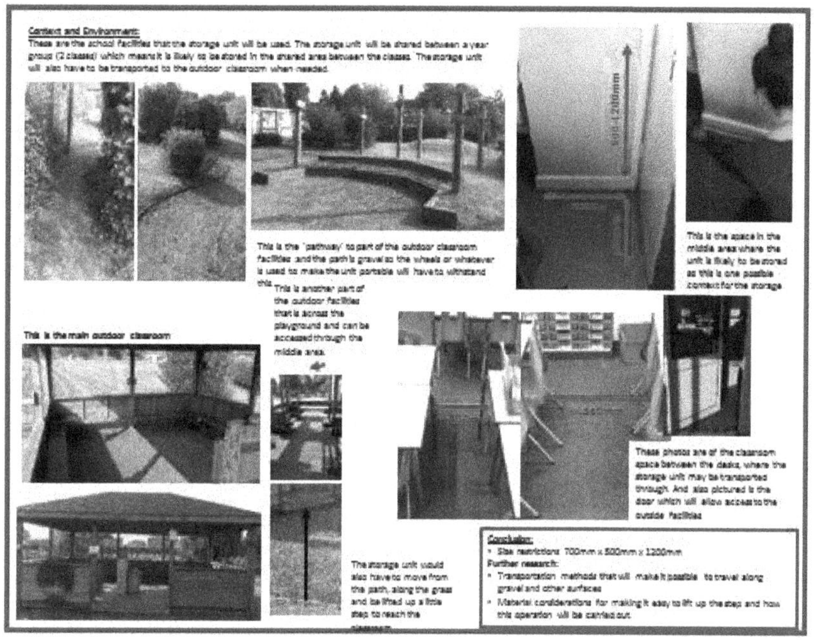

Figure 3.3.1 How a site visit might be used when investigating a primary school and the problem of moving teaching resources between classrooms.

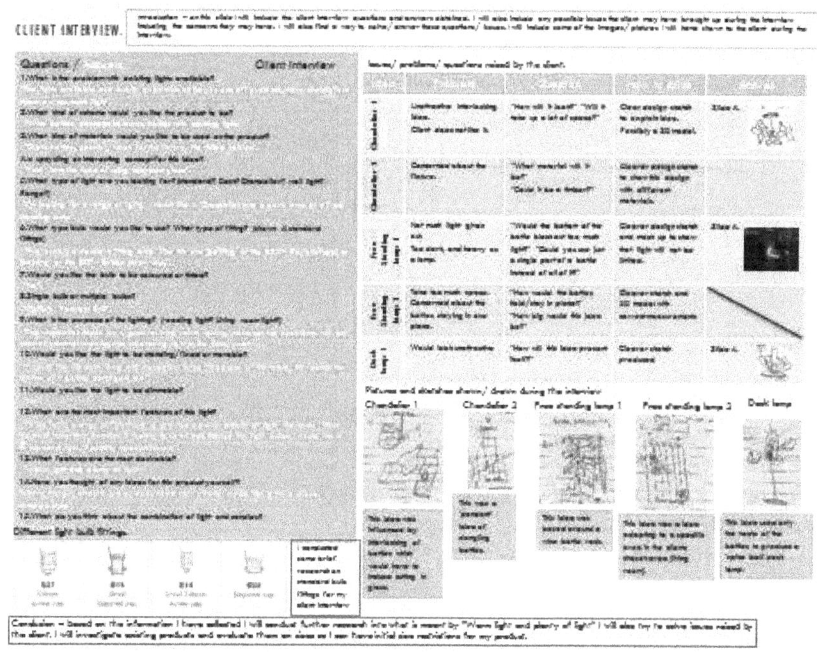

Figure 3.3.2 This is a good example of a client interview. The student is designing and making a light fitting that will use recycled materials. The student has also discussed points that the interview has raised with the client, including some initial ideas sketching.

Secondary techniques

Secondary techniques such as the internet, books and magazines are useful in identifying current fashions and trends. In addition to this, relevant anthropometric data can be obtained from books and the internet. This is essential when you are developing a prototype that must be ergonomic and 'fit' specific users. Specialist books and magazines can be useful in finding technical reports and comparing existing products and components.

Whatever methods of research you use, the material that you record in your folder must be relevant, and it must be used to help you formulate a design specification, and shape your design ideas and development.

At this stage, you should use your research to help produce your first design ideas. These are initial ideas that will be considered later in more detail. One will be developed into your prototype.

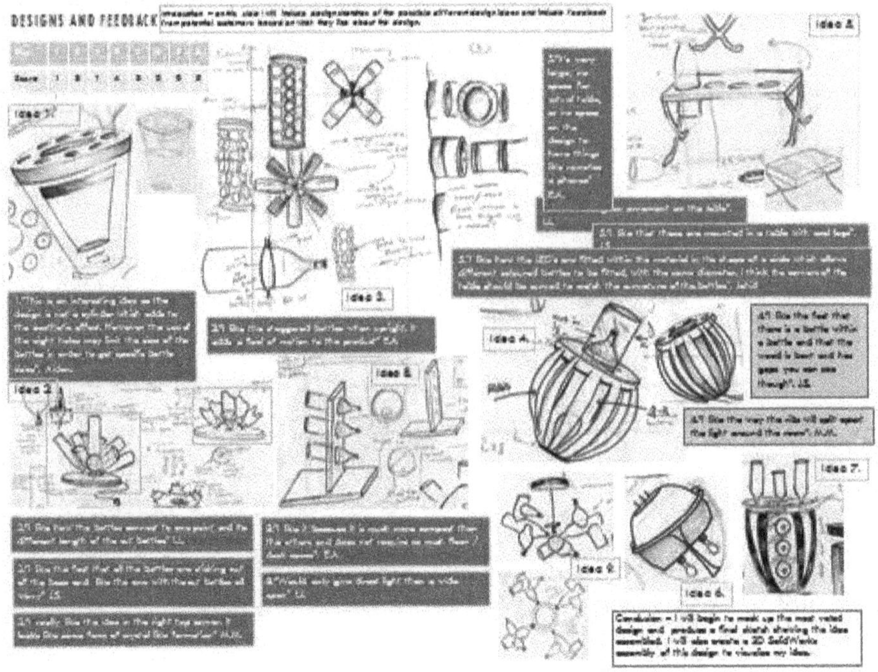

Figure 3.3.3 An example of initial concept ideas for a lighting project. The student has shown them to potential users, and obtained feedback which will help in developing the design further. The use of such feedback and analysis will score marks in the 'Analysing and evaluating' section.

Section B: Producing a design brief and specification

When you have selected a design context, you will write a design brief. This is a statement that explains what you will design and make to meet the design context, and will meet the needs of the intended user(s). An example AQA context is given below:

Animal care. Design and make prototype(s) for a product that could be used to assist in the care of animals or improve human interaction with animals.

An example of a student response to this context is given below:

The student has identified a problem at a dogs' home in that the dogs have a 2-hour-per-day exercise time, but they do not have much stimulation in the exercise area. The student wants to design and prototype a product that will provide stimulation in the exercise area, and provide a surface for the dogs to sit on other than the concrete floor of the exercise area.

The brief is given below:

I intend to design and make a dog exercise and relaxation station to be used by my client. The prototype I construct will be for the smaller sized dog such as a terrier. But it could be scaled up to for larger dogs. The product will continue to support the client's brand with its aesthetics. Dogs will be the users for this product. They will find it energising and relaxing in different areas of the structure because there will be different functions to make this possible. The dogs will also find interest in the product without human interaction or encouragement.

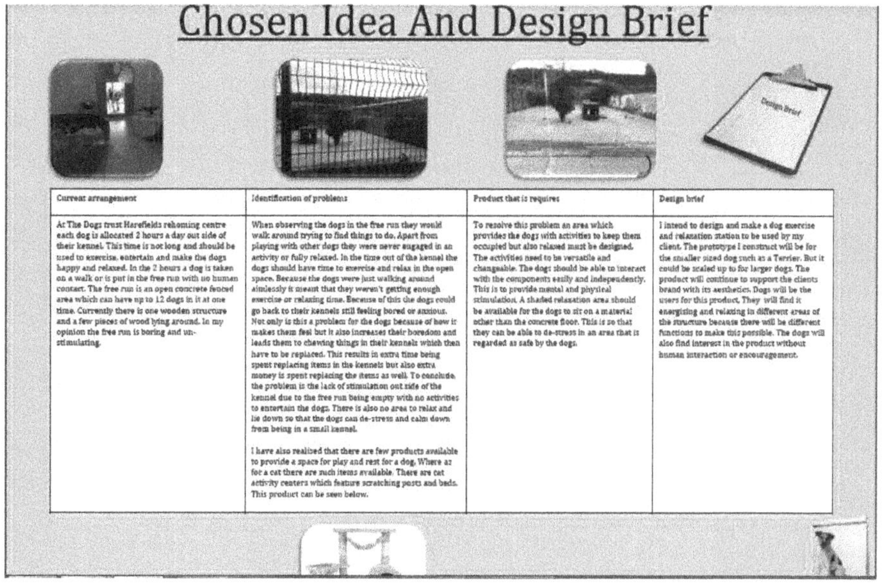

Chosen Idea And Design Brief

Current arrangement	Identification of problems	Product that is requires	Design brief
At The Dogs trust Harefields rehoming centre each dog is allocated 2 hours a day out side of their kennel. This time is not long and should be used to exercise, entertain and make the dogs happy and relaxed. In the 2 hours a dog is taken on a walk or is put in the free run with no human contact. The free run is an open concrete fenced area which can have up to 12 dogs in it at one time. Currently there is one wooden structure and a few pieces of wood lying around. In my opinion the free run is boring and un-stimulating.	When observing the dogs in the free run they would walk around trying to find things to do. Apart from playing with other dogs they were never engaged in an activity or fully relaxed. In the time out of the kennel the dogs should have time to exercise and relax in the open space. Because the dogs were just walking around aimlessly it meant that they weren't getting enough exercise or relaxing time. Because of this the dogs could go back to their kennels still feeling bored or anxious. Not only is this a problem for the dogs because of how it makes them feel but it also increases their boredom and leads them to chewing things in their kennels which then have to be replaced. This results in extra time being spent replacing items in the kennels but also extra money is spent replacing the items as well. To conclude, the problem is the lack of stimulation out side of the kennel due to the free run being empty with no activities to entertain the dogs. There is also no area to relax and lie down so that the dogs can de-stress and calm down from being in a small kennel. I have also realised that there are few products available to provide a space for play and rest for a dog. Where as for a cat there are such items available. There are cat activity centers which feature scratching posts and beds. This product can be seen below.	To resolve this problem an area which provides the dogs with activities to keep them occupied but also relaxed must be designed. The activities need to be versatile and changeable. The dogs should be able to interact with the components easily and independently. This is to provide mental and physical stimulation. A shaded relaxation area should be available for the dogs to sit on a material other than the concrete floor. This is so that they can be able to de-stress in an area that is regarded as safe by the dogs.	I intend to design and make a dog exercise and relaxation station to be used by my client. The prototype I construct will be for the smaller sized dog such as a Terrier. But it could be scaled up to for larger dogs. The product will continue to support the clients brand with its aesthetics. Dogs will be the users for this product. They will find it energising and relaxing in different areas of the structure because there will be different functions to make this possible. The dogs will also find interest in the product without human interaction or encouragement.

Figure 3.3.4 This example is set out well with a clear explanation of the context, a description of the problem and a design brief.

When you have completed your research, you will have enough information to produce a design specification. The specification is a list of detailed points that will be constantly referred to when developing your designs and making your prototype. Specifications include measurable criteria and qualitative points to which the design ideas and finished prototype can be compared.

Typical points that your specification could include are:

- Function: what will be the function of your prototype? What is the intended purpose?
- Target market: who is the product intended for?
- Anthropometrics: are there specific size ranges that should be considered?
- Ergonomics: are there specific ergonomic issues that should be considered?
- Size: are there any specific size constraints, for example a maximum size for the product?
- Aesthetics: are there any particular aesthetic requirements? These might have been specified by the client, for example colour preferences or styling preferences.
- Materials: there may be specific materials that you have identified in your research or your client may have specified (this may not be identified in your initial investigation but will be later in the development), which can be added to your specification at this point.
- Construction method: you may have identified a specific construction method in your investigation and can include this in your specification. (Again, this might be added later with ongoing investigation during the development and construction stages.)
- Finish: you may specify a specific finish that has been requested by the client or intended user(s).
- Safety: there may be specific safety requirements that you have identified such as British Standards.

- Cost: your investigation may have identified a maximum cost for your prototype, or a target retail price should the product be put into production.
- Environmental issues: your product might take into account specific environmental concerns, such as recyclability.

Figure 3.3.5 shows an example where a student has used a key on their specification: aesthetics (A), function (F), size (S) and materials (M). Each point of the specification under these criteria has been given a number: A1, A2, A3, F1, F2, M1 and so on. This specification is used later in the chapter when we discuss analysis and evaluating.

Specification

Essential:

Aesthetics:

- The product should represent the De Stijl design movement through use of primary colours and angular simple shapes - A2

Function:

- The switch must be accessible from outside the product – F1
- There must be two holes in the front to hold 50mm diameter K50 speakers – F3
- There should be an access panel in the back of the product to allow for the batteries to be removed and replaced – F4
- The battery pack should be fixed inside the product with a temporary fixing in order for it to be replaced – F5
- There should be a hole in the back for the jack cable to be connected – F6

Size Restrictions:

- There must be two 60mm diameter holes in the surface to allow for the 50mm speakers to fit in – S1
- The maximum width is 240mm and minimum width is 180mm in order to fit in the locker and fit all the components inside – S3
- The maximum height is 420mm and minimum height is 80mm in order to fit in the locker and fit all the components inside – S4
- The maximum depth is 420mm and minimum depth is 60mm in order to fit in the locker and fit all the components inside – S5
- There should be a 12mm hole for the switch in order for it to be attached to the speaker – S6

Material Properties:

Any material covering the speakers should be acoustically transparent so the sound is not affected – M2

Desirable:

Aesthetics:

- Speakers must be visible on the front of the product so that the sound is clear and directional - A1
- The switch must be on the front of the product A3

Function:

- Switch and speakers must be attached to product with a temporary fixing in order for it to be replaced if necessary – F2
- Must be able to hear it from 6metres away as this is the average distance in a room situation – F7
- There should be a 20mm space behind the speakers – F8

Size Restrictions:

- The access panel should be 85-95mm wide in order to fit your hand in to change the batteries – S2

Material Properties:

Material should be dense – M1

Any material covering the speakers should be acoustically transparent so the sound is not affected – M2

I have created a specification for my product to conform to. I have divided my specification points into **Essential** things that must be included for to product to be successful and **Desirable** things that can be compromised on if necessary.

8

Figure 3.3.5 An example of a specification for MP3 docking speakers.

Section C: Development of design proposals

To cover this section, you need to take your original concept designs and select one idea to develop. You will need to use your design brief and specification to help you to select an idea. You might also use feedback from your potential user(s) to help to select an idea to develop.

You will develop your designs through a series of sketches, test pieces and models. You might use scale or full-size models. At all times, refer back to your brief and specification to make sure that your final prototype will meet the needs of the user(s). It is important to make test pieces of parts of your prototype to experiment with making techniques. You should then evaluate these in your design folder. It is essential that you annotate your designs to explain what your thoughts are about them – suggestions for materials and construction method, how they meet the brief and specification criteria.

You might also practise finishing techniques to test out which is the most appropriate for your prototype. You should produce working drawings with dimensions to help you build your prototype to the correct size. CAD drawing is a communication technique that you could use in your project. For example, you could use CAD to produce 3D drawings of your designs, and 2D CAD drawings to produce working drawings. However, CAD should not be the only communication method you use. The development section should also include manufacturing plans that detail step by step how you will make your prototype. It is important that all through this process, you annotate your work, evaluate and explain your choices or decisions.

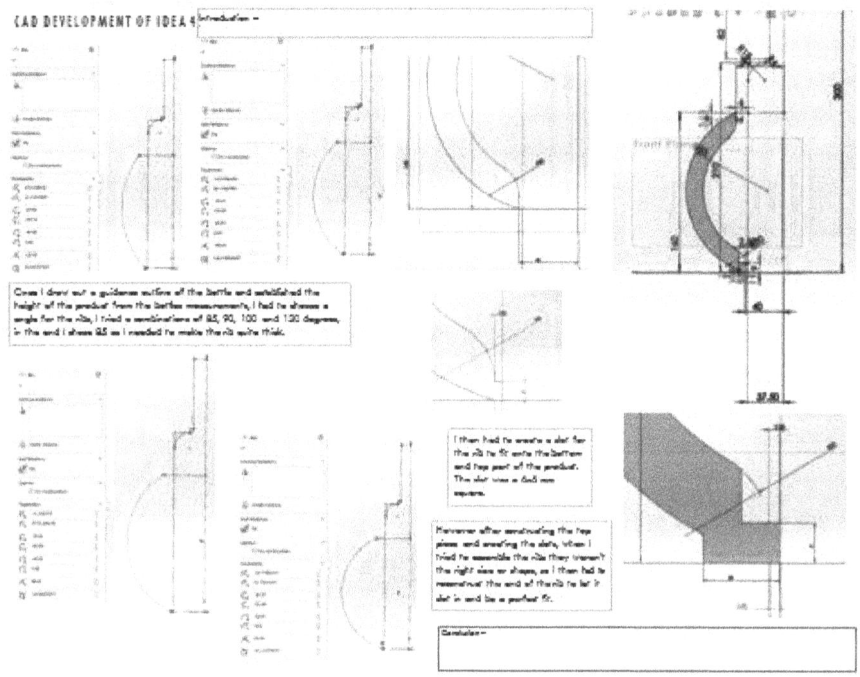

Figure 3.3.6 An example of 2D CAD working drawings: Such drawings have the necessary information that is needed to make the product. Critical sizes are given and the drawings can be used to compare the prototype against when it is being made.

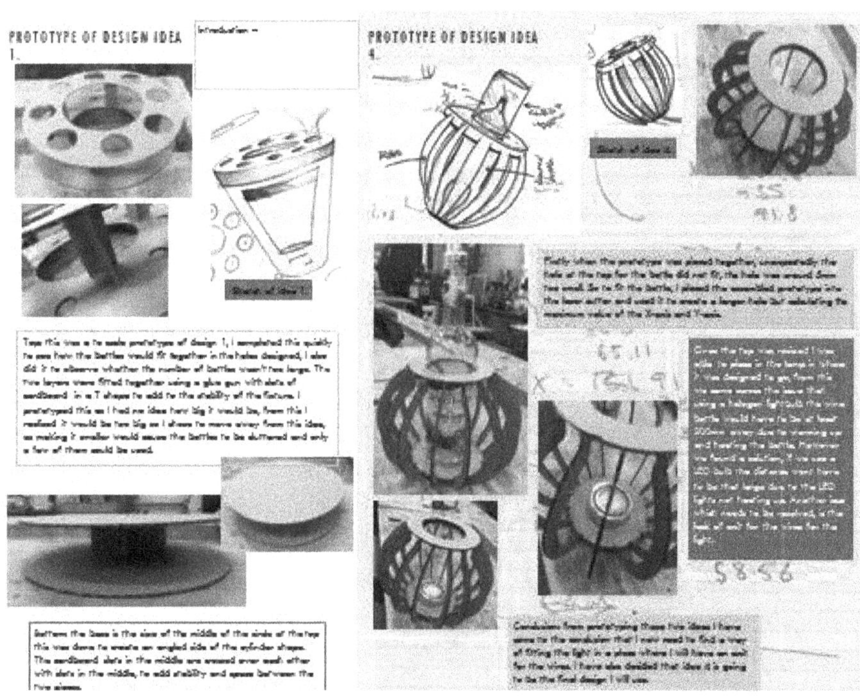

Figure 3.3.7 On this sheet, the student is using card modelling to develop a design proposal. Again, they have obtained feedback from potential users of the product to help them refine the prototype. On this sheet, they are making good use of 3D sketches and models, and you can see they are working out sizes. In this example, they have used a laser cutter for some of the card modelling which also helps to prove whether the construction technique works or not.

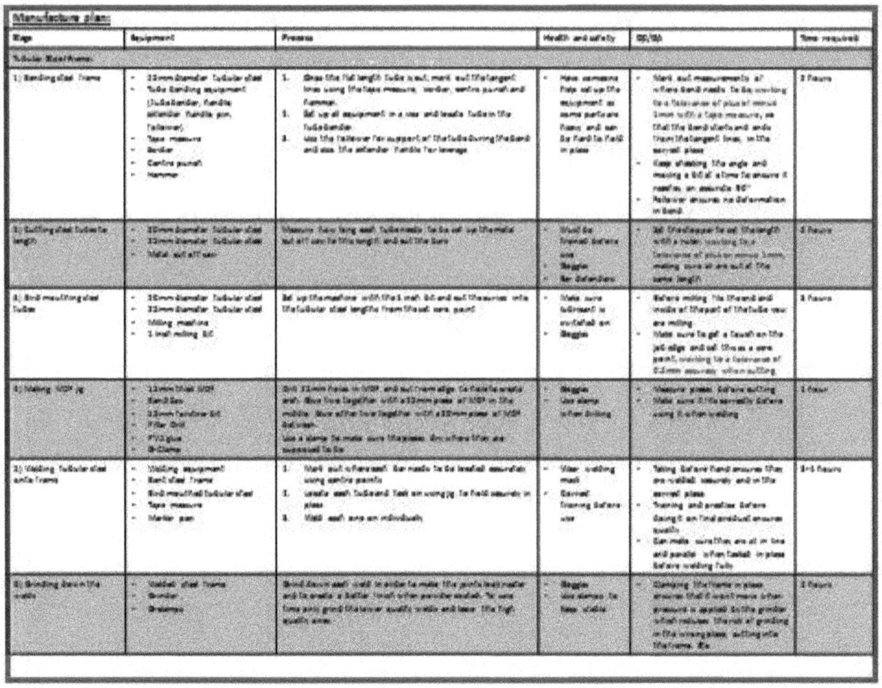

Figure 3.3.8 An example of part of a manufacturing plan. The student has detailed the making of their prototype, breaking it down into stages. This manufacturing plan details what equipment is to be used at each stage, what health and safety issues have to be considered, and how quality or accuracy will be maintained.

It is important to plan the manufacture of your prototype carefully before making it. These plans should be a 'live' document and adapted if you change your mind about how the product will be made. This is to be expected as you will often work out better or alternative methods during the manufacturing stages.

Section D: Development of design prototypes

For this assessment criterion, you can build a prototype and refine and develop it, or build several that lead to one high-quality finished prototype.

In this section of the NEA project, you should produce a prototype(s) and then test, evaluate, refine and develop it. As you build your prototype(s), it is expected that you will record any changes or refinements in your design folder using sketches, photographs and notes. It is critical that your prototype(s) directly links back to your developed designs, brief and specification.

It is crucial that you demonstrate a high level of skill in the making of your prototype(s). You will need to use a range of materials, tools, equipment and processes in making your prototype(s). You can use CAM equipment such as 3D printers, laser cutters, routers, milling machines, lathes and vinyl cutters, but you must not use CAM as the sole means of making your prototype(s). If your finished prototype is made using just CAM, it is vital that you make use of other making techniques in modelling and earlier development prototypes, to demonstrate hand skills.

On the next page is a good example of a developing design prototype. These images show the student developing a tripod stand for their lamp.

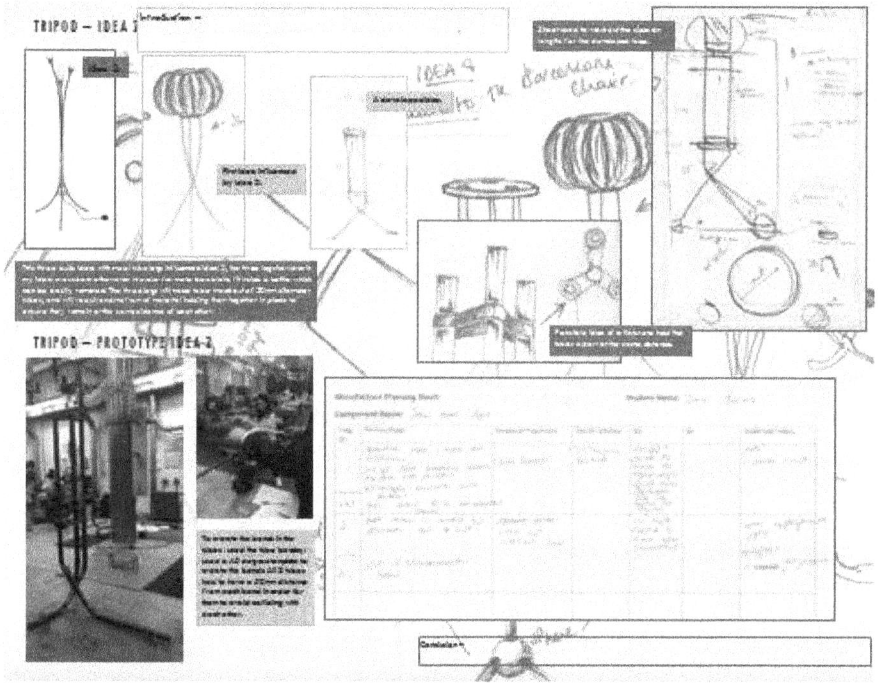

Figure 3.3.9 On this sheet, you can see how the student is sketching out ideas for how the tripod will be made, and they are trying out an idea with practical test pieces. It is a very busy sheet which also includes a section of a step-by-step making plan.

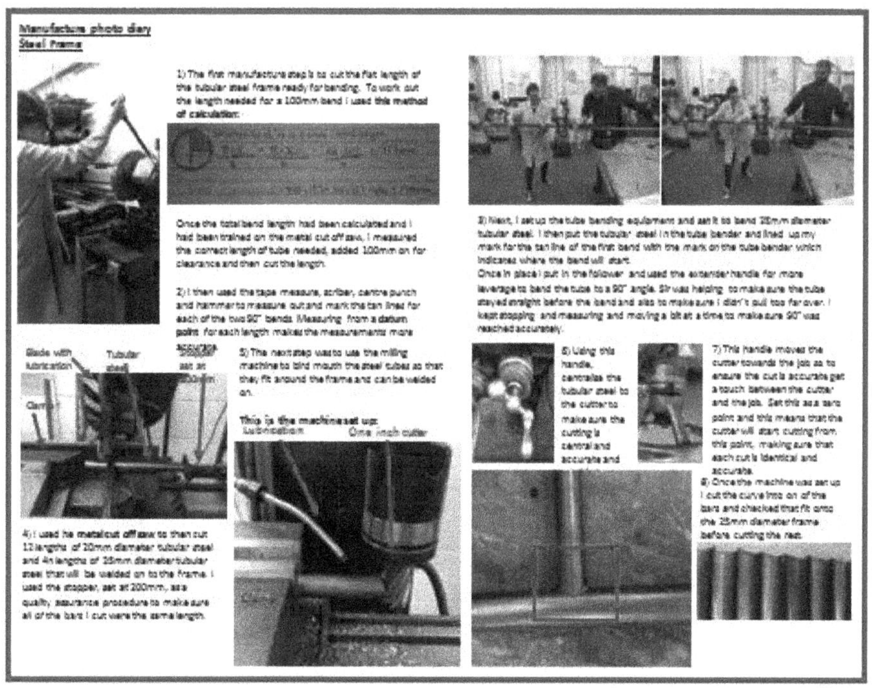

Figure 3.3.10 Here, the student is making a prototype trolley from tubular steel and manufactured board. They make good use of notes and photographs to explain how the prototype is made. This kind of photographic evidence is important because it shows the skills and level of accuracy in your work.

This sheet also shows how the student is using maths skills to work out the total length of the bends required in the steel. This is obviously important in this project.

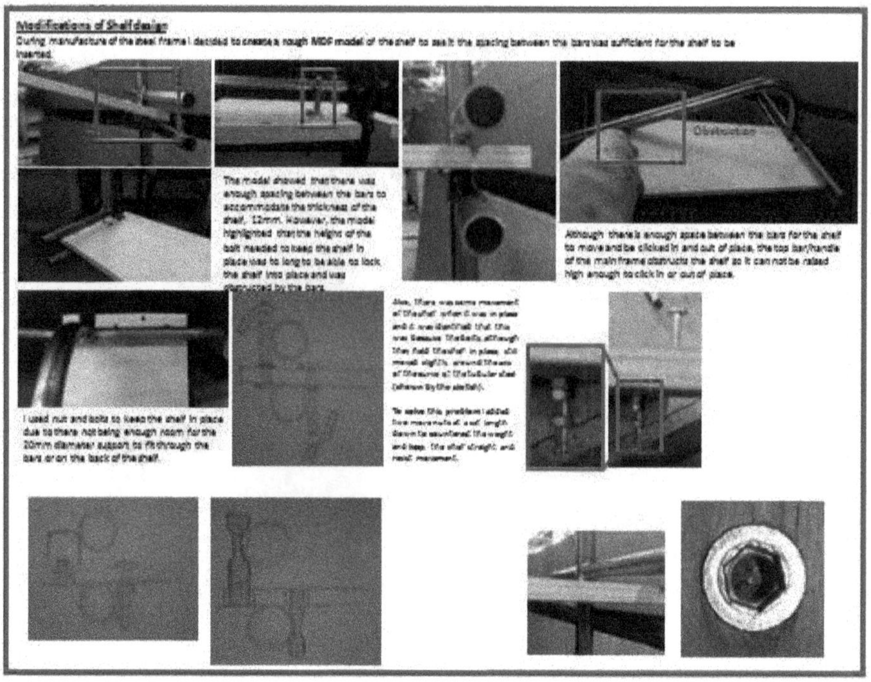

Figure 3.3.11 Here, the student is making excellent use of photographs and sketching to communicate some of the problems, and to show how the prototype needs to be developed further.

Section E: Analysing and evaluating

You can generate evidence for this part of the NEA at any point of the project. You will analyse and evaluate your research work, your design ideas and development drawings, your models and prototype.

An important part of getting good marks for this assessment criterion is to involve your client or target user(s) in the design and development process. By doing this, you will make sure that the prototype will meet the needs of the client or target user(s).

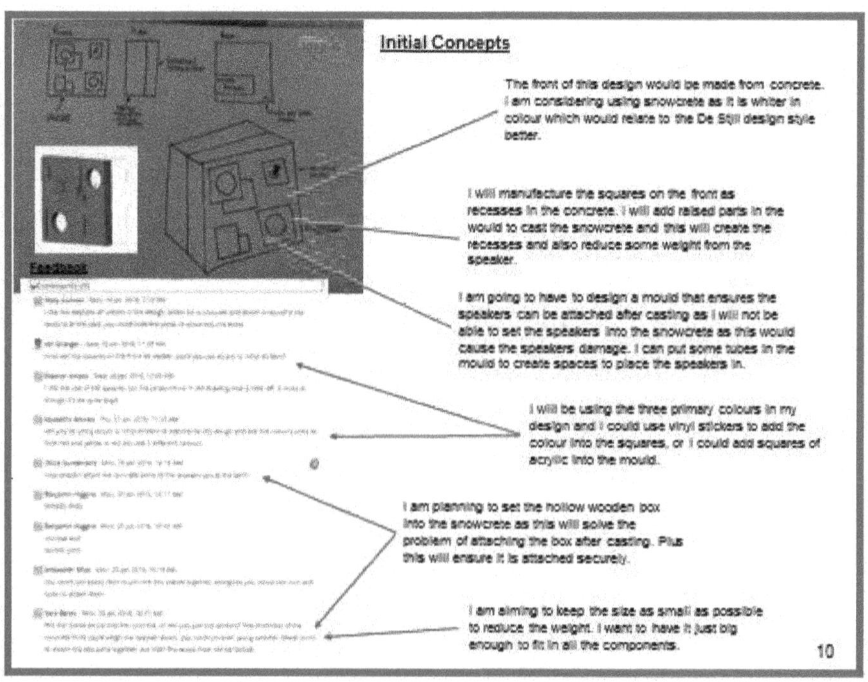

Figure 3.3.12 This example shows the student has obtained feedback about one of their design concepts by using social media. Here you can see that the student has responded to the feedback by suggesting how they will develop their design.

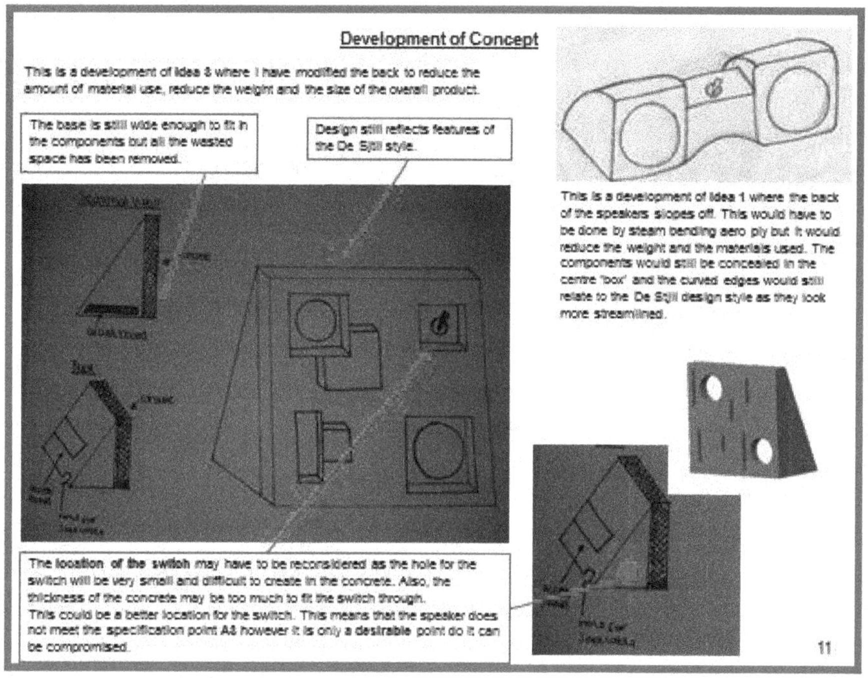

Figure 3.3.13 Here you can see ongoing evaluation in part of the student's development work. They are making critical/evaluative comments about their design and explaining decisions made. You can see they are referring back to their specification, with 'A3' meaning a point about aesthetics.

For this assessment criterion, you will need to:

- test your prototype(s) and use the information from this testing to refine your prototype during the development stage. You will gain credit for making use of your client or intended user(s) in testing the prototype(s) if it is done correctly
- test your final prototype and use this information to suggest improvements to the product that would be required if it were to go into production. Again, you will gain credit for testing the final prototype with your client or intended user(s)
- explain how the design would need to be modified if it were to be made using batch and mass production. The best way to do this is through a series of notes and sketches or CAD drawings.

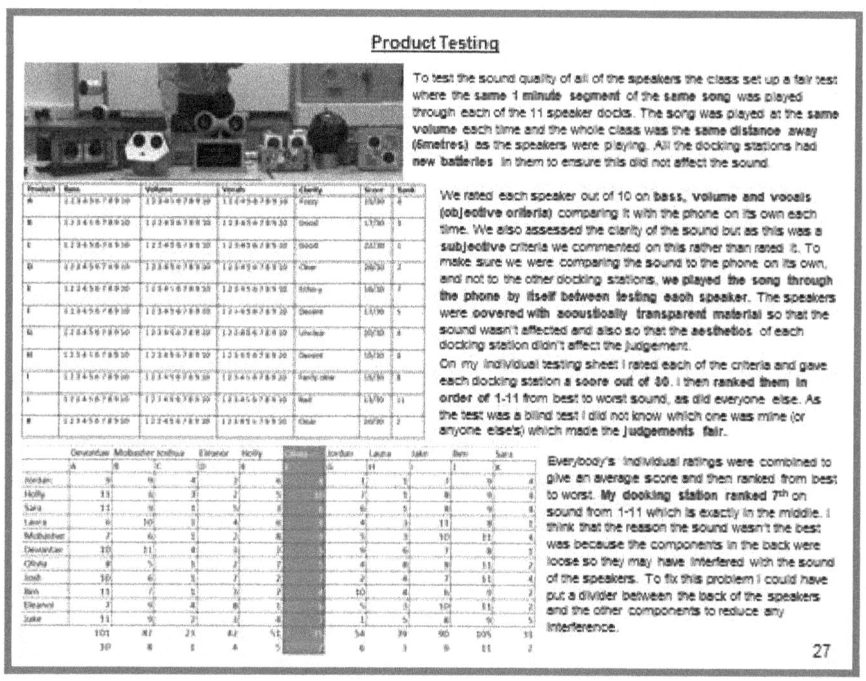

Figure 3.3.14 This is an example of a student who carried out comparative testing of prototype MP3 speakers that have been designed and made by a class of students. The student is testing their docking speaker on bass, volume and vocals. To make the testing objective, each student rated all of the speakers using the above criteria, and the speakers were covered so that students would not be swayed by the aesthetic appearance.

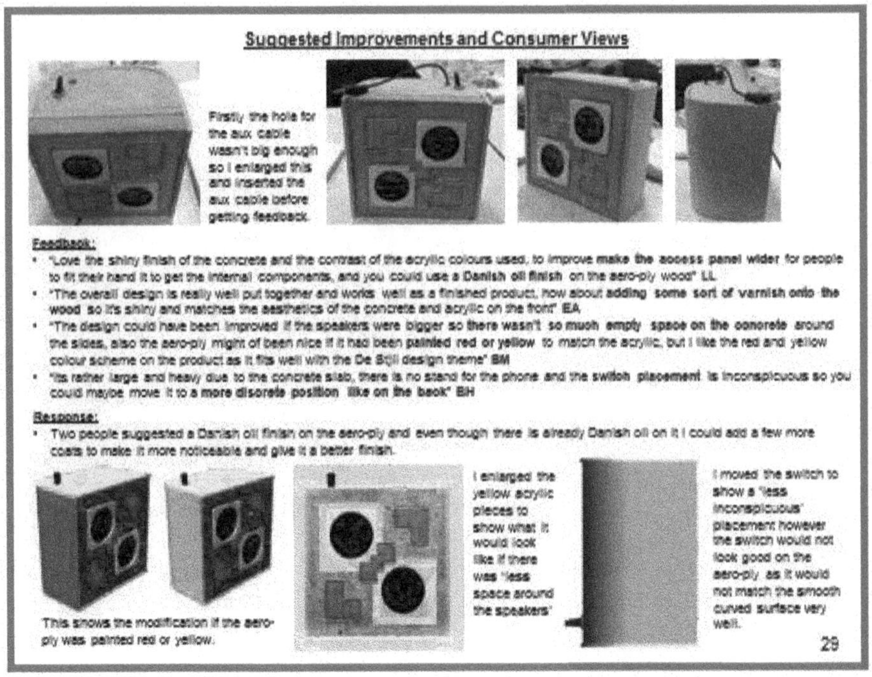

Figure 3.3.15 Here the student has shown more photos of the prototype, and they have obtained third party feedback. They have then used this feedback to suggest some simple modifications about the finish, perhaps using more yellow acrylic around the speaker.

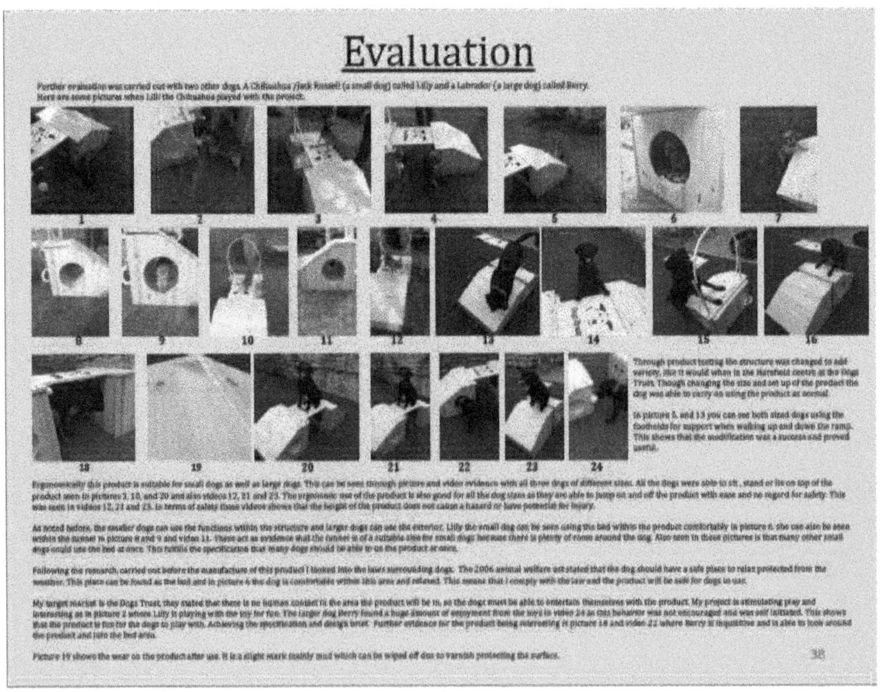

Figure 3.3.16 This is an example of testing a prototype piece of dog exercise equipment. It was produced in response to the context and brief that we saw earlier in this section of the book.

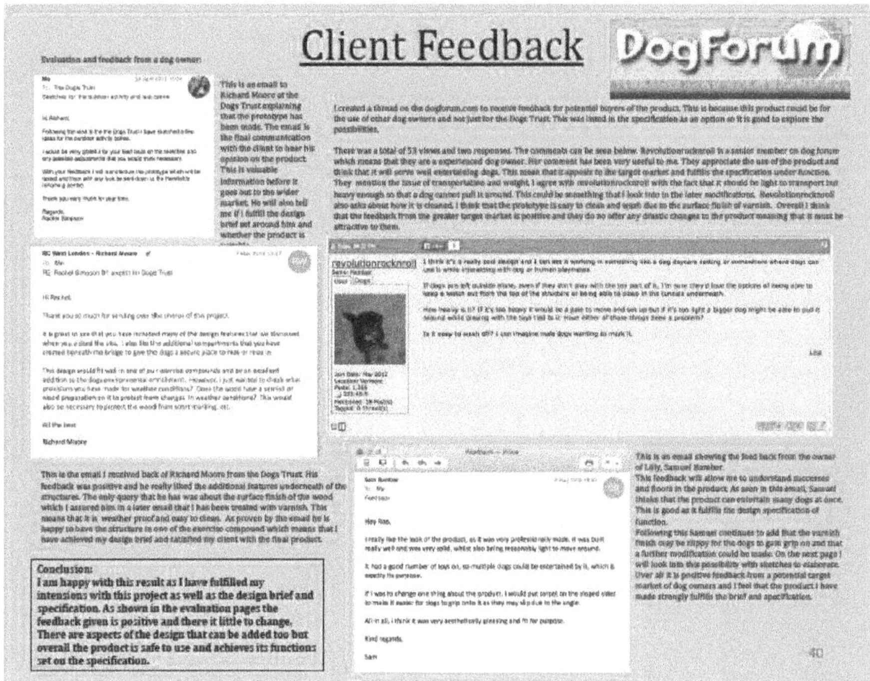

Figure 3.3.17 Here the student uses a dog owner's forum to obtain feedback from potential customers for the product. They also contact members of the Dogs Trust to obtain 'expert opinion'.

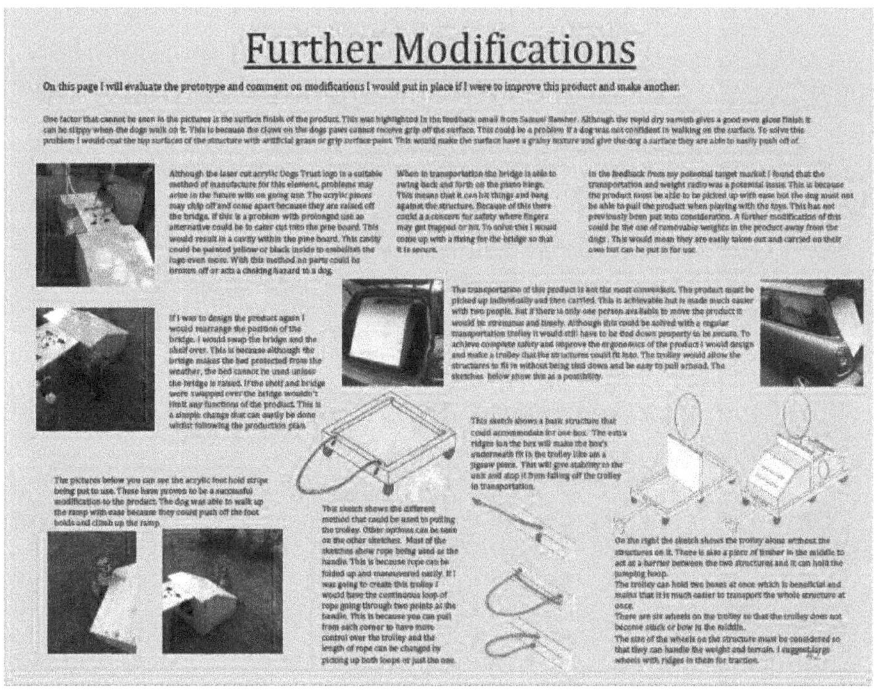

Figure 3.3.18 On this sheet, the student suggests modifications they would make to their prototype in response to the feedback that they have had about their prototype.

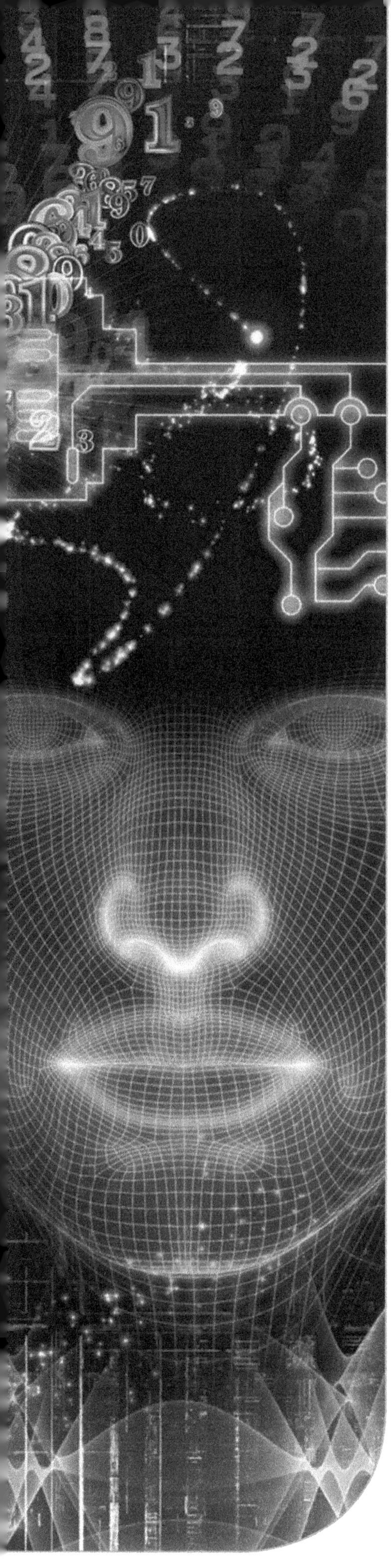

Section 4 Written exams

In this section you will learn about the following:

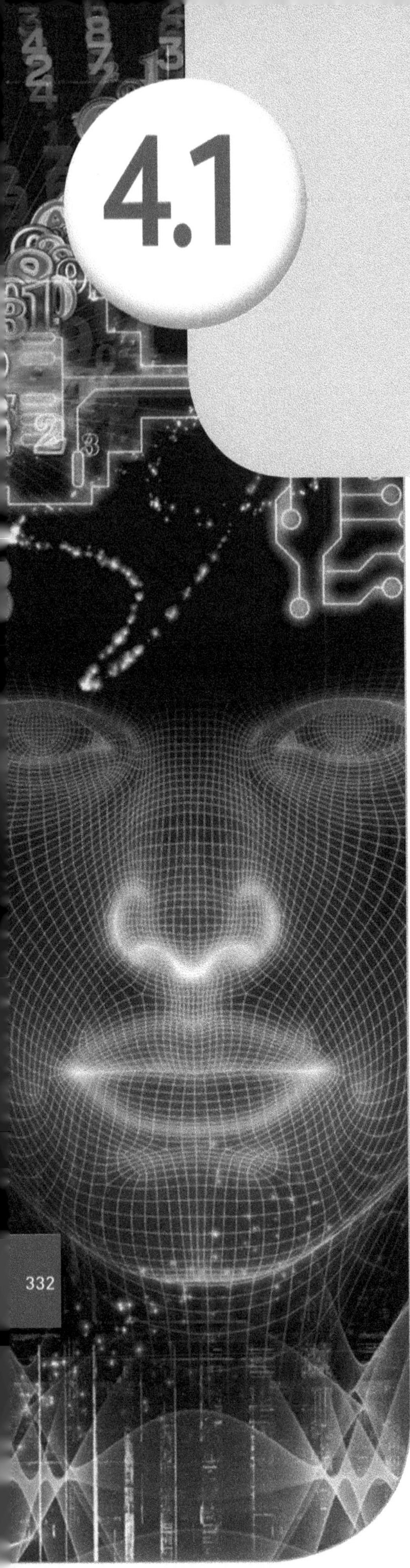

4.1 The AS written exam

Introduction to the exam

In the written exam, you will be tested on:

- technical principles
- designing and making principles.

These topics are covered in Chapters 1 and 2 of this book. The written exam is worth 80 marks and 50 per cent of your total AS mark.

When will the written paper be taken?

The paper will be available in the summer exam series in May/June.

How long will I have?

The exam will last 1 hour and 30 minutes.

What format will the written paper take?

There will be a mixture of short and longer answer questions. Multiple choice questions could also be included. Questions will cover technical principles and designing, as well as making principles, to cover the full range of the specification.

Mathematics and science skills in a product design context will also be examined as part of the paper. Mathematics questions may be short answer or multiple choice or may form part of a longer question.

How will I be assessed?

You will be assessed against the two assessment objectives (AOs) shown in the table below.

Table 4.1.1 Assessment objectives.

Assessment objective (AO)	Weighting (%)
AO3: Analyse and evaluate: • design decisions and outcomes, including for prototypes made by themselves and others • wider issues in design and technology.	15
AO4: Demonstrate and apply knowledge of: • technical principles • designing and making principles.	35
Overall weighting	50

General advice on answering questions

You should aim to answer ALL questions in your written examination. Remember to:

- read the questions carefully in order to understand exactly what is needed
- look for the command words – identify the key words before answering the question
- use the allocated marks to guide how much time to spend on each answer
- use the answer book to draw a quick plan or write down key points or ideas
- write clearly
- include clear diagrams with annotation when asked
- cross out any draft work that is not part of your answer; crossed out work will not be marked
- allow time to check back over your work.

Questions give you the opportunity to:

- demonstrate knowledge, for example giving the definition of a non-ferrous metal, naming a non-ferrous metal and giving an example of a product made from the named non-ferrous metal
  - common commands words may include: define, name, which
- demonstrate understanding, for example giving reasons why the non-ferrous metal you named is suitable for the product – factual recall alone is not sufficient, you need to show understanding
  - common commands words may include: explain, give reasons, suggest why
- apply your knowledge, analyse and evaluate, for example applying your subject knowledge to a real-life context such as selecting and applying given anthropometric data to a seating product or evaluating a product in terms of function, materials, ease of use and manufacture
  - common commands words may include: analyse, apply, compare, contrast, evaluate, justify.

Table 4.12 lists the command words you are likely to encounter.

Table 4.1.2 Command words.

Command word	Meaning	Example question
Analyse	Separate information into components and identify their characteristics.	Analyse the data from the table to select the most suitable height for an office desk.
Apply	Put into effect in a recognised way.	Apply the anthropometric data in the table shown to give the most suitable height range for an adjustable office chair.
Calculate	Work out the value of something.	Calculate the number of products which need to be sold to make a 5 per cent profit.
Comment	Present an informed opinion.	Comment upon the suitability of ABS for a remote-control casing.
Compare	Identify similarities and/or differences.	Compare the material properties of aluminium and PP. Which is the most appropriate material for a lemon juicer?
Consider	Review and respond to given information.	Consider the manufacturing processes and costs shown in the table, and state which process would be most economical to use for a batch of 50 products.
Contrast	Identify differences.	Contrast the environmental impact of the lemon juicers.
Criticise	Assess worth against explicit expectations. 'Criticise' is often used along with the terms 'analyse' or 'evaluate'.	Critically evaluate the ergonomic features of the toothbrush shown.
Define	Specify meaning.	Define the term 'ergonomics'.
Describe	Set out characteristics.	Describe how the properties of phosphorescent pigment could enable energy to be saved.
Evaluate	Judge from available evidence.	Evaluate the environmental issues associated with the use of hardwoods for product manufacture.
Explain	Set out purposes or reasons.	Explain why beech is a suitable material for a cooking spoon.
Give	Produce an answer from recall.	Give an example of a bio-degradable polymer.
Identify	Name or otherwise characterise.	Identify the composite material from a given list.
Justify	Support a case with evidence.	Select a suitable material for a laptop casing and justify your choice.
Select	Choose as being the best or most suitable.	Select the most appropriate manufacturing method (from a given list) for a carrier bag made from LDPE.
Sketch	Use drawings to present clarifying examples.	Use notes and sketches to describe the blow moulding process.
State	Express clearly and briefly.	State the meaning of the term 'elastomer'.
Suggest	Present a possible case/solution.	Suggest a reason why aluminium is suitable for a window frame.

Types of questions

The paper will include the question types listed below. Familiarise yourself with question wording, question types and how answers are assessed.

Multiple choice questions

Multiple choice questions require the correct answer to be selected from a list of potential answers. There may be a single question with a list of potential answers to choose from, or a question that requires one or more answers to be matched or selected (such as the question shown below).

Question

Match the process description shown below, to the correct workshop manufacture technique. **[3 marks]**

A Wasting
B Addition
C Redistribution

You should use each letter once only.

Manufacture technique	**Process description**
MIG welding	☐
Drilling a hole	☐
Sand casting	☐

Answer

MIG welding	B	(Addition)
Drilling a hole	A	(Wasting)
Sand casting	C	(Redistribution)

Short answer questions

Short answer questions range from 1 to about 6 marks. Short answer questions worth 1 or 2 marks may form part of a series of questions on a topic or theme.

Question

Name a specific 'composite material' and a sports product that it is used in. **[2 marks]**

Answer

Carbon-fibre reinforced polymer (CFRP) used in Formula 1 racing car bodies.

Maths questions

For more guidance on mathematical skills in design and technology, see Chapter 4.4.

Product analysis questions

These questions usually show a photograph of a product that must be analysed. Analysis could be on a range of topics such as environmental impact, suitability for the end user, ergonomic or anthropometric evaluation, or analysis of the suitability of a manufacturing method.

Question

Analyse the environmental impact at the end of life for the office chair shown below. **[6 marks]**

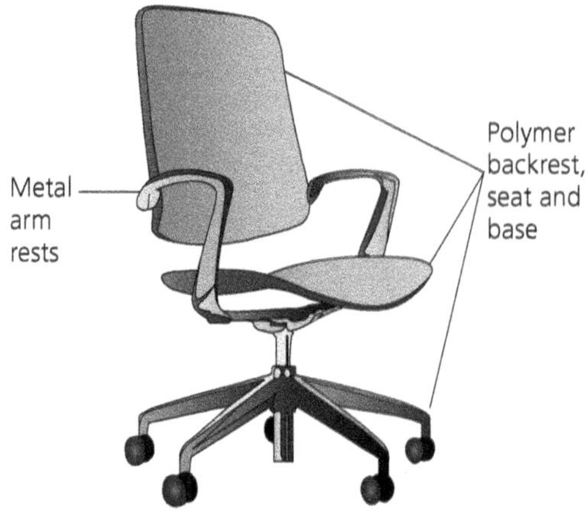

Figure 4.1.1 Office chair.

Extended response questions

Extended response questions may include a source photograph or may be on a specific topic from the specification. Knowledge of appropriate methods, legislation, examples and justification will be expected in an extended response.

Question

What can a manufacturer do to ensure the safety of products for the consumer? **[8 marks]**

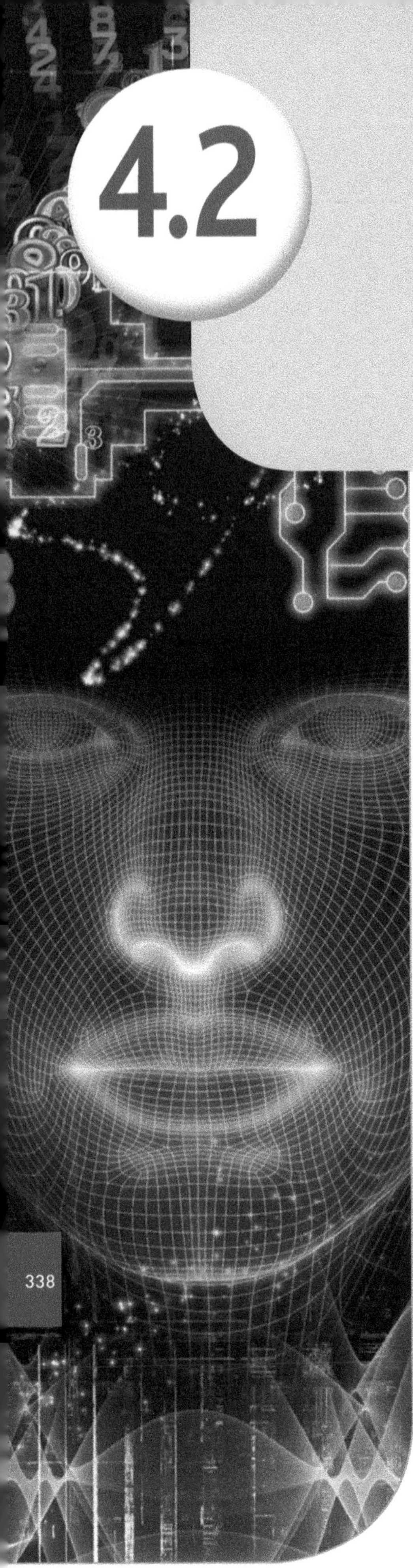

4.2 A-level paper 1

Introduction to the paper

In paper 1, you will be tested on:

- technical principles.

This topic is covered in Chapter 1 of this book. The written is worth 120 marks and 30 per cent of your total A-level.

When will the written paper be taken?

The written paper will be taken in the summer exam series in May/June.

How long will I have?

The exam will last 2 hours 30 minutes.

What format will the written paper take?

There will be a mixture of short and longer answer questions. Multiple choice questions could also be included. Questions will cover technical principles from across the full range of the specification.

Mathematics and science skills in a product design context will also be examined as part of the paper. Mathematics questions may be short answer or multiple choice or may form part of a longer question.

How will I be assessed?

You will be assessed against the two assessment objectives (AOs) shown in the table below.

Table 4.2.1. Assessment objectives.

Assessment objective (AO)	Weighting (%)
AO3: Analyse and evaluate: • design decisions and outcomes, including for prototypes made by themselves and others • wider issues in design and technology.	7.5
AO4: Demonstrate and apply knowledge of: • technical principles.	22.5
Overall weighting	30

General advice on answering questions

You should aim to answer all questions in the A-level paper 1 written exam. Remember to:

- read the questions carefully in order to understand exactly what is needed
- look for the command words – identify the key words before answering the question
- use the allocated marks to guide how much time to spend on each answer
- use the answer book to draw a quick plan or write down key points or ideas
- write clearly
- include clear diagrams with annotation when asked
- cross out any draft work that is not part of your answer; crossed out work will not be marked
- allow time to check back over your work.

Questions give you the opportunity to:

- demonstrate knowledge, for example giving the definition of a non-ferrous metal, naming a non-ferrous metal and giving an example of a product made from the named non-ferrous metal
  - common commands words may include: define, name, which
- demonstrate understanding, for example giving reasons why the non-ferrous metal you named is suitable for the product – factual recall alone is not sufficient, you need to show understanding
  - common commands words may include: explain, give reasons, suggest why
- apply your knowledge, analyse and evaluate, for example applying your subject knowledge to a real-life context such as selecting and applying given anthropometric data to a seating product or evaluating a product in terms of function, materials, ease of use and manufacture
  - common commands words may include: analyse, apply, compare, contrast, evaluate, justify.

Table 4.2.2 lists command words.

Table 4.2.2 Command words.

Command word	Meaning	Example question
Analyse	Separate information into components and identify their characteristics.	Analyse the data from the table to select the most suitable height for an office desk.
Apply	Put into effect in a recognised way.	Apply the anthropometric data in the table shown to give the most suitable height range for an adjustable office chair.
Calculate	Work out the value of something.	Calculate the number of products which need to be sold to make a 5 per cent profit.
Comment	Present an informed opinion.	Comment upon the suitability of ABS for a remote-control casing.
Compare	Identify similarities and/or differences.	Compare the material properties of aluminium and PP. Which is the most appropriate material for a lemon juicer?
Consider	Review and respond to given information.	Consider the manufacturing processes and costs shown in the table, and state which process would be most economical to use for a batch of 50 products.
Contrast	Identify differences.	Contrast the environmental impact of the lemon juicers.
Criticise	Assess worth against explicit expectations. 'Criticise' is often used along with the terms 'analyse' or 'evaluate'.	Critically evaluate the ergonomic features of the toothbrush shown.
Define	Specify meaning.	Define the term 'ergonomics'.
Describe	Set out characteristics.	Describe how the properties of phosphorescent pigment could enable energy to be saved.
Evaluate	Judge from available evidence.	Evaluate the environmental issues associated with the use of hardwoods for product manufacture.
Explain	Set out purposes or reasons.	Explain why beech is a suitable material for a cooking spoon.
Give	Produce an answer from recall.	Give an example of a bio-degradable polymer.
Identify	Name or otherwise characterise.	Identify the composite material from a given list.
Justify	Support a case with evidence.	Select a suitable material for a laptop casing and justify your choice.
Select	Choose as being the best or most suitable.	Select the most appropriate manufacturing method (from a given list) for a carrier bag made from LDPE.
Sketch	Use drawings to present clarifying examples.	Use notes and sketches to describe the blow moulding process.
State	Express clearly and briefly.	State the meaning of the term 'elastomer'.
Suggest	Present a possible case/solution.	Suggest a reason why aluminium is suitable for a window frame.

Types of questions

The paper will include the question types listed below. You should aim to familiarise yourself with question wording, question types and how answers are assessed.

Multiple choice questions

Multiple choice questions require the correct answer to be selected from a list of potential answers. There may be a single question with a list of potential answers to choose from, or a question that requires one or more answers to be matched or selected (such as the question shown below).

Question

Match the product to the most appropriate manufacturing process. **[3 marks]**

A Injection moulding
B 3D printing
C CNC plotter cutter
D Vacuum forming

Product	**Manufacturing process**
Vinyl lettering for a sign	☐
Polylactic acid (PLA) prototype of a design	☐
PP (polypropylene) yoghurt pot	☐

Answer

Vinyl lettering for a sign	C	(CNC plotter cutter)
PLA prototype of a design	B	(3D printing)
PP (polypropylene) yoghurt pot	D	(vacuum forming)

Short answer questions

Short answer questions are usually worth up to 6 marks. Short answer questions worth 1 or 2 marks may form part of a series of questions on a topic or theme. An example question is given below:

The photograph below shows a toy scooter. The body of the scooter is made from aluminium.

Question

State one physical and one mechanical property of aluminium, and explain why these properties make aluminium suitable for the body of the scooter. **[4 marks]**

Figure 4.2.1 Toy scooter.

Maths questions

For more guidance on mathematical skills in design and technology, see Chapter 4.4.

Product analysis questions

These questions usually show a photograph of a product that must be analysed. Analysis could be on a range of topics such as environmental impact, suitability for the end user, ergonomic evaluation, anthropometric evaluation or analysis of the suitability of a manufacturing method.

The photograph below shows a ski pole handle.

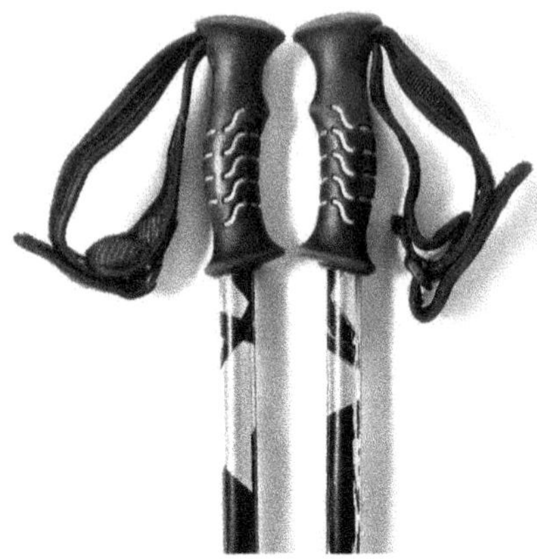

Figure 4.2.2 Ski pole handle.

Question

Explain the implications of using CAD in the design of the ski pole handle. **[6 marks]**

Extended response questions

Extended response questions may include a source photograph or may be on a specific topic from the specification. Knowledge of appropriate methods, legislation, examples and justification of your thoughts will be expected in an extended response.

Question

Explain how designs need to be modified to make them more efficient to manufacture. **[12 marks]**

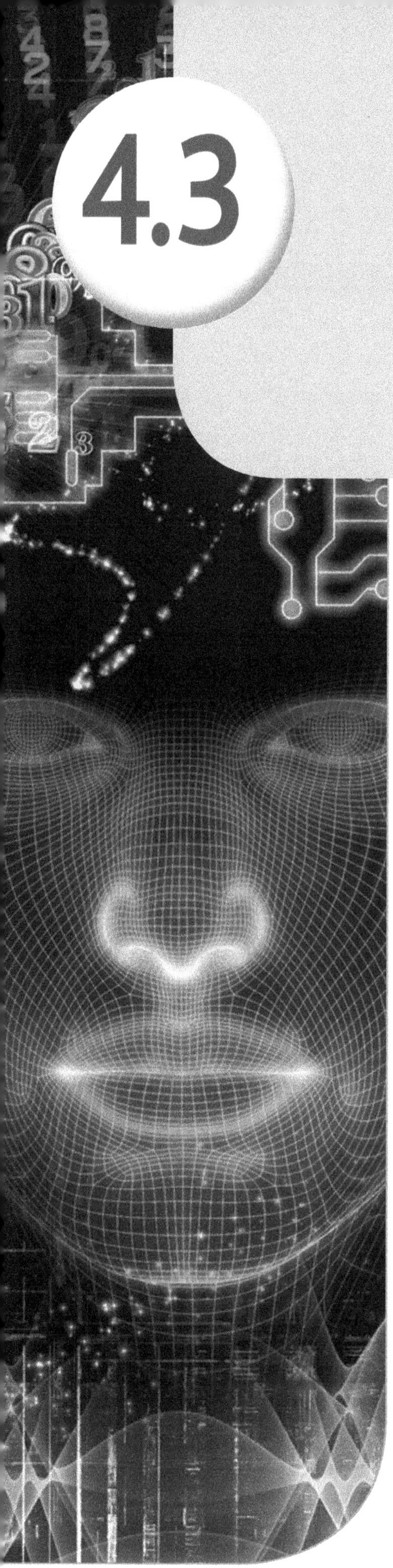

4.3 A-level paper 2

Introduction to the paper

In paper 2, you will be tested on:

- designing and making principles.

This topic is covered in Chapter 2 of this book. The written exam is worth 80 marks and 20 per cent of your total A-level.

When will the written paper be taken?

The written paper will be taken in the summer exam series in May/June.

How long will I have?

The exam will last 1 hour 30 minutes.

What format will the written paper take?

There will be a mixture of short and longer answer questions. Questions will cover designing and making principles from across the full range of the specification.

The paper will be divided into two sections:

- Section A: Product analysis
  - 30 marks are available for this section.
  - There will be up to six short answer questions based on the visual stimulus of product(s).
- Section B: Commercial manufacture
  - 50 marks are available for this section.
  - There will be a mixture of short and extended response questions.

Mathematics and science skills in a product design context will also be examined as part of the paper.

Mathematics questions may be short answer or multiple choice or may form part of a longer question.

How will I be assessed?

You will be assessed against the two assessment objectives (AOs) shown in the table below.

Table 4.3.1 Assessment objectives.

Assessment objective (AO)	Weighting (%)
AO3: Analyse and evaluate: • design decisions and outcomes, including for prototypes made by themselves and others • wider issues in design and technology.	7.5
AO4: Demonstrate and apply knowledge of: • technical principles • designing and making principles.	12.5
Overall weighting	20

General advice on answering questions

The same advice provided for A-level paper 1 applies to paper 2. Refer back to page 339 to remind yourself of this advice when preparing for Paper 2.

You should also remind yourself of the command words which are provided on pages 339–40.

Types of questions

The paper will include the question types listed below.

Multiple choice questions

Multiple choice questions require the correct answer to be selected from a list of potential answers. There may be a single question with a list of potential answers to choose from, or a question which requires one or more answers to be matched or selected.

Question

The length of a cardboard box for a product's packaging is specified as 500 ± 10 mm. Calculate the percentage tolerance which would be acceptable on this dimension. Tick the box with the correct answer. **[1 mark]**

A 2% ☐
B 4% ☐
C 5% ☐
D 10% ☐

Short answer questions

Short answer questions are usually be worth up to 6 marks.

Question

Second World War rationing led to the development of 'utility' products. Use a specific product example to explain what is meant by this. **[3 marks]**

Maths questions

For more guidance on mathematical skills in design and technology, see Chapter 4.4.

Product analysis questions

These questions usually show a photograph of a product or products that must be analysed. Analysis could be on a range of topics such as materials or design processes used; how the design is influenced by design styles, movements or other designers; the influence of technological developments; or socio-economic, moral, ethical or environmental influences.

Question

The images on this and the following page show two dining room tables.

For each table, compare the materials used and their suitability. **[8 marks]**

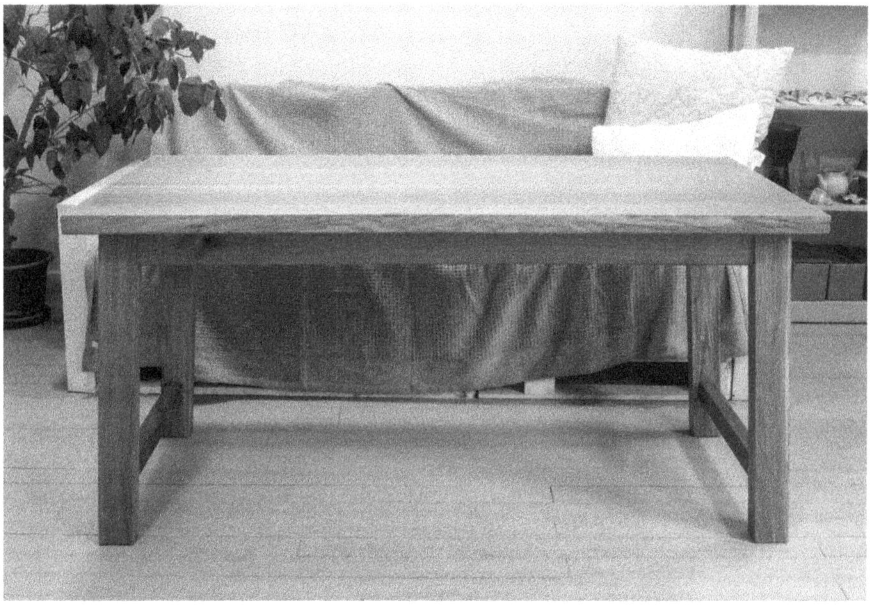

Figure 4.3.1 An oak table.

Figure 4.3.2 A MDF table.

Extended response questions

Extended response questions may include a source photograph or may be on a specific topic. Knowledge of appropriate methods, legislation, technologies, examples and justification of your thoughts will be expected in an extended response.

Question

Using specific examples, analyse the impact that product packaging has on the environment and how it can be minimised. **[12 marks]**

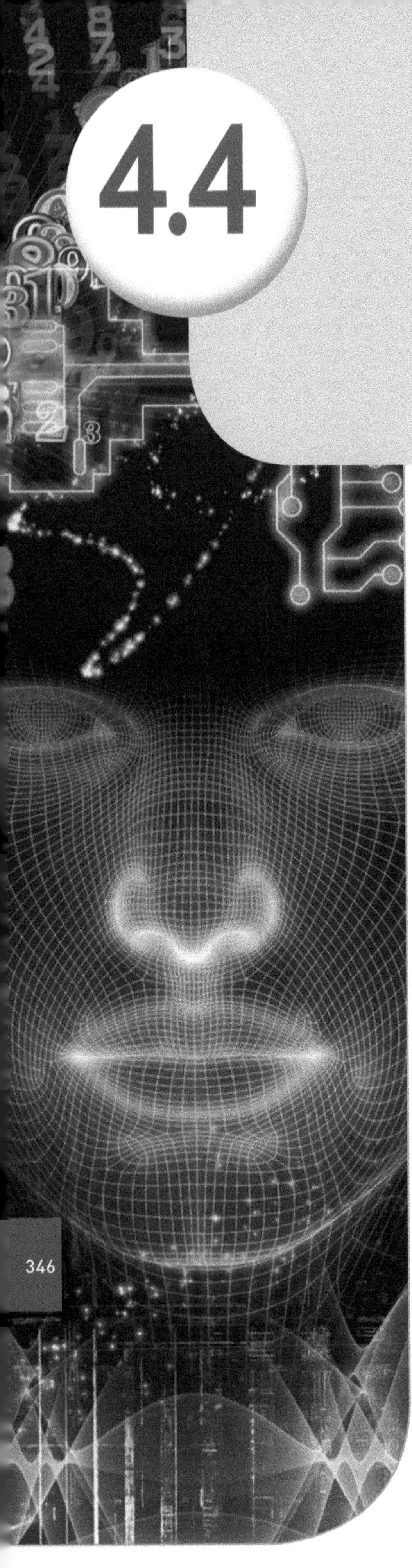

4.4 Mathematical skills for the written exams

This chapter covers the key mathematical elements that appear within the written papers, and how they relate to product design.

You should aim to follow the rules outlined below.

- 15% of the examination must be from the application of mathematical skills.
- The level of mathematical skill expected is Level 2 grade C (4 +) and above.
- You will be able to use a scientific calculator within the exam.
- All of the AS and A-level exams will include mathematics content.
- When undertaking mathematics questions students must use the full calculator value throughout the calculation only 'rounding' the final figure to the accuracy required.
- All working steps are expected to be shown (method marks are awarded).
- The use of π within a question should be the calculator value. (Although the value of π should be the calculator value all mark schemes for calculations using p will accept a range of values indicated as: [minimum acceptable value, maximum acceptable value]. This is shown in worked examples within this chapter.)

Number and percentages

A percentage (% or per cent) is a fraction recorded as a number out of 100. This means that 10 per cent is 10 out of 100. A percentage can also be converted to a decimal. This is done by dividing the percentage by 100; this means that 10 per cent becomes 0.1 as a decimal. By multiplying a decimal by 100, you can convert the decimal back to a percentage; this means that 0.05 is equivalent to 5 per cent.

Percentages are often used when comparing quantifiable parameters and tolerances such as:

- costings: materials, production and wastage
- measurements: linear, area, volume, mass and density
- variables: speed, time, temperature and force.

You may need to express or convert a percentage with reference to an increase, decrease or as an acceptable range, which could be classed as a tolerance.

Worked example – material use and production costs

A printing company uses 11,000 litres of ink per day for one week. The ink is used in the following percentages:

Table 4.4.1 Ink percentages.

Colour	Monday	Tuesday	Wednesday	Thursday	Friday
Cyan	20%	40%	10%	25%	15%
Magenta	5%	10%	15%	20%	10%
Yellow	35%	20%	15%	15%	20%
Black	40%	30%	60%	40%	55%
TOTAL	100%	100%	100%	100%	100%

Question

How many litres of black are used during the week?

Answer

Step 1: Calculate the total amount of ink used over one week.

5 × 11,000 = 55,000 litres

Step 2: Calculate the amount of black ink used for the week.

This calculation is:

(40 + 30 + 60 + 40 + 55 divided by 100 + 100 + 100 + 100 + 100)

all multiplied by the total litres

= 225/500 × 55,000

= 24,750 litres.

Question

Express the amount of yellow ink to the amount of magenta ink used over the week as a ratio (see the next section) in its simplest form.

Answer

Total percentage of yellow used over the week is:

(35 + 20 + 15 + 15 + 20)/500 = 21%

Total percentage of magenta used over the week is:

(5 + 10 + 15 + 20 + 10)/500 = 12%

The ratio of yellow to magenta is 21:12. You should always express ratios in their simplest form, so this would be 7:4. (3 is a factor of 21 and 12 so the ratio simplifies to 21/3:12/3 = 7:4.)

Question

The firm is looking at a process to reduce ink usage by 10 per cent. If all ink colours cost £1.25 per litre what would be their total saving over a four-week period?

Answer

In four weeks the firm uses:

55,000 × 4 = 220,000 litres of ink

10 per cent of this would be:

220,000 × 0.1 = 22,000 litres

The cost saving would therefore be:

22,000 × £1.25 = £27,500

Worked example – volume

Question

An injection mould for a dice has a cavity 20 mm × 20 mm × 20 mm. Once cooled, the component shrinks along its linear length by 10 per cent. Calculate the percentage reduction in volume between the mould and the component.

Answer

The cavity volume is:

$20 \times 20 \times 20 = 8{,}000\ \text{mm}^3$

The reduced linear length is:

$20 \times 0.9 = 18\ \text{mm}$

(The linear length becomes 10 per cent smaller, so when cooled, it is 90 per cent or 0.9 of the original size.)

Component size after cooling (therefore component volume) is:

$18 \times 18 \times 18 = 5{,}832\ \text{mm}^3$

The volume reduction is:

$8{,}000 - 5{,}832 = 2{,}168\ \text{mm}^3$

The percentage volume reduction is therefore:

$(2{,}168/8{,}000) \times 100 = 27.1\%$

Worked example – tolerance

Question

The ideal speed of rotation for a specific drill is 800 rpm, although it is acceptable to run it between 780 – 820 rpm. Express the acceptable tolerance as a percentage.

Answer:

Maximum rpm – minimum rpm = acceptable tolerance:

$820 - 780 = 40$

As a percentage, the tolerance is:

$(40/800) \times 100 = 5\%$

This tolerance would be expressed as ± 2.5%, because 2.5 per cent less than 800 or 5 per cent more than 800 are both acceptable.

Ratios

A ratio is a way of saying how much of one thing there is compared to another thing. Ratios can be written in various ways, but it is most common to show a ratio as a series of numbers separated by colons.

For example, if you had 10 lengths of timber, 7 of which are pine and 3 of which are oak, the ratio of pine to oak is 7 to 3 and could be written as 7:3. Seven tenths of the timber is pine; three tenths of the timber is oak.

If a ratio is shown as 8:4 then it should be displayed in its simplified form. As both sides can be divided by 4, the simplified form would therefore be 2:1. Always try to divide by the highest common factor.

When a ratio is in its simplest form, all of the numbers are whole numbers. If you have a ratio with decimal numbers, then both sides of the ratio should be multiplied by the same value to create whole values on both sides of the colon.

When working with mixing liquids, it may be that quantities are given as ratios. For example, when diluting paint, you may mix a 2.5 litre tin of paint with 0.5 litres of water.

The ratio of 2.5:0.5 should be multiplied by 2 to give the whole numbers 5:1.

Ratios are often used to divide measurements into set proportions. This may be applied to:

- mixing constituent parts: paints, adhesives, composites (such as concrete)
- dividing sizes: linear, area, volume and angular
- dividing costings: material, production and wastage.

You may need to use or determine a ratio from given data.

Worked example – mixing constituent parts

Question

A concrete casting is produced using the following ratios:

Sand: 3 litres
Cement: 1 litre
Water: 0.5 litres
Aggregate: 1.5 litres

Express the ratio using only whole numbers.

Answer

Ratio: 3:1:0.5:1.5

Multiplying by 2 throughout the ratio to remove the decimals gives:

Ratio: 6:2:1:3

Question

The casting requires 30 litres of concrete. How much aggregate is required?

Answer

In our whole numbers ratio, there are 3 units of aggregate in the total (6 + 2 + 1 + 3) 12 units of liquid.

This is 3/12 or 1/4 of the liquid.

If the total liquid is 30 litres, to find the proportion that is aggregate, you need to multiply the proportion with the total:

$3/12 \times 30 = 1/4 \times 30 = 7.5$ litres

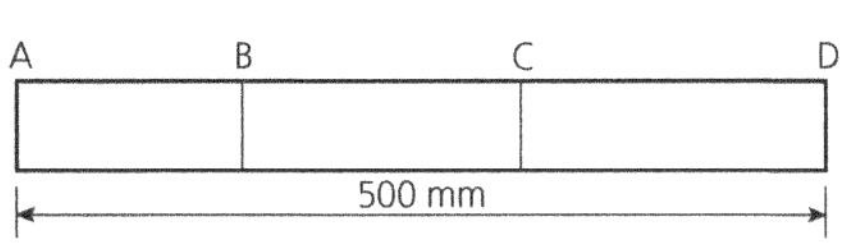

Figure 4.4.1 Aluminium strip bent along B and C.

Worked example – dividing sizes

Question

The diagram shows a strip of aluminium that is to be bent along bend lines B and C. The lengths AB, BC, CD are in the ratios of 3:4:5.

The overall length of the strip (AD) is 500 mm. Calculate the length of each line:

AB
BC
CD

Answer

The overall length is split into (to 2 decimal places)

AB = 3/12 × 500 = 125 mm
BC = 4/12 × 500 = 166.66 mm
CD = 5/12 × 500 = 208.33 mm

We can also use ratios to scale drawings up or down (by multiplying or dividing).

$P = 4S$

$A = S^2$

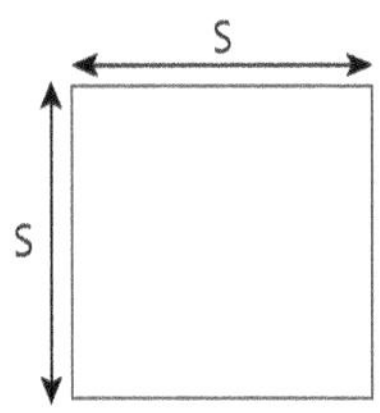

Figure 4.4.2 Perimeter and area of a square.

$P = 2a + 2b$

$A = ab$

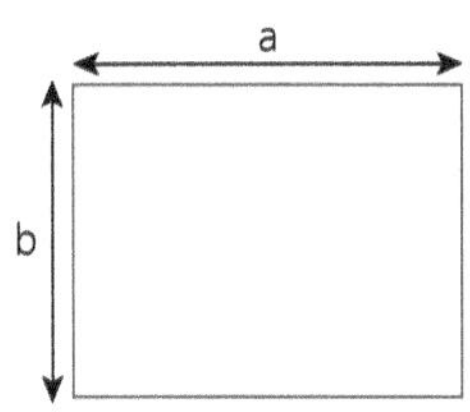

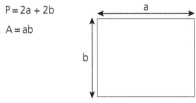

Figure 4.4.3 Perimeter and area of a rectangle.

$P = a + b + c$

$A = \frac{1}{2}bh$

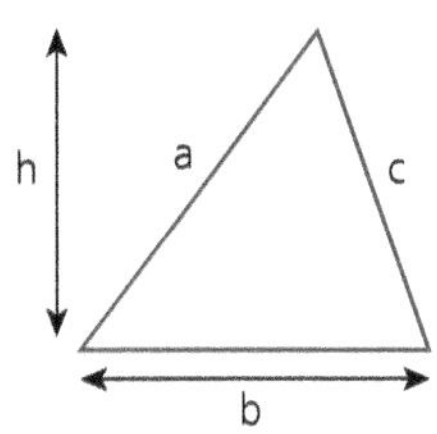

Figure 4.4.4 Perimeter and area of a triangle.

$C = 2\pi r$

$A = \pi r^2$

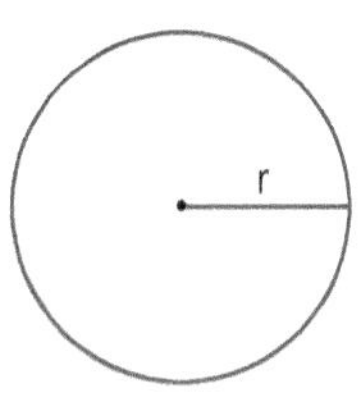

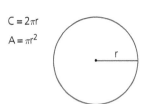

Figure 4.4.5 Circumference and area of a circle.

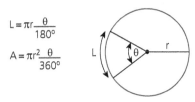

$L = \pi r \frac{\theta}{180°}$

$A = \pi r^2 \frac{\theta}{360°}$

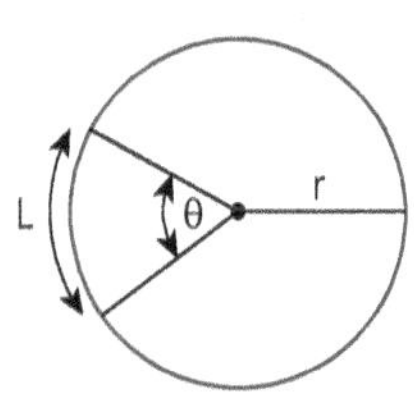

Figure 4.4.6 Circular sector.

Calculating surface area and volume

Calculating surface area

Calculating surface areas is required when ordering the correct amounts of materials and/or finishes, and to calculate effective usage of materials with minimum wastage.

Standard mathematical formulae for surface area can be used and combined to provide the required information.

Areas of 2D geometric shapes include rectangles, triangles, circles (and parts of) and shapes constructed from these.

All 2D shapes have a perimeter and area. The perimeter (P) is defined as the total length of all edges, and the area (A) is the space enclosed by the edges.

To calculate the area of a square or rectangle, you must multiply the height by the width. The standard units of measurement used within product design are millimetres for length, and for area the units are mm^2.

The formula for the area of a triangle is 0.5 × base × vertical height.

The circumference (C) of a circle is defined as:

π × diameter or 2 × π × radius.

The area of a circle is defined as:

πr^2.

Using parts of a circle

It is often necessary to work out lengths of arcs (L) for sheet or tube bending. This can be done using the formula for the circumference of a circle × the angle of the sector/360°.

Length of arc = $2\pi r \times \theta/360$.

When calculating the length of tube required for a tube bend, you should use the mean radius of the bend (the radius from the centre line of the tube to the centre of the arc).

Figure 4.4.7 Wassily chair tube bend.

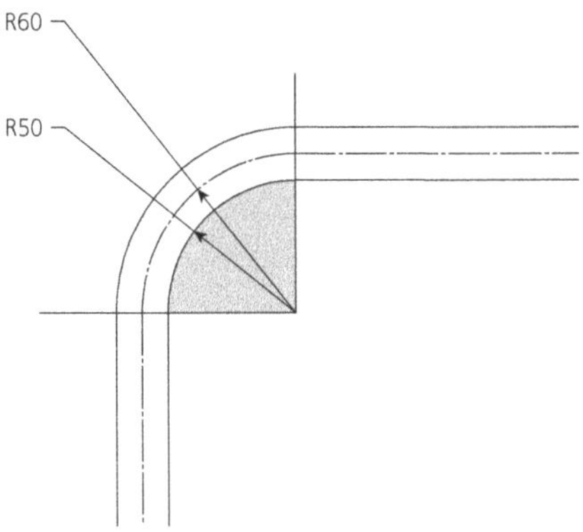

Figure 4.4.8 Tube bend diagram.

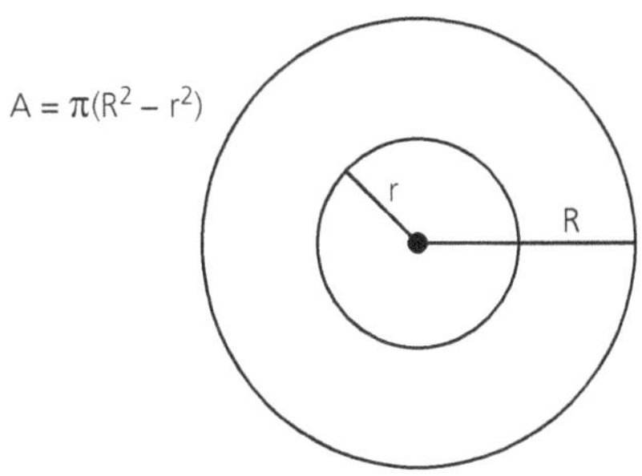

Figure 4.4.9 Circular ring.

Worked example

Question

The arm of a B3 'Wassily chair' by Marcel Breuer has been bent through 90° around a former with a 50 mm radius. The tube diameter is 20 mm.

What is the length of the tube required to form this bend (assuming no excess is needed)?

Answer

Step 1: Find the mean radius of the bend:

radius + 0.5 × tube diameter

$50 + 0.5 \times 20 = 60$ mm

Step 2: Find the length of the tube (arc):

$2 \times \pi \times$ mean radius $\times$ angle of arc/360

$2 \times \pi \times 60 \times (90/360) = [94.2, 94.26]$ mm

This shows the range of acceptable answers based on the acceptable value of π being 3.14, 3.142.

When calculating the cross-sectional area of a hollow object, the area of the hollow shape must be subtracted from the area of the outer shape. This is shown in Figure 4.4.9, with the cross section of a circular ring.

Worked example

Question

If the outer circle has a radius of 30 mm and the inner hole has a radius of 20 mm, what is the cross-sectional area of the ring?

Answer

Area of the ring = area of circle without hole – area of the hole

Step 1: Find the area of the circle without hole:

$\pi \times 30^2 = [2826, 2827.8]$ mm^2

Step 2: Find the area of the hole:

$\pi \times 20^2 = [1256, 1256.8]$ mm^2

Area of the ring:

$[2826, 2827.8] - [1256, 1256.8] = [1569.2, 1571.8]$ mm^2

This situation may occur when calculating the volume of a tube.

Calculating surface areas and volumes of 3D forms

You may find you need to calculate surface areas and volumes of 3D forms to ensure correct material finishing and machining costings, ordering of sufficient materials and tool life expectancy. It is also essential for the accurate design of moulds, formers, jigs and templates.

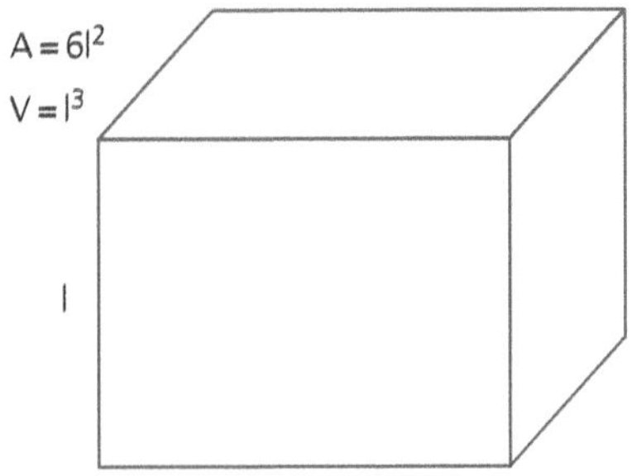

Figure 4.4.10 Cube area and volume.

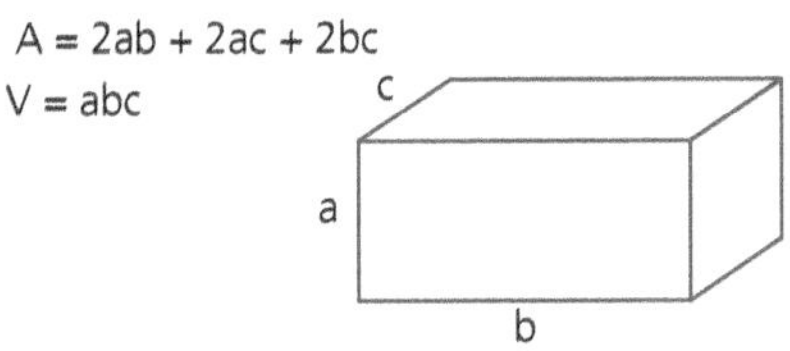

Figure 4.4.11 Rectangular prism area and volume.

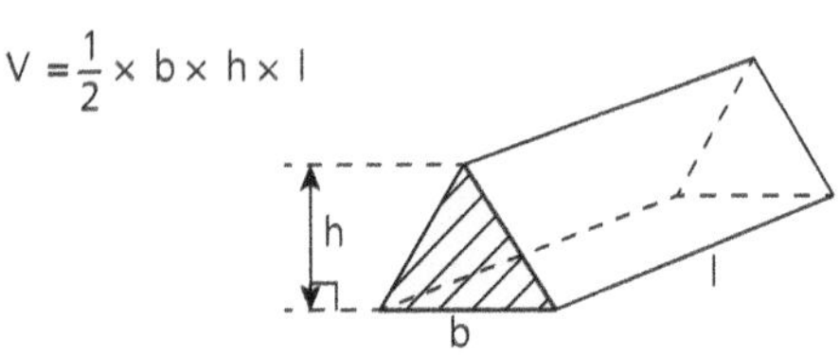

Figure 4.4.12 Triangular prism volume.

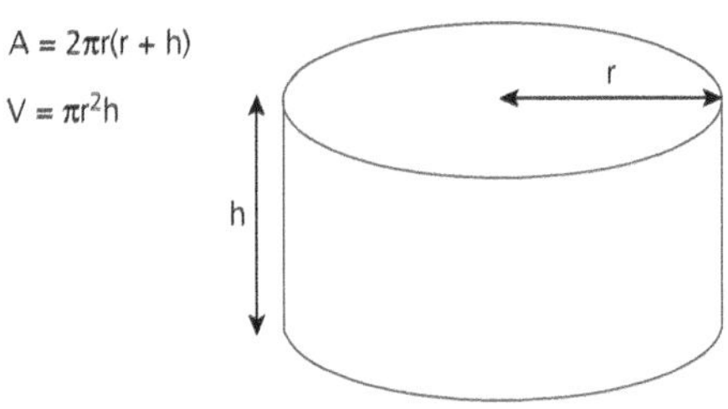

Figure 4.4.13 Cylinder area and volume.

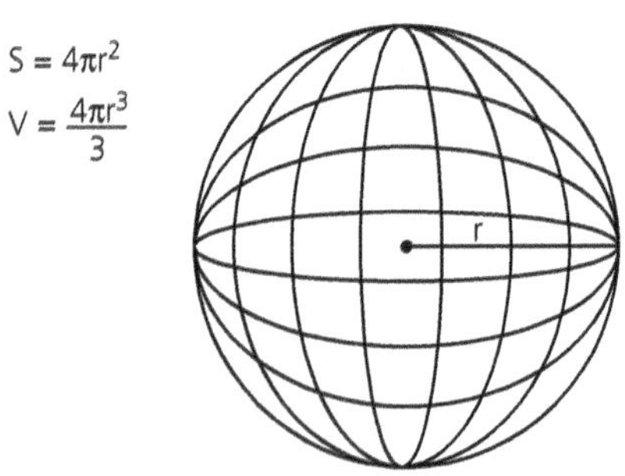

Figure 4.4.14 Sphere surface area (S) and volume.

Volumes of standard geometric forms

Cuboids, cylinders, cones, prisms, spheres (and parts of) are known as simple geometric forms.

The standard units of measurement used within product design are millimetres for length, mm^2 for area and mm^3 for volume.

Although the standard dimensions you will come across are mm^3 you may experience situations where cm^3 and m^3 are used. The table below shows the relationship between the three most common units and how to convert between these.

Table 4.4.2 The relationship between common units and how to convert.

	mm^3	cm^3	m^3
1 mm^3	1	0.001	0.000,000,001
1 cm^3	1,000	1	0.000,001
1 m^3	100,000,000	1,000,000	1

To calculate the volume of a cube or cuboid you multiply: height × width × depth.

When calculating the volume of any prism – triangular, rectangular, hexagonal and cylindrical – the calculation has two stages.

Stage 1: Calculate the cross-sectional area (as seen in the previous section).

Stage 2: Multiple the cross-sectional area by the height/length of the prism.

Worked example

Question

Find the volume of a cylinder with a radius of 25 mm and a height of 50 mm.

Answer

Step 1: Find the cross-sectional area.

The cross-sectional area of a cylinder is a circle, therefore the formula is:

πr^2

Calculation of cross sectional area:

$\pi \times 25^2 = [1962.5, 1963.75]\ mm^2$

The volume of a cylinder is cross-sectional area × height.

Calculation of volume:

$[1962.5, 1963.75] \times 50 = [98125, 98187.5]\ mm^3$

Calculating the volume of a sphere uses the formula: volume = $(4\pi \times r^3)/3$. Although this is useful, the more common situation you will find will use a hemisphere, requiring you to divide the volume in half.

Calculating both the volume of a pyramid or cone uses the same formula:

(area of the base shape × vertical height)/3

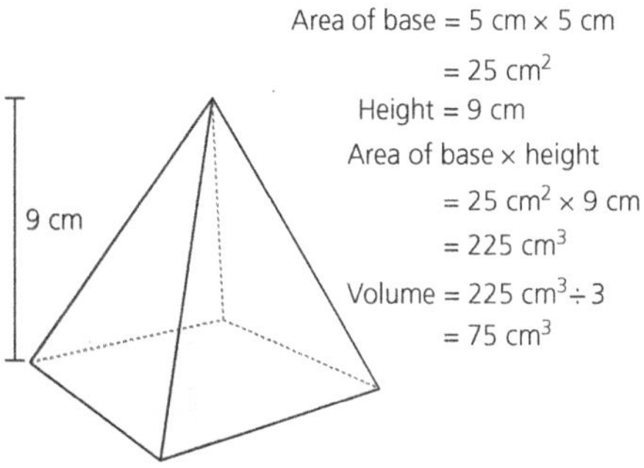

Figure 4.4.15 Square-based pyramid

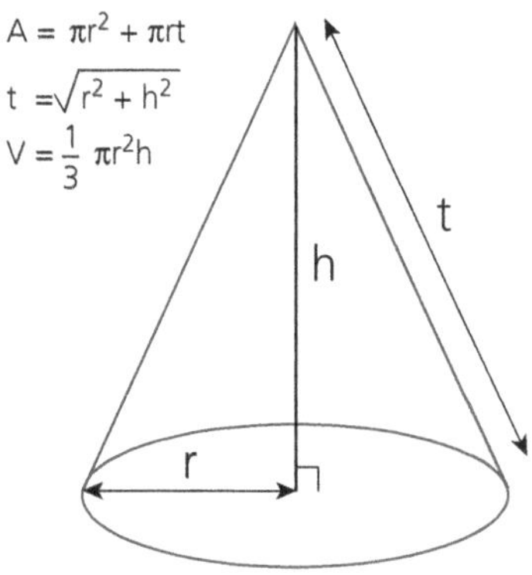

Figure 4.4.16 Cone surface area and volume.

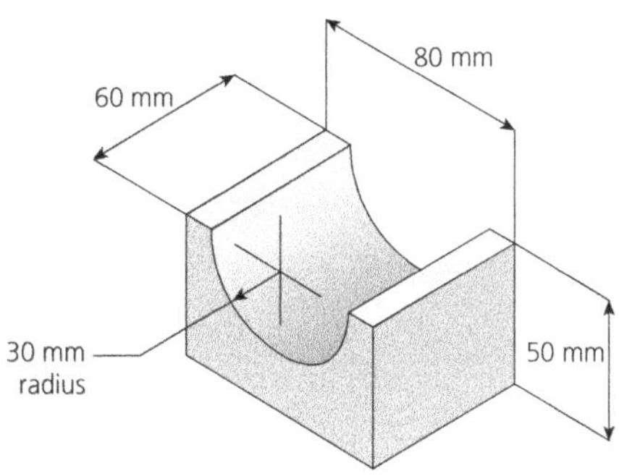

Figure 4.4.17 Dimensioned support block.

Worked example

Question

Calculate the volume of a cone with a base radius of 20 mm and a vertical height of 60 mm.

Answer

Step 1: Calculate the area of the base:

$\pi \times 20^2 = [1256, 1256.8]$ mm^2

Step 2: Multiply the circular base by the vertical height:

$[1256, 1256.8] \times 60 = [75360, 75408]$

Step 3: Divide by 3.

$[75360, 75408] / 3 = [25120, 25136]$ mm^3

Combining forms

The majority of products will be a combination of geometric forms, such as shown in Figures 4.4.10–4.4.16. To work out volumes in these situations requires several steps.

Worked example

Question

Figure 4.4.17 shows a cast metal block used for supporting a bicycle handle bar. Calculate the volume of the casting.

Answer

Step 1: Determine the geometric forms used to construct the object.

The object is produced from a cuboid and a half cylindrical channel.

Step 2: Calculate the volume of the cuboid (before the channel is removed).

Volume of a cuboid = height × width × depth

$50 \times 80 \times 60 = 240{,}000$ mm^3

Step 3: Calculate the cross-sectional area of the semicircle channel.

$(\pi \times 30^2) / 2$

$[2826, 2827.8] / 2 = [1413, 1413.9]$ mm^2

Step 4: Calculate the volume of the channel.

Volume of half a cylinder = cross-sectional area of semicircle × length

$[1413, 1413.9] \times 60 = [84780, 84834]$ mm^3

Step 4: Subtract the channel volume from the cuboid volume

$240{,}000 - [84780, 84834] = [155166, 155220]$ mm^3

Comparing weights using density

Density is defined as mass per unit volume, and is used in design calculations to compare suitable materials for components with weight restrictions. Density data will often be given for a specific material. This can then be combined with the calculated volume to compare component weights.

Table 4.4.3 Densities of common non-ferrous metals.

Metal	Density: (g/cm^3)	Density (g/mm^3)
Aluminium	2.7	0.0027
Copper	8.96	0.00896
Zinc	7.13	0.00713
Iron	7.87	0.00787
Gold	19.32	0.01932
Lead	11.36	0.01136
Silver	10.49	0.01049

Due to the vast number of different steels available, it is not possible to give a definitive value for density, but the value stated below is sufficiently accurate for comparison at this level.

Table 4.4.4 Steel density.

Metal	Density: (g/cm^3)	Density (g/mm^3)
Steel (variable)	7.85	0.00785

Worked example

Question

The dimensioned support block shown in Figure 4.4.17 is to be cast twice – once in aluminium and once in zinc. Calculate the mass of each component. Which should be used if the casting had to weigh less than 1 kg?

Answer

Step 1: Calculate the mass of the component when cast from aluminium.

(volume of component in mm^3) × (density of aluminium in g/mm^3)

155,190 × 0.0027 = 419 grams

Step 2: Calculate the mass of the component when cast from zinc.

(volume of component in mm^3) × (density of zinc in g/mm^3)

155,190 x 0.00713 = 1,106.5 grams

If the casting had to weigh less than 1 kg, then aluminium should be used.

Area and volume scale factors

The use of proportion is key and scaling products up or down in size may be required. You may be required to increase or decrease a length, area or volume.

The scaling of a length is covered in the ratios section, however scaling area and volume are more problematic. The mathematic skills for scaling area and volume at this level will therefore only cover rectangular and cuboid forms.

Scaling area

When changing surface area of a rectangle, you may be given a percentage increase or decrease in area and be asked to calculate the change in side length.

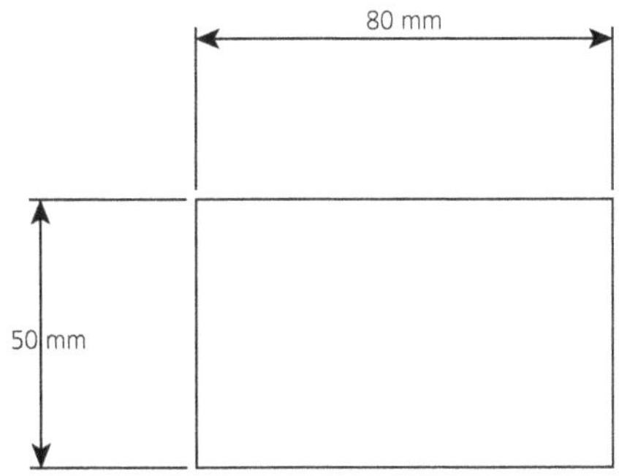

Figure 4.4.18 Rectangular surface area.

Worked example

Question

The rectangular label in Figure 4.4.18 is 50 mm × 80 mm, giving a surface area of 4,000 mm^2. If I wish to reduce label surface area by 50 per cent, how long would each side be?

Common mistake

If we reduce each side length by 50 per cent, the new rectangle will be 25 × 40 mm, giving a surface area of 1,000 mm^2 which is a reduction in surface area of 75 per cent. This is, of course, wrong.

Therefore, to calculate the reduced side length when reducing surface area, we must follow the steps below.

Answer

Step 1: Calculate the area scale factor.

The area scale factor is expressed as a decimal, which is converted from the new size as a percentage:

The area scale factor = 0.5 (50 per cent of the original)

Step 2: Calculate the length scale factor.

As we have an area scale factor and an area is measured in mm^2, but we need a length scale factor that is measured in mm, we must apply a square root to the area scale factor

The length scale factor:

$\sqrt{0.5} = 0.707...$

Step 3: Calculate the new length of each side.

original length × length scale factor

$50 \times 0.707... = [35.35, 35.36]$

$80 \times 0.707... = [56.56, 56.57]$

Step 4: Check new surface area:

$[35.35, 35.36] \times [56.56, 56.57] = [1999.396, 2000.3152]\ mm^2$

When changing volume of a cuboid or prism, you may be given a percentage increase or decrease in volume and need to calculate the change in side length.

Worked example

Question

The manufacturers of a chocolate bar shown in Figure 4.4.19 wish to reduce the volume of chocolate used by 30 per cent, while keeping the same proportions. Calculate the new lengths of each side.

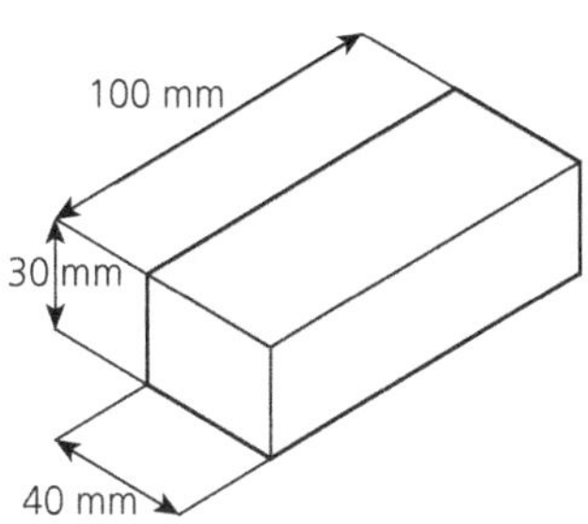

Figure 4.4.19 Chocolate bar.

Answer

Step 1: Calculate the original volume.

$30 \times 40 \times 100 = 120{,}000\ mm^3$

Step 2: Calculate the volume scale factor.

The volume scale factor is expressed as a decimal, which is converted from the new size as a percentage:

The volume scale factor = 0.7 (70 per cent of the original)

Step 3: Calculate the new volume required.

Original volume × volume scale factor

$120{,}000 \times 0.7 = 84{,}000 \text{ mm}^3$

Step 4: Calculate the length scale factor.

As we have a volume scale factor and volume is measured in mm^3, but we need a length scale factor that is measured in mm, we must apply a cube root to the volume scale factor.

The length scale factor = $\sqrt[3]{0.7} = 0.887 \ldots$

Step 5: Calculate the new length of each side.

original length × length scale factor

$30 \times 0.887 \ldots = [26.61, 26.637] \text{ mm}$

$40 \times 0.887 \ldots = [35.48, 35.516] \text{ mm}$

$100 \times 0.887 \ldots = [88.7, 88.8] \text{ mm}$

Step 6: Check new volume.

$[26.61, 26.637] \times [35.48, 35.516] \times [88.7, 88.8] = [83743.69, 84008.32] \text{ mm}^3$

Trigonometry

Trigonometry can be applied to determine the distances between points and angles between lines and surfaces. It is often used for accurate marking out, modelling, prototyping and testing.

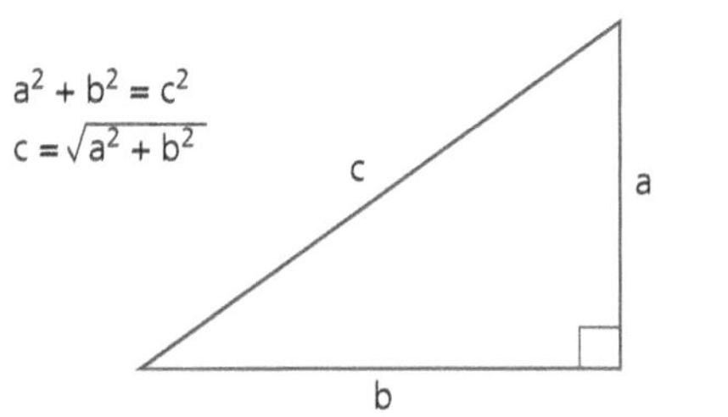

Figure 4.4.20 Pythagoras's theorem.

Pythagoras' theorem is used to calculate the length of the third side of a triangle when the other two lengths have been given.

The theorem states:

The square of the longest side = the sum of the squares of the other two sides.

This only applies to right-angled triangles.

Applying trigonometry within product design

Worked example

Question

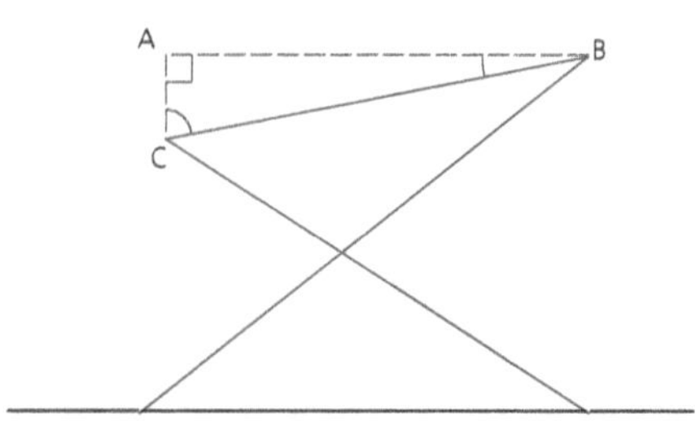

Figure 4.4.21 Chair frame.

When designing a chair, a frame has been produced out of steel tube (see Figure 4.4.21). A seat is to fit along BC. The designer must calculate the length of board needed for this. They know that the depth of the chair AB is 400 mm and that the seat should drop 100 mm along AC.

Express your answer to the nearest millimeter.

Answer

Step 1: Select Pythagoras' theorem and substitute values.

The square of the longest side = the sum of the squares of the other two sides

$BC^2 = AB^2 + AC^2$

$BC^2 = 400^2 + 100^2$

$BC^2 = 160{,}000 + 10.000$

$BC^2 = 170{,}000$

$BC = \sqrt{170{,}000}$

$BC = 412$ mm

When calculating angles or lengths (if only one side length is provided), you can use one of the following three formulae:

sin x = opposite/hypotenuse

cos x = adjacent/hypotenuse

tan x = opposite/adjacent

All of these formulae apply to right-angled triangles only. The formula you select will depend on the information you have been given. Note that trigonometry is used for the three calculations above, not Pythagoras' theorem.

Worked example

Question

If, for the chair frame in Figure 4.4.21, you wish to find the angle of the seat from the horizontal line AB, you must find angle ABC. To calculate this angle, you require two side lengths.

You are given the following data:

AB = 400

AC = 100

Answer

Stage 1: Select the correct formula.

We have the length of sides opposite and adjacent to the angle we require, we have not been given the hypotenuse. This means we must select the tan formula.

Stage 2: Substitute values.

tan ABC = opposite (AC)/adjacent (AB)

tan ABC = 100/400

tan ABC = 0.25

Stage 3: Use the inverse tan function on a calculator to find the angle.

ABC = $\tan^{-1}$ (0.25)

ABC = 14.04°

Construction, use and analysis of charts and graphs

Charts and graphs are often used to display data from research activities such as questionnaires, surveys and focus groups. A range of methods can be used, including histograms, pictograms, bar charts and pie charts.

Technical data may also be presented for interpretation when selecting materials, finishes, components or production processes, or when designing for inclusivity (use of anthropometric and ergonomic data).

Figure 4.4.22 Exemplar infographic.

The term 'infographic' refers to the presentation of information in a graphical format to improve understanding. This form of graphical presentation often makes it easier to interpret results and to visualise trends and variations.

The appropriate selection of communication method is dependent on the data set, the intention of the analysis and the audience.

Co-ordinates and geometry

Co-ordinates and geometry include methods of determining the position of points, lines and planes in relation to each other or datums.

When using co-ordinates, it is essential that you understand the concept of using a datum. A datum is a start point or a location/position from which measurements are taken (see figure 4.4.23). An ideal location for a datum would be a corner, where horizontal and vertical dimensions can be taken from.

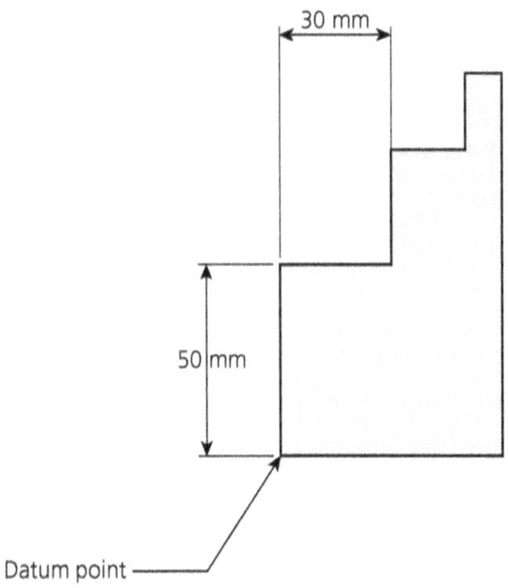

Figure 4.4.23 Measuring from a datum.

All points on a 2D shape have a horizontal and vertical co-ordinate in relation to the set datum, and these co-ordinates are displayed as dimensions in standardised metric units.

If you dimension without using the same datum for each measurement, you risk inaccuracy in measurements. This is because each dimension has an acceptable tolerance, which is the level of accuracy possible with the machining processes available.

If the tolerance on each dimension is ± 0.2 mm then by dimensioning from a datum (as shown on the left of Figure 4.4.24), the overall height of a product will have a tolerance of ± 0.2 mm. If you fail to use the same datum for all measurements (as shown on the right in Figure 4.4.24), you could end up with a cumulative error. As there are three dimensions, each with a tolerance of ± 0.2 mm, the combined error could be as much as ± 0.6 mm.

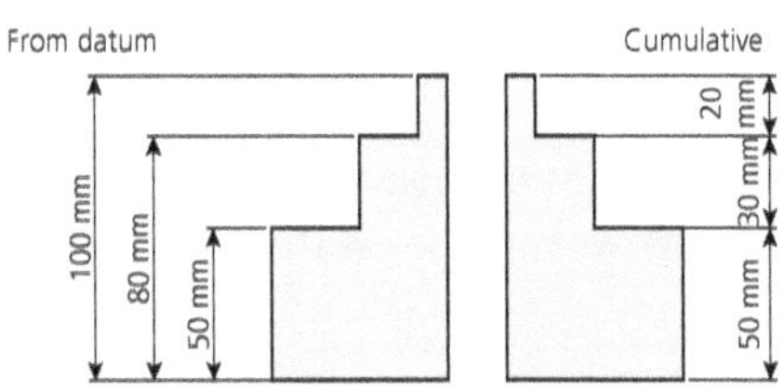

Figure 4.4.24 Comparing dimensioning strategies.

When dimensioning radii or diameters of circles you must always reference the centre point of the circle.

In Figure 4.4.25, the datum is set as the centre of the circle, and the position of the centre of each hole is specified using a separation angle and a radial length from the datum (centre of circle). This is defined as a pitch circle diameter (PCD).

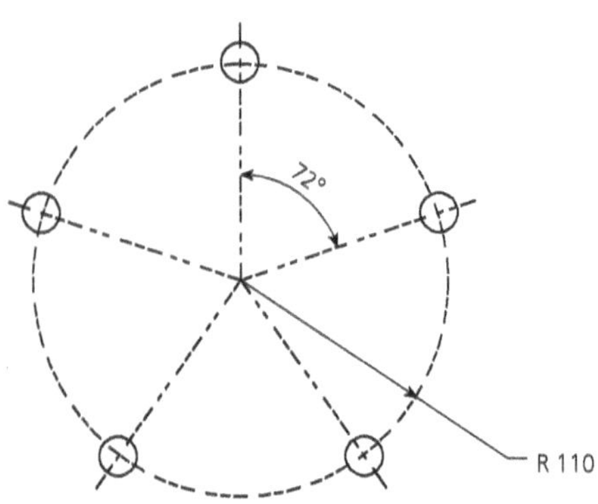

Figure 4.4.25 Pitch circle diameter dimensioning.

Statistics and probability

Probability refers to the likelihood of an outcome. This may be used within product design for a variety of reasons, such as predicting product sales, failure rates and sales trends.

Feedback from focus groups and consumers can be used in the prediction of product sales, and data from QC testing can be used to calculate the likelihood of product failures and plan aftersales customer support requirements.

Worked example

Question

An electrical manufacturer tests 1,000 identical washing machines. The motors on 2 machines fail after 500 washes and the drive belts of 25 machines fail after 500 washes. Both outcomes were independent of each other. What is the probability of a machine suffering from both issues?

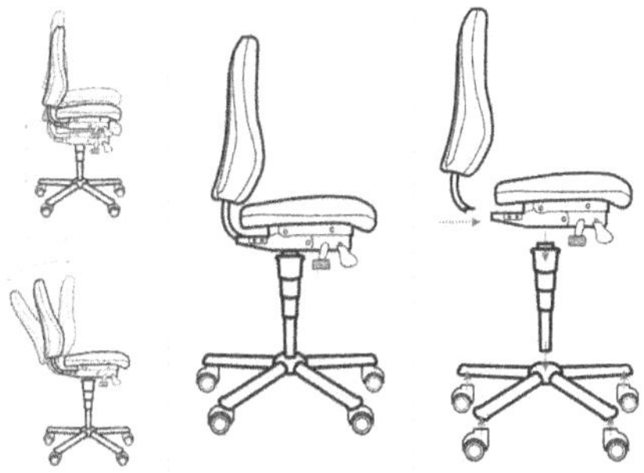

Figure 4.4.26 A side elevation of an adjustable office chair.

Answer

Probability of motor failure: 2/1,000

Probability of drive belt failure: 25/1,000

Probability of both failures:

$(2/1{,}000) \times (25/1{,}000) = 5/100{,}000 = 1/20{,}000$

When undertaking a product design project, it is often necessary to refer to collected data, so as to design products suitable for the largest number of users. The data collected could be anthropometric measurements to assist in the production of ergonomic forms. It could also be results of market research or focus group feedback.

The worked example below shows how you may interpret data to help make design decisions.

Worked example

Question

When designing an office chair, the seat height must be adjustable to suit a wide range of users. Figure 4.4.27 and Table 4.4.5 show the comfortable sitting heights of a sample of office workers.

Calculate in mm the range of adjustment required for an office chair suitable for the 5th to 95th percentiles of the sample.

Table 4.4.5 Comfortable sitting heights of a sample of office workers.

Seat height (to nearest mm)	Number in sample
300–320	4
321–340	7
341–360	19
361–380	24
381–400	45
401–420	65
421–440	87
441–460	98
461–480	87
481–500	65
501–520	45
521–540	24
541–560	19
561–580	7
581–600	4

Answer

Calculate sample size:

$4 + 7 + 19 + 24 + 45 + 65 + 87 + 98 + 87 + 65 + 45 + 24 + 19 + 7 + 4 = 600$ people

Calculate 5th percentile of sample:

$600 \times 0.05 = 30$ people

Calculate 95th percentile of sample:

$600 \times 0.95 = 570$ people

Question

If the pneumatic cylinder used to adjust the height is restricted to 140 mm within any height range, what is the maximum percentage of the sample that could be comfortably accommodated?

Answer

Step 1: As each row of the table is 20 mm of movement, you must find the seven adjacent rows with the largest number of workers within them.

380–520 mm

Step 2: Calculate the total number of workers within those rows.

$98 + 87 + 87 + 65 + 65 + 45 + 45 = 492$

Step 3: Calculate the percentage of the sample included in the rows.

$492/600 \times 100 = 82\%$

82 per cent of the sample would be accommodated.

Index

Page numbers in **bold** indicate key terms.

Photo credits

All photos that have not been listed below have been kindly supplied by the author team.

p.6 *c* © Fotogal/stock.adobe.com, *b* © Monticellllo/stock.adobe.com; **p.7** © Peter Hermes Furian/Shutterstock.com; **p.8** © Feng Yu/Shutterstock.com; **p.9** *t* © Jan Hetman/Alamy Stock Photo, *c* © APStudio 2015/Shutterstock.com, *b* © Photographee.eu/stock.adobe.com; **p.10** © Mark Wherry/Alamy Stock Photo; **p.11** *tl* © BESTWEB/Shutterstock.com, *b* © Nadezda Murmakova/Shutterstock.com; **p.12** *t* © Eshma/Alamy Stock Photo; **p.16** © Ampol sonthong/123RF; **p.25** © Stocksolutions/Shutterstock.com; **p.26** © Andy Wilkinson/123 RF; **p.31** © Stocksolutions/Shutterstock.com; **p.33** © Roger Ressmeyer/Corbis/VCG/Corbis Documentary/Getty Images; **p.34** © Konrad Mostert/Shutterstock.com; **p.35** © TLaoPhotography/Shutterstock.com; p.42 *tl* © Sanddebeautheil/Shutterstock.com, *tc* © Daniel Hopkinson/Arcaid Images/Alamy Stock Photo, *tr* © John Edward Linden/Arcaid Images/Alamy Stock Photo, *c* © Industria/Alamy Stock Photo, *b* © Craig Russell/Shutterstock.com; **p.47** © Oleksandr Rybitskyi/123RF; **p.54** *tl* © Gino Santa Maria/123RF, *tr* © Radovan1/Shutterstock.com, *cl* © Texelart/123RF, *cr* © Photka/123RF, *bl* © Eugene Sergeev/Alamy Stock Photo, *br* © Pro3DArtt/Shutterstock.com; **p.58** *l* © Jeffrey B. Banke/Shutterstock.com, *r* © Vvoe/Shutterstock.com; **p.67** © Tcsaba/Shutterstock.com; **p.69** *uc* © BART photography/Shutterstock.com, *lc* © Tad Denson/Shutterstock.com, *lb* © Khongdet Khumsri/Shutterstock.com; **p.71** © Stepan Popov/123RF; **p.75** © SNEHIT/Shutterstock.com; **p.76** *tl* © Grigvovan/Shutterstock.com, *bl* © Steve Cordory/Shutterstock.com; **p.77** *t* © Valzan/Shutterstock.com, *uc* © Olivierl/123RF, *lc* © MG Still Life/Alamy Stock Photo, *bl* © IB Photography/Shutterstock.com; **p.78** *tl* © Vadim Orlov/Shutterstock.com, *ul* © Sheila Quattrocchi/Alamy Stock Photo, *lb* © Oleksandr Chub/Shutterstock.com, *rb* © Stanislav Lazarev/Shutterstock.com; **p.80** *tl* © Hugh Threlfall/Alamy Stock Photo, *cl* © Tony Watson/Alamy Stock Photo, *bl* © Richard Heyes/Alamy Stock Photo; **p.81** © Andrey Ruzaev/Shutterstock.com; **p.82** *tl* © Sauletas/Shutterstock.com, *cl* © DIY Disc Sander made by FloweringElbow" by Stephen John Saville; **p.83** *ut* © Skdesign/123RF, *uc* © Alexeysun/Shutterstock.com, *lc* © Chones/Shutterstock.com, *ub* © Leungchopan/Shutterstock.com, *lb* © WIRACHAIPHOTO/Shutterstock.com; **p.86** *tl* © Vereshchagin Dmitry/Shutterstock.com, *cl* © Ed Phillips/Shutterstock.com; **p.87** © Tamisclao/Shutterstock.com; **p.89** © Moreno Soppelsa/Shutterstock.com; **p.90** © Stephen mulcahey/Shuttetstock.com; **p.91** © Cristian M. Vela/Alamy Stock Photo; **p.92** © BC Photo/Alamy Stock Photo; **p.93** *tl* © Oramstock/Alamy Stock Photo, *tr* © Natallia Khlapushyna/Alamy Stock Photo, *cl* © Showcake/stock.adobe.com; **p.100** *tl* © Freeprod/123RF, *bl* supplied by Dave Sumpner; **p.101** © Photo 12/Archives Snark/Alamy Stock Photo; **p.103** *tl* © Porsche AG, *cl* © Xieyuliang/Shutterstock.com, *bl* © John Henderson/Alamy Stock Photo; **p.104** *tl* © Monty Rakusen/Cultura Creative (RF)/Alamy Stock Photo, *cl* © Stanislav Lazarev/Shutterstock.com, *bl* © Paul Bunch/Alamy Stock Photo; **p.105** © Alan Piscaglia/123RF; **p.106** *tl* © Ergochair Ltd/Adapt®, *bl* © Uli nusko/Alamy Stock Photo; **p.107** *cl* © Jim West/Alamy Stock Photo, *bl* © Keystone Pictures USA/ZUMA Press/Alamy Stock Photo; **p.108** © Everett Kennedy Brown/EPA/Shutterstock/Rex Features; **p.109** © Bossemeyer,Klaus/TravelCollection/Alamy Stock Photo; **p.110** © Dmitry Kalinovsky/Shutterstock.com; **p.112** *t* © Grethe Ulgjell/Alamy Stock Photo, *b* © Anthony Collins Cycling/Alamy Stock Photo; **p.116** © Iain Masterton/Alamy Stock Photo, **p.117** *tl* © Vladimir Nenezic/Shutterstock.com, *cl* © Art Directors & TRIP/Alamy Stock Photo, *bl* © oYOo/Shutterstock.com; **p.118** *tl* © David J. Green - industry/Alamy Stock Photo, *cl* © Jiř Zuz nek/123RF, *bl* © Iain Masterton/Alamy Stock Photo; **p.119** © Italyan/Shutterstock.com, *lb* © Mathew Alexander/Shutterstock.com; **p.121** *tl* © Patthana Nirangkul/Shutterstock.com, *cl* © Science Picture Co/Alamy Stock Photo; **p.122** © Xieyuliang/Shutterstock.com; **p.126** © Gilles BASSIGNAC/Gamma-Rapho/Getty Images; **p.128** *cl* © VW Pics/ Universal Images Group/Getty Images, *bl* © Egon Bömsch/imageBROKER/Alamy Stock Photo; **p.129** *tl* © Stephen Chung/Alamy Stock Photo, *cl* © DmyTo/Shutterstock.com; **p. 130** *cl* © Goddard Automotive/Alamy Stock Photo, *bm* © Funkyfood London - Paul Williams/Alamy Stock Photo, *br* © Lasse_Sven/Shutterstock.com; **p.131** *tc* © Chris Willson/Alamy Stock Photo, *cl* © Matthew Richardson/Alamy Stock Photo; **p.132** *tl* © Marco Uliana/123RF, *tr* © Rclassenlayouts/123RF; **p.133** *cr* © Juice Images249/Alamy Stock Photo; **p.134** *b*© Christian Rolfes/Ford of Europe, *b* © Christian Rolfes/Ford of Europe; **p.145** *tl* © Yaoinlove/Shutterstock.com, *bl* © Kitch Bain/123RF; **p.146** *bl* © Philippa Willitts/123RF; *tr* © V_ctoria/Shutterstock.com; **p.147** *tl* © Reed Saxon/AP/REX/Shutterstock.com, *bl* © The British Toy & Hobby Association;**p.148** *tl* © The British Toy & Hobby Association; **p.151** *tl* © Rachel Griffith/Alamy Stock Photo, *cl* © David Kilpatrick/Alamy Stock Photo; **p.152** *cl* © Granger Historical Picture Archive/Alamy Stock Photo, *bl* © Bettina Strenske/Alamy Stock Photo; **p.153** *tl* © Carolyn Jenkins/Alamy Stock Photo, *bl* © Houdek Martina/CTK/Alamy Stock Photo; **p.154** *cl* © Michael Cyran/Shutterstock.com; **p.158** *br* © Andersonrise/123RF; **p.160** © DPD ImageStock/Alamy Stock Photo; **p.161** *tl* © Peter Essick/Aurora Photos/Alamy Stock Photos; **p.162** *cl* © Elena Elisseeva/Alamy Stock Photo, *bl* © Praethip

Docekalova/Alamy Stock Photo; **p.163** *tl* © Fairphone, *bl* © Robert Stainforth/Alamy Stock Photo; **p.164** *tl* © PBimages/ Alamy Stock Photo, *bl* © Redorbital Photography/Alamy Stock Photo; **p.166** © KBImages/Alamy Stock Photo; **p.167** *tl* © Rainer Unkel/vario images GmbH & Co.KG/Alamy Stock Photo, *cl* © JIPEN/Alamy Stock Photo; **p.171** *tl* © Cheskyw/123RF, *bl* © Tatiana Popova/123RF; **p.172** *tl* © Kostsov/123RF, *cl* © Nataliia Natykach/123RF, *bl* © Auremar/123RF, *br* © Auremar/123RF; **p.176** *tl* © Brian Jackson/123RF, *cr* © Alona Stepaniuk/123RF; **p.179** © Adrian Sherratt/Alamy Stock Photo; **p.186** *tc* © Stockphoto mania/Shutterstock.com, *bc* © Attaphong/Shutterstock.com; **p.187** *cl* © Andy Dean Photography/Shutterstock.com, *bl* © AnnaKu/Shutterstock.com; **p.197** *bl* © Heritage Image Partnership Ltd/Alamy Stock Photo, *br* © John Hammond/The National Trust Photolibrary/Alamy Stock Photo; **p.198** *bl* © Francois Roux/Alamy Stock Photo, *bc* © Nikreates/Alamy Stock Photo, *br* © Marc Tielemans/Alamy Stock Photo; **p.199** *tl* © Picture Partners/Alamy Stock Photo, *cr* © Julian Castle/Arcaid Images/Alamy Stock Photo; **p.200** *cl* © Juergen Hanel/ Alamy Stock Photo, *cr* © Ros Drinkwater/Alamy Stock Photo; **p.201** *cr* © Arne Jacobsen/Fritz Hansen, *bl* © INTERFOTO/Alamy Stock Photo, *bc* © Elizabeth Leyden/Alamy Stock Photo, *br* © National Motor Museum/Motoring Picture Library/Alamy Stock Photo; **p.202** *bl* © The Zanone Collection/Coutesy MEMPHIS MILANO, *br* © The Zanone Collection/Coutesy MEMPHIS MILANO; **p.203** *cr* © Courtesy of Alessi; **p.206** *bl* © Interfoto/Alamy Stock Photo; **p.207** *l* © V&A Images/Alamy Stock Photo, *r* © Marc Tielemans/Alamy Stock Photo; **p.208** *tl* © The Zanone Collection/ Coutesy MEMPHIS MILANO, *tr* © Nikreates/Alamy Stock Photo; **p.211** *tl* © D Guest Smith/Alamy Stock Photo, *bl* © Juergen Hanel/Alamy Stock Photo, *br* © INTERFOTO/Alamy Stock Photo; **p.212** *cl* © War Archive/Alamy Stock Photo; **p.213** *tl* © Hulton Deutsch/Corbis Historical/Getty Images, *bl* © Hulton Deutsch/Corbis Historical/Getty Images, *br* © Johannes Simon/Getty Images Entertainment/Getty Images; **p.214** *cl* © Sciencephotos/Alamy Stock Photo; **p.215** *cl* © Andrew Paterson/Alamy Stock Photo, *cr* © Sashkin/Shutterstock.com; **p.217** *tl* © Heinz Zinram Collection/Mary Evans, *tr* © Monty Rakusen/Cultura Creative (RF)/Alamy Stock Photo, *cl* © Borthwick Institute/Heritage Image Partnership Ltd/Alamy Stock Photo, *bl* © Kwangmoo/123 RF, *bc* © Ilan Amihai/PhotoStock-Israel/Alamy Stock Photo, *br* © Hugh Threlfall/Alamy Stock Photo; **p.218** *tl* © Shaun Finch - Coyote-Photography.co.uk/Alamy Stock Photo, *tr* © Olekcii Mach/Alamy Stock Photo, *cl* © Trinity Mirror/Mirrorpix/Alamy Stock Photo, *cr* © Chris Willson/Alamy Stock Photo, *bl* © INTERFOTO/Alamy Stock Photo; **p.219** *ul* © Bettmann/Getty Images, *uc* © PHILIPS, *ur* © Chris Willson/Alamy Stock Photo, *bl* © INTERFOTO/Alamy Stock Photo, *bc* © Chris Willson/Alamy Stock Photo; **p.220** *bl* © Hufton and Crow/VIEW Pictures Ltd /Alamy Stock Photo; **p.221** *cl* © MASASHI ARAI/123 RF, *bl* © NCP Images/Greenshoots Communications/Alamy Stock Photo; **p.222** *cl* © Juice Images 264/Alamy Stock Photo; **p.223** *bl* © DeAngelo Lee/Alamy Stock Photo; **p.224** *tl* © Alexander Tolstykh/Alamy Stock Photo; **p.225** *tl* © Lucadp/123 RF, *bl* © Wavebreak Media ltd PH78/Alamy Stock Photo; **p.227** *tl* © Stocktrek Images, Inc./Alamy Stock Photo, *bl* © Agencia Brasil/Alamy Stock Photo, **p.228** *bl* © Doughoughton/Alamy Stock Photo; **p.229** *tl* © Per Andersen/Alamy Stock Photo, *bl* © ACORN 1/Alamy Stock Photo; **p.230** *tl* © GeoPic/Alamy Stock Photo, *bl* © Geoff Moore/REX/Shutterstock; **p.231**© Practical Action/ Uttam Kumar Saha; **p.232** *tl* © Sandy young/Alamy Stock Photo, *bl* © Piero Cruciatti/Alamy Stock Photo; **p.233** *tl* © Europa Newswire/Alamy Stock Photo, *bl* © Hospitainer; **p.234** *tl* © Watchtheworld/Alamy Stock Photo; **p.235** *bl* © Heinrich/Agencja Fotograficzna Caro/Alamy Stock Photo; **p.252** *bl* © Monty Rakusen/Cultura Creative (RF)/Alamy Stock Photo, *br* © VW/Dpa picture alliance/Alamy Stock Photo; **p.253** *bl* © Peter Jurik/123 RF; **p.254** *tl* © Prisma by Dukas Presseagentur GmbH/Frischknecht Patrick/Alamy Stock Photo, *bl* © Alberto Giacomazzi/Alamy Stock Photo; **p.255** *tl* © Stephen Grant/Alamy Stock Photo, *tr* © Robert hyrons/Alamy Stock Photo; **p.256** *tl* © Sciencephotos/Alamy Stock Photo, *bl* © BSI; **p.262** *cl* © FABIAN BIMMER/AP/REX/Shutterstock; **p.267** *tl* © Rob Winter, *tc* © Science & Society Picture Library/Getty Images, *tr* © Zoonar/Alfred Hofer/Zoonar GmbH/Alamy Stock Photo, *bl* © Roman Gorielov/123 RF, *br* © Pukhov K/Shutterstock.com; **p.268** *tl* © Chromorange/Alfred Hofer/Alamy Stock Photo, *tr* © Mark Humphreys/123RF, *bl* © Alfred Hofer/123 RF, *br* © Bigalbaloo/YAY Media AS/Alamy Stock Vector; **p.269** *tc* © David Bleeker Photography/Alamy Stock Photo, *ul* © Kemphoto artistic works/Alamy Stock Photo, *bl* © Vvoennyy/123 RF; **p.270** *cl* © Brännhage Bo/Prisma by Dukas Presseagentur GmbH/Alamy Stock Photo, *cr* © IS098Q1ML/Image Source/Alamy Stock Photo; **p.271** *tl* © Lumi Images/Alamy Stock Photo, *bl* © Richard Ellis/Alamy Stock Photo; **p.272** *tl* © SoloStock Industrial/Alamy Stock Photo, *cl* © Javier Larrea/Age fotostock/Alamy Stock Photo, *bl* © Oprea Florin/ Alamy Stock Photo; **p.273** *bl* © Trevor Smith/Alamy Stock Photo; **p.277** *cl* © Ange/Alamy Stock Photo; **p.282** *ul* © Martin Lee/Alamy Stock Photo, *ur* © David Lee/Alamy Stock Photo, *bl* © KeepCup; **p.299** *tl* ©Richard McDowell/Alamy Stock Photo; **p.300** *tr* © Aumm graphixphoto/Shutterstock.com; **p.301** *tl* © Monty Rakusen/Cultura Creative/Alamy Stock Photo; **p.307** *tl* © Standard Studio/Alamy Stock Photo, *cl* © David J. Green/Alamy Stock Photo, *bl* © Forest Stewardship Council®, *cr* © Eltoro69/Shutterstock.com; **p.308** *tl* © Simon Belcher/Alamy Stock Photo, *br* © EU Ecolabel; **p.309** © Forest Stewardship Council®; **p.341** *tl* © Coprid/Shutterstock.com, *bl* © Frantic00/Shutterstock.com; **p.347** *tl* © Vadim Orlov/Shutterstock.com, *bl* © Ewastock diningrooms/Elizabeth Whiting & Associates/Alamy Stock Photo; **p.362** *tl* © Rawpixel Ltd/Alamy Stock Photo; **p.363** *tl* © Mingis/Shutterstock.com.